Housing Policies in Eastern Europe and the Soviet Union

A commitment to providing an acceptable level of accommodation for all has meant that housing has enjoyed a high place on the agendas of most socialist countries. However, this place has not been undisputed and at various times in the different countries of the eastern bloc, housing has had considerably differing degrees of priority.

This book examines the issues related to housing in eastern Europe. It describes the broad distinctions and similarities in the situation in east and west Europe, outlines trends in housing conditions and construction since the war and discusses the relevance of factors external to the housing market. The book also places an emphasis on the various eastern bloc economic and social systems and discusses likely future trends.

Its geographical coverage is broad and it provides a wealth of information that will be invaluable for anyone working in this area.

The Author
J.A.A. Sillince is Senior Lecturer in the Department of Computer Science, Coventry Polytechnic.

Housing Policies in Eastern Europe and the Soviet Union

Edited by J.A.A. Sillince

ROUTLEDGE
London and New York

First published 1990
by Routledge
2 Park Square, Milton Park, Abingdon, Oxon, OX14 4RN

Simultaneously published in the USA and Canada
by Routledge
270 Madison Ave, New York NY 10016

Transferred to Digital Printing 2007

British Library Cataloguing in Publication Data

Housing policies in Eastern Europe and the Soviet Union.
 1. Communist countries. Housing. Policies on governments
 I. Sillince, John, *1946*–
 363.5'56'091717

 ISBN 0-415-02134-0

Library of Congress Cataloging-in-Publication Data

Housing policies in Eastern Europe and the Soviet Union / edited by
 J.A.A. Sillince.
     p.   cm.
   Includes bibliographical references.
   ISBN 0-415-02134-0
   1. Housing policy — Europe, Eastern. 2. Housing policy — Soviet
 Union. 1. Sillince, John, 1946–
 HD7332.H59   1990
 363.5'8'0947–dc20   89-77046
                                                                 CIP

Publisher's Note
The publisher has gone to great lengths to ensure the quality of this reprint
but points out that some imperfections in the original may be apparent

Printed and bound by CPI Antony Rowe, Eastbourne

CONTENTS

Contents

Contents

Contents

LIST OF TABLES

List of Tables

List of Tables

LIST OF FIGURES

List of Figures

INTRODUCTION

This book seeks to investigate one major
aspect of social and economic policy in Eastern
Europe and the Soviet Union - that of housing
policy. In many countries of the region there
has been a severe cutback since 1975 in state
resources available for housing, and a consequent
promotion of policies which use private or co-
operative funds. This change has been part of a
wider change in political ideas in many parts of
the region towards a greater role for the
'market' and a reduced role for the administrator.
This change has included a revaluation of
traditional ideas towards the maintenance of low
rents and low prices. These changes are
interesting because they are a barometer of other
changes, and because it is possible to see, in the
detailed chapters that follow, policy being
formulated. The studies that follow reveal policy
formulation in Eastern bloc countries to be
hesitant and subject to rapid revision.
    Is there a housing crisis in Eastern Europe
and the Soviet Union? This is the question which
Chapter 1 attempts to answer, by means of
investigating the extent of the housing shortage
there, using official comparative statistics and
national press reports. The definition of housing
shortage involves considering quantitative aspects
(numbers, dwelling size, average living space
etc.) and qualitative aspects, such as quality of
finish. Chapter 1 suggests that a housing
shortage does exist, and proceeds to investigate
its causes, whether due to underinvestment, low
productivity, poor quality materials and so on.
Also, the effects are analysed, whether
homelessness (a poor guide), frustration with
allocation rules, illegal sublets, over-crowding,

1

Introduction

or low birth rates. The chapter concludes with a brief survey of postwar housing history. Immediately after the war there was a system of state housing construction, with housing allocated by administrators. This allocation system led to inequalities, though probably the inequalities were less than would have arisen under a market system. After 1960 the state monopolies were ended as new sectors began - private and co-operative construction, and new finance arrangements (notably mortgages) began. Housing construction suffers from considerable organisational problems. Also, completed houses have noticeably poor quality finishes. The chapter finishes by hinting at the kinds of new policies being discussed in many countries in the region. These policies form a radical break with the past.

Poland (Chapter 2) has suffered from the problems of extensive war damage, fast population growth, migration to the towns, and rising expectations. A relatively large part of the houses constructed (75 per cent) are built by co-operatives for sale, although recently the private sector has been promoted. Problems of the house construction industry include slow rates of construction, poor finish, and a large black market. There is a large difference in housing quality between towns (good) and villages (bad) and between previously German areas (good) and other parts of the country. Rents have been rising relative to income.

Czechoslovakia (Chapter 3) suffered relatively little war damage, and complacency played a part in its being the only part of Europe where housing conditions deteriorated between 1950 and 1961. Since then construction rates have risen, and demolition rates have also been high, at least before 1975. Since 1980 the state sector has declined and the co-operative and private sectors have grown. Rents have remained low and virtually unchanged since the war. An interesting development is that some state flats are now being offered for sale.

Romania (Chapter 4) is an area where housing policy is influenced by ideas of social engineering and limiting the independence of the individual. A sign of the ideological "purity" of the government is that, because of concern over speculation and inequality in housing, there has

2

been a cutback in sales since 1986. High densities and high-rise housing designs are partly motivated by social engineering ideas, but also partly by concern about the use of agricultural land. There is a patent lack of interest by the government in the conservation of architecture and minority culture. The government's housing policy rejects the counter urbanisation trends of Albania without allowing scope for the private sector.

Bulgaria (Chapter 5) has now withdrawn from a commitment to monolithic high-rise housing placed insensitively in its environment, and now rigid height limits are being imposed, mainly for aesthetic reasons. During the 1970s the co-operative and private sectors grew until they now, with the state sector, each account for a third of current construction. There has been a policy during the 1980s of shifting the cost burden from the state.

The USSR (Chapter 6) has two main housing goals, that every family should have a flat, and that rents should be low. These goals are probably in conflict. A Soviet commentator has named housing as the main Soviet social problem. Allocation practices have been criticised openly as unfair and in some instances, corrupt. A reform of prices is needed, with many experts arguing they should be raised steeply, others saying they should reflect size and quality. The existence of the command economy in housing has been criticised for discouraging democracy. Despite a high rate of housing construction, quality of finish is poor, and demolition rates are high. The building industry is inefficient, and quality and quantity of building materials too low. Economic enterprises have too much say in housing construction and allocation, while local government is weak and inexperienced, as well as being resistant to new policies such as the encouragement of the private and co-operative sectors. All these problems are being discussed more openly now, but as yet no firm policy has emerged. Many questions exist about policy. How will the government reform local government? How can the quality of building materials be improved? If it is not possible to effect a sharp change to private construction in the towns, how can this become a significant source of new housing (since the great majority of the population is urban)? Will the population accept rent rises? What kind of persuasion will be required by politicians? How can the black market and illegal building workers be regulated?

The German Democratic Republic (Chapter 7) suffered relatively little war damage, and so there are a large number of poor quality old dwellings. There is an ambitious rehabilitation programme to modernise this older stock. House construction suffers from poor quality and a lack of skilled labour. There has been a limited increase in private construction since 1973 but materials, plots, and permits are hard to get. Housing is used as a means of directing labour to areas where it is scarce, as a means of regional planning, and as a means of increasing densities and hence efficient utilization of infrastructure.

Albania (Chapter 8) was, until recent leadership changes, strongly "Stalinist" in its interpretation of communist ideology, and therefore politically and commercially isolated from the other Eastern European countries, as well as from the West. It is the poorest country in Europe, and generally lacks sewerage, piped water, and piped gas. There is a high birthrate. New housing blocks are spartan, usually in the form of walk-up, 6 storey buildings. Housing is allocated by workplace trade unions. Often state housing is built by voluntary labour organised into brigades within the workplace. Although private house ownership is fairly high in rural areas, there is no trend towards the state selling off flats. There is strong government control of migration and the rural population, uniquely within Eastern Europe, is growing.

Yugoslavia (Chapter 9) suffered huge wartime housing losses, and administrative allocation was instituted after 1945. In 1966 self-managed housing enterprises were established, and ten years later local self-management committees were set up to liaise with planners, workplaces, and construction companies. Despite high investment, the rate of construction has been disappointing. Problems include fast urbanisation, uncoordinated development and poor organisation resulting from a lack of hierarchical structure and clear accountability, low building productivity, the monopoly position of building enterprises, and irrational credit policies. However, 70 per cent of all housing is post-1945. Home-ownership is high and rising - although not openly encouraged by government policies. Economic enterprises control some housing construction, though most house building is private. Co-operatives have had problems of quality and corruption in the past. In the 1987 housing reform, the government introduced economic rather than social need criteria to decide credit-worthiness, measures to transfer costs from the state to the private sector, and increased credit facilities for private building. Rents will rise, state housing is being sold off, rebates to the poor are being increased, and (by 1990) rents will reach "economic" levels.

Hungary (Chapter 10) has made great strides in housing provision since the end of the war, during a time when war damage and migration resulting from rapid industrialisation created enormous difficulties and shortages. Initially new housing was not of high quality, but later quality improved, as did the amount of demolition of old stock. The first big surge in housing construction came during the 1960s with the development of industrialised methods of building. This was followed by a vast increase in mortgage funds for owner occupation in the second acceleration in the 1970s, during which period shortages were reduced considerably, allowing very progressive allocation and subsidy policies to be introduced in 1971 for the needy. These policies were partially withdrawn in 1983 as a result of the country's economic difficulties. Since then state funding for housing has been reduced, along with encouragement of the private sector. Further means of reducing the state burden of housing were announced at the end of 1988 involving considerable rent rises, and a raise in the mortgage rate (for new borrowers) from the previous 6 per cent to 18 per cent now.

Chapter One

**HOUSING POLICY IN EASTERN EUROPE
AND THE SOVIET UNION**

John Sillince

        In this chapter some attempt is made to
consider housing problems and policies in Eastern
Europe and the Soviet Union as a whole. Such an
overview with independent data and information
sources is necessary for two reasons. First,
reading about housing in individual countries may
give hints of similarities and differences, but
international comparative data are required to
ensure that such apparent relationships are not
spurious. Second, despite having a common brief
as to what aspects of housing to include, it is
inevitable that writers from different subject
backgrounds will emphasise or ignore different
things, so that an overview is needed to smooth
out these differences.
        The impression gained by even a superficial
acquaintance with this topic is that housing in
Eastern Europe and the Soviet Union is in a state
of deepening crisis. We shall have to investigate
to what extent this impression is an accurate one,
by attempting to define what we mean by a housing
shortage, by considering the available statistics,
and by comparing them with similar data for
Western countries. We shall also consider the
possible causes of the housing problem, and also
the effects. Finally, the relative importance of
housing compared to other social and economic
issues will be dealt with, together with an
attempt to understand continuities and recent
innovations in housing policy.

THE HOUSING SHORTAGE - QUANTITATIVE ASPECTS

        The most obvious definition of housing
shortage is a purely quantitative one. If we
take, for any period such as a year, the number of

6

newly constructed dwellings and subtract demolitions, marriages, and divorces from this figure, we arrive at a crude indicator. But this calculation assumes that all marriages are by young people leaving the parental home rather than their own dwelling. It also assumes that there are no divorced people, living alone, who remarry. For this reason the figure for shortages in Table 1.1 excludes divorces for the upper bound and includes them for the lower bound. Furthermore, Table 1.1 excludes the effect of rural-urban migration, mortality or births, events which would form important dimensions in any more sophisticated calculation of housing need. This method is essentially that used by Morton (1979) although he excluded divorces from his calculations. Missing data have been dealt with by interpolation except in Romania's case, where regional average rate of demolitions to completions has been used: all substitutions for missing data are in parentheses.

There are several striking features to Table 1.1. First, in most countries completions (new dwellings constructed) rose to a high point between 1975 and 1980 and then fell, as a result presumably of worsening international economic conditions. This was true for Bulgaria, Hungary, the GDR, Poland, Romania (with an earlier peak in 1960 also), Czechoslovakia, and Yugoslavia. The exception was the USSR with peaks in 1960 and 1970. The second interesting feature is the relatively low rate of demolitions to completions - in 1986 it was 12.3 per cent compared with 25.3 per cent for the 11 rich Western countries for which data were available (United Nations, 1986, Table 4.) Third, apart from Bulgaria in 1977 and 1980, and Romania in 1960, the regional norm has been one of chronic housing deficit.

The quantitative deficit and its relative significance can be more clearly seen in Table 1.2. Column 6 shows the total accumulated shortage up to 1986. Relative to population, Bulgaria and the USSR have the worst problem, Hungary's problem is relatively small, and the other countries' are somewhere in between. One in ten of the population in Bulgaria and the USSR are without the accommodation they need, while for Hungary only one person in forty is in this position. However, the problem takes on a greater significance if one considers the 1986 shortage as percentage of the 1986 stock (Column 8).

Table 1.1: Housing shortage (thousands)

		1960	1965	1970
Bulgaria	A	49.8	45.2	45.7
	B	(4.5)	(4.5)	(4.5)
	C	69.6	65.8	73.2
	D	7.1	9.0	10.2
	Housing Shortage	− 24.3 to − 31.4	− 25.1 to − 34.1	− 32.0 to − 42.2
Hungary	A	58.1	54.6	80.3
	B	(18.3)	(18.3)	21.6
	C	89.1	89.5	96.3
	D	17.0	20.3	22.8
	Housing Shortage	− 49.3 to − 66.3	− 53.2 to − 73.5	− 37.6 to − 60.4
GDR	A	80.5	68.2	76.1
	B	(29.6)	(29.6)	(29.6)
	C	166.7	129.5	131.4
	D	24.1	27.3	27.3
	Housing Shortage	−115.8 to −139.9	− 90.9 to −118.2	− 84.9 to −112.2
Poland	A	142.0	171.0	194.0
	B	(21.2)	(21.2)	(21.2)
	C	244.3	198.8	277.6
	D	14.9	23.7	34.6
	Housing Shortage	−123.5 to −138.4	− 49.0 to − 72.7	−104.8 to −139.4

Notes: A: Dwellings completed during year;
B: Dwellings demolished or use changed during year
C: Marriages during year;
D: Divorces during year; Housing Shortage = A−B−C to A−B−C−D is a measure of that year's shortage (−) or surplus (+) of dwellings.

8

Table 1.1 (continued)

		1975	1977	1980	1986
Bulgaria	A	57.2	75.9	74.3	56.0
	B	(4.5)	5.9	4.0	3.5
	C	74.9	75.0	70.1	65.4
	D	11.3	13.0	13.3	9.8
	Housing Shortage	- 22.2 to - 33.5	+ 5.0 to - 8.2	+ 0.2 to - 13.1	- 12.9 to - 22.7
Hungary	A	99.6	93.4	89.1	69.4
	B	(18.3)	20.4	(18.3)	14.0
	C	103.6	97.1	80.3	72.0
	D	26.4	27.7	27.8	29.7
	Housing Shortage	- 22.3 to - 48.7	- 23.7 to - 51.4	- 9.5 to - 37.3	- 16.8 to - 46.5
GDR	A	141.0	117.4	102.2	121.6 [2]
	B	(29.6)	29.6	(29.6)	(29.6)
	C	141.3	147.5	133.9	138.1
	D	42.0	43.6	45.2	53.2
	Housing Shortage	- 29.9 to - 71.9	- 59.7 to -103.3	- 61.3 to -106.5	- 46.1 to - 99.3
Poland	A	264.0	276.0	217.1	184.9
	B	(21.2)	(21.2)	26.2	16.2
	C	331.6	327.6	307.3	259.2
	D	41.4	41.8	39.3	52.6
	Housing Shortage	- 88.4 to -130.2	- 72.8 to -114.6	-116.4 to -155.7	- 90.5 to -143.1

Sources: A and B from United Nations 1980 and 1986 Table 4;
C and D from Statisticheski Ezhegodnik 1987 Tables
4 and 6, and United Nations 1985
Notes: (1) 1985 data; (2) 1984 data; (3) Calculation based
on new total floor space divided by useful square
metres per dwelling for USSR (52.3 in 1980, 56.8 in
1985); (4) UN and CMEA data conflict for this item
and so UN data have been presented; (5) 1966 data

Table 1.1 (continued)

		1960	1965	1970
Romania	A	253.0	192.0	159.0
	B	(47.3)	(35.9)	(29.7)
	C	197.9	164.1	146.6
	D	37.2	37.0	7.9
	Housing	+ 7.8 to -	8.0 to -	17.3 to
	Shortage	- 29.4 -	45.0 -	25.2
USSR	A	2591.0	2227	2266.0
	B	277.3	(348.1)	418.8
	C	2617.1	2020.5	2365.7
	D	277.3	371.6	634.1
	Housing	-303.4 to -	141.6 to -	518.5 to
	Shortage	-508.7 -	513.2 -	1152.6
Czechoslovakia	A	73.8	77.8	122.6 [4]
	B	(26.6)	(26.6)	33.7
	C	106.8	112.1	126.3
	D	15.3	18.7	24.9
	Housing	- 59.6 to -	60.9 to -	37.4
	Shortage	- 74.9 -	79.6 -	62.3
Yugoslavia	A	75.7	122.0	128.8
	B	(8.85)	10.0	15.1
	C	168.1	173.5	183.6
	D	21.5	21.6	19.5
	Housing	-101.3 to -	61.5 to -	69.9 to
	Shortage	-122.8 -	83.1 -	89.4

Notes: A: Dwellings completed during year;
B: Dwellings demolished or use changed during year;
C: Marriages during year;
D: Divorces during year; Housing Shortage = A-B-C
to A-B-C-D is a measure of that year's shortage
(-) or surplus (+) of dwellings.

Table 1.1 (continued)

		1975	1977	1980	1986
Romania	A	165.0	145.0	197.8	105.6[1]
	B	(30.9)	(27.1)	(37.0)	(19.7)
	C	190.0	200.2	182.7	161.3[1]
	D	34.6	25.9	45.2	31.8[1]
	Housing Shortage	− 55.9 to − 90.5	− 82.3 to − 108.2	− 21.9 to − 67.1	− 75.4to −107.2
USSR	A	2228.0	2111.0	2004.0 [3]	1991.0
	B	(405.4)	391.9	229.4 [3]	(227.9)
	C	2735.0	2783.5	2745.9	2760.5
	D	792.3	910.5	933.1	957.7
	Housing Shortage	−912.4 to −1704.7	−1064.4 to −1974.9	−964.1 to −1897.8	−997.4to −1955.1
Czechoslovakia	A	147.0	144.0	128.8	78.7
	B	(26.6)	25.0	29.0	18.9
	C	141.1	137.2	117.7	119.4
	D	32.4	31.2	33.4	38.8
	Housing Shortage	− 20.7 to − 53.1	− 18.2 to − 49.4	− 17.9 to − 51.3	− 59.6to − 98.4
Yugoslavia	A	145.5	142.5	136.8	127.6[1]
	B	7.1	7.5	8.0	5.4[1]
	C	180.9	178.8	171.4	167.9[2]
	D	25.0	23.7	22.6	21.2[2]
	Housing Shortage	− 42.5 to − 67.5	− 43.8 to − 67.5	− 42.6 to − 65.2	− 45.7to − 66.9

Source: A and B from United Nations 1980 and 1986 Table 4;
C and D from Statisticheski Ezhegodnik 1987 Tables
4 and 6, and United Nations 1985

Notes: (1) 1985 data; (2) 1984 data; (3) Calculation based
on new total floor space divided by useful square
metres per dwelling for USSR (52.3 in 1980, 56.8 in
1985); (4) UN and CMEA data conflict for this item
and so UN data have been presented: (5) 1966 data

Table 1.2: Total housing shortages and total housing stock,
around 1970 and 1986 (thousands)

	1	2	3	4
	Initial shortage (year)	Initial stock (year)	Initial shortage as % of initial stock	Initial shortage as % of initial population
Bulgaria	− 472 (1965)	2055 (1965)	23.0	5.7
Hungary	− 6 (1973)	3346 (1973)	0.2	0.1
GDR	− 340 (1971)	6068 (1971)	5.6	2.0
Poland	− 1357 (1974)	8548 (1974)	15.9	4.2
Romania	− 575 (1966)	5250 (1966)	11.0	3.0
USSR	− 13690 (1970)	59201 (1970)	23.1	5.7
Czechoslovakia	− 438 (1970)	4410 (1970)	9.9	3.2
Yugoslavia	−	−	−	−

Notes: (1) Author's own calculation, from Column 2 and
Table 1 above
(2) Calculated on assumption that demolitions are
regional average
(3) Calculated from 1960 on assumption that there
was no shortage prior to 1960

Table 1.2 (continued)

	5	6	7	8	9
	Subsequent Accumulated Shortage to 1986	Total 1986 Shortage	Total 1986 Stock	1986 Shortage as % of 1986 Stock	1986 Shortage as % of 1986 Population
Bulgaria -	408.4 -	880.4	3216	27.4	9.8
Hungary -	251.0 -	257.0	3891	6.6	2.4
GDR -	841.7 -	1181.7	6914	17.1	7.1
Poland -	1217.8 -	2574.8	10788	23.9	6.9
Romania -	582.9[2] -	1157.9	8287[1]	14.0	5.0
USSR	- 12972.4	-26662.4	88163[1]	30.2	9.5
Czechoslovakia -	439.6 -	877.6	5746	15.3	5.7
Yugoslavia-	1634.7[3] -	1634.7[3]	6841	23.9	7.0

Source: Columns 1 and 2 : Morton (1979) p. 302-303;
   Column 5: Table 1.1 above, by multiplying lower
             bound for given year by 5 to approximate
             shortage for subsequent 5 years;
   Column 7: United Nations (1986)

13

In order to get rid of its housing deficit, the USSR needs 30.2 per cent more dwellings, Bulgaria needs 27.4 per cent more dwellings, Poland and Yugoslavia both need 23.9 per cent more dwellings, and so on. Again Hungary, while in deficit, has a noticeably smaller problem. Also, the problem has worsened in the last fifteen years. Comparing Columns 3 and 4 with Columns 8 and 9 the worsening shows across all countries in the region. The greatest worsening was in the GDR (more than a trebling of the deficit) and in Hungary (from a very small problem initially).

It is interesting that according to United Nations (1980; 1986, Table 4) this worsening took place alongside an increase in the quantity of dwelling stock relative to population from 1970 to 1986 - from 300 to 359 per thousand population (Bulgaria - first figure for 1977), from 298 to 370 (Czechoslovakia), from 355 to 416 (GDR), from 310 to 366 (Hungary), from 255 to 288 (Poland), from 243 to 315 (USSR), and from 245 to 294 (Yugoslavia - latter figure for 1985). But of course these two trends are not incompatible. The proportion of a dwelling available per person is gradually growing. But the percentage of households who have to share accommodation (mostly young newlyweds sharing with one or both parents) is also growing. It is this sharing of dwellings which is recorded as housing shortage in Tables 1.1 and 1.2.

Another aspect of the quantitative situation is average dwelling size. The size of new dwellings has gradually risen in all countries - in 1986 it ranged from 2.5 (Bulgaria) to 4.0 habitable rooms (Poland). More detailed data on size of newly constructed dwellings are shown in Table 1.3, Column 1. Space standards vary considerably - the average new dwelling in the USSR in 1986 was only 68 per cent the size of its equivalent in Hungary. Apart from exceptional cases (GDR in 1980-1986, Bulgaria in 1970-1980) space standards have been rising in newly-built dwellings. However, in 1986 the average size of a new dwelling in the region was only 61.3 square metres compared with an average of 113.5 square metres in the 10 rich Western countries for which data were available (United Nations 1986, Table 4.)

Table 1.3: Space and facilities in new dwellings

	Useful floor space per dwelling (square metres) in newly built dwellings			Percentage dwellings with piped water in newly built dwellings			Percentage dwellings with fixed bath or shower			Percentage dwellings with central heating in newly built dwellings			People per dwelling (all dwellings)		GNP per capita (£ sterling) (3)
	1970	1980	1986	1970	1980	1986	1970	1980	1986	1970	1980	1986	1970	1986	
Bulgaria	63.7	59.0	66.9	75.7	94.1	98.8	63.0	88.5	91.4	28.8	35.6	33.1	3.8	2.8	1591 (1983)
Czech'kia	67.2	73.8	81.8	88.8	99.4	99.3	80.6	98.2	100.0	85.6	96.4	97.3	3.4	2.7	3515 (1982)
GDR	55.0	62.7	61.2	100.0	100.0	100.0	99.3	100.0	100.0	73.8	96.0	98.5	2.9	2.4	4848 (1983)
Hungary	61.5	67.0	83.0	70.2	98.1	99.6	74.9	98.8	99.6	38.2	50.9	75.1	3.4	2.7	2533 (1983)
Poland	54.3	64.0	71.0 (1)	-	96.3	98.3	-	92.9	96.2	-	91.4	95.0	4.2	3.5	1667 (1984)
Romania	44.9	57.0	57.5 (1)	-	- (2)	-	-	- (2)	-	-	- (2)	-	3.6	2.8	3182 (1984)
USSR	46.8	52.3	56.8	-	89.9	- (1)	-	86.9	- (1)	-	86.6	- (1)	4.1	3.2	1576 (1984)
Yugo'via	59.2	70.9	72.5	61.3	90.4	94.4	59.3	85.1	89.8	23.4	34.1	33.6	-	3.4	1884 (1984)
Albania	-	-	-	-	-	-	-	-	-	-	-	-	-	-	503 (1981)

Source: United Nations 1980; 1986, Table 4

Notes: (1) 1985 data; (2) Urban areas only;
(3) Some 1983 Western figures comparable with Column 6 are USA 8152, UK 4373, Austria 5018, West Germany 5727, Sweden 8982, Switzerland 8732

The result has been that the region has begun to meet internationally accepted norms for living space. The average number of persons per habitable room in all dwellings was 1.0 in Bulgaria (1975), 0.9 in Czechoslovakia (1980), 1.0 in Hungary (1980) and 1.2 in Poland (1978) according to data from the United Nations (1985, pp. 241-250.) In the USSR, the "minimum sanitary living norm" is 9 square metres of "living space" per person, (living space is calculated as 70 percent of total housing space), a norm which many Soviet citizens have yet to enjoy. In 1964 the USSR figure was 6.5 square metres of living space per person, compared with the Western European average of 13.4 square metres (Hanson 1968, p. 65). By 1980, however, average living space (9.3 square metres) slightly exceeded the norm for the first time (Warwick Statistics Service 1983, pp. 13-15). By contrast, average living space in Yugoslavia has increased from 8.7 square metres (1951), to 9.9 square metres (1961), to 12.2 square metres (1971), showing Yugoslavia to be relatively much better off (Sirc 1979, p. 144).

Thus quantitative changes have been complex. On four indicators, persons per room, persons per dwelling, average dwelling size, and rooms per dwelling, the trends have all been favourable. Unfortunately, marriages and divorces have led to falling numbers per household and to a vastly expanded number of households. This has overwhelmed the favourable changes, so that, despite the fact that fewer persons were living together in larger dwellings, they did so as part of a far greater number of households. The consequence was a rise in the number of households who were forced to share a dwelling.

Does the above quantitative assessment accord with other estimates? One internal USSR estimate states that "over 80 per cent of the urban population live in separate apartments" (Narkhoz 84 1985, p .400). This has been used as the basis of an estimate of 9 million for the USSR urban shortage (Trehub 1986a) smaller than the estimate of urban shortage of 9,562,106 for 1970 in Morton (1979) despite the fact that Trehub accepts and even provides evidence for a growing USSR housing deficit (Trehub 1986a, p. 5). Moreover, as Voznesenskaya (1986, p. 5) points out, "Communal living quarters for young families are still being built. Moreover, there is no explanation of how many years a family remains 'young', of how many years its 'temporary' accommodation in such quarters may last, how many families altogether live in youth housing complexes, and whether or not they are included among the 80 per cent of the population that are claimed to be living in separate apartments."

Indeed, the 1988 official estimate of housing shortage is now 13 million units (Izvestia 1988a) and even that seems on the low side since the Soviet economist Aganbegyan has estimated that by the end of the century, a mere twelve years away, 40 million units will need to be built (Radio Budapest 1987).

Another method of independently checking comes from a report of a meeting on 30th January 1988 at the Moscow City Soviet, where a figure of 1,800,000 was given as Moscow's current housing shortage (British Broadcasting Corporation 1988a). If we divide Morton's (1979, Table 1) figure for total 1970 USSR shortage by his figure for the urban 1970 USSR shortage, and multiply this by the 1986 USSR urban population (186 million) divided by the 1986 Moscow population (8 million) and multiply this by 1,800,000 (Moscow's shortage currently) we obtain 59,917,900 as the estimate of current USSR shortage. Now, the figure in Table 1.2 above of 26,662,400 as the estimated 1986 USSR housing deficit does not take account of rural-urban migration. Yet we know, for example, that during the 1970-1979 period alone 15.6 million migrated from the countryside to the towns, a figure rather larger than the 12 million natural increase of that period (Warwick Statistics Service 1983, p. 1). So our figure of 26,662,400 is almost certainly an underestimate.

For Poland there is an independent estimate for the 1987 shortage of 2 million units (Trybuna Ludu 1987), although another source states that more than 3 million are on waiting lists for co-operative flats, the predominant form of residential construction in urban areas (Zycie Warszawy 1984). Table 1.2 above gives our 1986 estimate as 2,574,800.

In Albania between 1950 and 1970 there was an extra 916,700 population and 190,000 dwellings were built (Marmullaku 1975 p. 107, Appendix Table 1). Even assuming there was no additional need due to migration, or demolition, this rate of house-building appears insufficient compared with population growth.

17

To summarise, our estimates of housing shortage given in Table 1.2, while imperfect, are probably the best guide to the quantitative problems of housing in the region. The underlying data are sometimes incomplete and interpolation and other calculations have had to be resorted to. Moreover sometimes the official statistics do not report such basic elements as numbers of households and numbers of dwellings. Nevertheless the problem does not seem to be one of falsification (Hanson 1986 Chapter 2; Andrusz 1988).

THE HOUSING SHORTAGE - QUALITATIVE ASPECTS

Table 1.3 gives some quality indicators for new housing construction, and shows that standards have risen in all countries. However, Table 1.4 shows that sanitary conditions are still far from adequate, a factor (among others) which relates to the worsening life expectancy in the region (see Table 1.11). Some of the underlying factors are given in Table 1.5: although these data are old and incomplete it gives a stark picture of housing quality in the region. There are of course much higher standards in new construction than in existing housing and conditions have undoubtedly improved - see the data in Table 1.5 for Poland in 1984. All countries for which data existed had between 1966 and 1975 more than 60 per cent of their dwellings with a density of greater than 1 person per room; the current situation has seen an improvement to a norm of about 1 person per room for the countries for which data exist (see above). The norm in the majority of rich Western countries is about 0.5 persons per room (Organisation for Economic Co-operation and Development 1986, p. 129).

Problems are aggravated by poor quality finishes on new dwellings (Mieczkowski 1975, p. 29). Often new occupants are expected to do a certain amount of finishing work or repairs when shoddy construction is carried out. A large proportion of property repairs are carried out by people doing second jobs (Trehub 1986b, p. 3). In the USSR, dormitories and youth housing complexes are often of a poor quality, and many families live in poor quality temporary accommodation (Voznesenskaya 1986).

18

Table 1.4: Population coverage with adequate sanitary
facilities and safe water

	Percentage of population with adequate sanitary facilities (year)		Percentage of population with safe water (year)	
Albania	–		92.0	(1980)
Bulgaria	–		96.0	(1980)
Czechoslovakia	60.5	(1983)	74.5	(1983)
GDR	70.0	(1985)	90.0	(1985)
Hungary	60.0	(1984)	84.0	(1984)
Poland	50.0	(1980)	67.0	(1980)
Romania	50.0	(1980)	77.0	(1980)
USSR	50.0	(1980)	100.0	(1983)
Yugoslavia	58.4	(1981)	67.8	(1981)

Source: World Health Organisation 1986 pp. 51-55

Table 1.5: Housing quality, all dwellings

Country (year)	Percentage of all dwellings built after 1945	Percentage of all dwellings with piped water	Percentage of all dwellings with central heating	Percentage of all dwellings with toilet inside dwelling	Percentage of all dwellings with density greater than 2 persons per room	Percentage of all dwellings with more than one household	Percentage of all dwellings with density greater than 1 person per room
Albania	-	-	-	-	-	-	-
Bulgaria (1975)	70.4	66.1	7.5	28.0	12.4	10.6	60.2
Czechoslovakia (1970)	38.3	75.3	30.9	52.4	10.4	8.2	67.9
GDR (1971)	20.8	82.1	-	43.4	-	-	-
Hungary (1970)	37.5	36.1	-	-	8.5	9.6	64.4
Poland (1970)	43.2	47.3 (92.7, 54.0)	22.2 (67.3, 30.5)	33.4 (77.0, 40.2)	27.6	12.9	83.0
Romania (1966)	-	12.3	-	-	27.8	-	81.5
USSR	-	-	-	-	-	-	-
Yugoslavia (1971)	59.1	33.6	4.5	29.2	30.8	5.7	83.1

Source: Shoup (1981, pp. 385-421)
For Poland, figures in brackets refer to 1984 data: first figure refers to urban households, second figure refers to rural households (Radio Free Europe Research 1986a). These data were based on a sample survey by the Polish Central Statistical Office of 1 million households

Housing Policy in E. Europe and the USSR

There has, since 1986, been much open criticism in the USSR of the State Committee for Civil Construction and Architecture, which has provided a convenient focus for the frustrations created by the housing problem (Trehub 1986c, British Broadcasting Corporation 1987a). In Poland some building materials containing toxic substances are still used in housing construction (Swidlicka 1986). To discuss and publicise Poland's housing problems a "Congress of People without Housing" was held on 16th January 1988 in Warsaw (Radio Free Europe Research 1988a). A special investigative commission in Warsaw discovered that many buildings registered as ready for occupancy in 1985 were far from finished (Polish Television 1986, Slowo Powszechne 1986). This frustration with poor quality has also appeared in Hungary, where resources are slowly being withdrawn from the state sector, and where repairs and maintenance of state housing are nonexistent or far less adequate than in private sector housing (Lengyel 1985). In Romania one important dimension of housing quality is domestic heating and electricity use, where cuts have begun to really affect quality of life. There have been five cuts since 1979, although until the one in November 1987 these had applied to the cold season only. Since November 1987, however, the government has provided only a minority of domestic electricity needs throughout the year. For example, for a one room dwelling only 17.4 kilowatt hours of electricity per month is provided rather than the government's estimated minimum requirement of 64.4 kilowatt hours (Romania Libera 1987a). Supplies of coal, oil, and wood have also been progressively cut. The cuts have been greatest in rural areas, where it is presumably hoped by the government that firewood can be acquired outside of the state's supply system. Besides this, the hostels for young people in Romania, which in 1983 housed 523,000 in 2344 buildings (Directia Generala de Statistica 1986 p. 235) have only limited amenities and the older ones are in very poor repair (Scinteia Tineretului 1986a, 1986b, 1986c, 1986d) while noise, drunkenness, and prostitution have been found to be associated problems (Scinteia Tineretului 1985a).

THE CAUSES OF THE HOUSING SHORTAGE

Table 1.6 shows how much of total capital investment has been devoted to housing. Morton (1979 pp. 314-315) surveying the years 1950 to 1975 concluded that in all countries (except Poland) the proportion had fallen. By 1970 for example the share of investment given to housing in the USSR was 17.1 per cent (Table 1.6) comparing unfavourably with the 20.3 per cent in the United States (Campbell, 1974 p. 101). The extended data up to 1986 suggest a reversal of that trend - five of the seven countries show a rise between 1975 and 1986. Romania and Czechoslovakia - perhaps the least sensitive to popular feelings - are the exceptions to the general rise. In Poland and Hungary, where public discussion and criticism of government policies has some relative freedom, more than a fifth of all investment went into housing in 1986.

However the startling thing about Tables 1.6 and 1.7 is the fact that while five out of the seven countries show an increased proportion of their investment going into housing in the period 1975-86, all countries (except the GDR) show a fall in house-building rate per head. We can see an explanation for this by consulting Table 1.8 which shows total national investment volumes at stationary prices (i.e. real values unaffected by inflation). In particular the last columns of Tables 1.7 and 1.8 give 1980 to 1986 changes in house-building rates (Table 1.7) and total investment (Table 1.8.) Hungary, Poland, and Czechoslovakia all had a fall in building rates together with falling or stationary investment in housing in real terms (discounting the effect of inflation). Romania's total investment remained constant in real terms, but the proportion of that investment devoted to housing fell, as the government concentrated on repaying its foreign debt. The only country which had a higher house-building rate in 1986 than in 1980 was the GDR, and that country expanded its total investment the most as well as increasing the proportion of that investment devoted to housing.

In Yugoslavia, the late 1970s saw a high level of investment in housing and industry which was a major factor in fuelling inflation and the unfavourable trade balance. Housing construction funded through the banks was the fastest growing sector of investment.

Table 1.6: Investment in housing as a percentage of total capital investment outlays in national currency, 1950-1986

	1950 capital investment	%	1960 capital investment	%	1965 capital investment	%	1970 capital investment	%	1975 capital investment	%	1980 capital investment	%	1986 capital investment	%
Bulgaria	78	21.1	193	14.1	238	12.0	344	9.7	509	9.5	955	13.3	1200	13.0
Hungary	4180	20.0	6628	15.6	8194	16.8	15547	15.3	28681	17.6	37965	17.8	53549	20.7
GDR	-	-	1911	11.9	1596	7.8	3248	7.4	5468	9.8	7584	11.5	8617	13.1
Poland	4538	11.7	24159	21.7	28872	18.7	48566	17.4	82773	13.2	371857	24.0	442923	23.2
Romania	708	11.2	4340	15.7	5053	10.7	7859	9.8	13362	9.7	21930	10.4	20063	8.1
USSR	2300	18.0	9456	22.5	9638	16.9	15793	17.1	19215	15.0	21123	14.0	30875	15.9
Czechoslovakia	3368	17.0	8916	15.3	9445	14.7	15631	17.3	20235	15.2	20127	13.2	17413	1.9
Albania	-	8.0	-	8.0	-	8.0	-	7.0	-	5.0	-	-	-	-

Source: Morton (1979 Table 4);
Statisticheski Ezhegodnik Stran-Chlenvov Soveta
Ekonomicheskoy, 1987 Tables 52 and 57, pages 143 and 149;
Marmullaku 1975 Appendix Table 12

Notes: (1) 1985 data

Housing Policy in E. Europe and the USSR

Much of the housing financed by bank loans was for private ownership. In the state sector housing investment also rose fast. Between 1977 and 1978 the share of total investment devoted to housing rose from 16.7 to 17.2 per cent (Singleton 1982, pp. 288-289). These proportions compare favourably with most of the countries in Table 1.6 for the late 1970s. However, the proportion of investment devoted to the non-productive sector (housing and social services) fell from 36 per cent in 1962-1964 to 27.4 per cent in 1967-1969, rising again to 32 per cent in 1972-1974. Throughout that period the share taken by the private sector (mainly private housing) rose steeply from 7.5 through 14.5 to 17.3 per cent (Schrenk, Ardalan, and El Tatawy 1979, p. 167).

Another problem has been that labour costs have been particularly high and rising in the region's construction industries compared with other industrial sectors (Vienna Institute for Comparative Economic Studies 1986, pp. 70-108) resulting in a smaller number of dwellings for the same given investment. The inefficiency of the USSR house-building industry has in recent years been much discussed by the Communist Party's Central Committee and the government, and the same kinds of inefficiencies probably exist also elsewhere in the region. For example, in Poland and Hungary there is much open criticism of the monopoly position and inefficiency of the state house- building industry.

During the period 1960 to 1975 the housing shortage got worse because too much state investment went into industry rather than into housing. Since 1975 a noticeable switch has been made away from industry towards housing. Yet the problem has again worsened because this larger percentage for housing was coming from a static or decreasing total investment fund. Also, house-building costs were rising faster than other costs.

For this reason the problems endemic to the house-building industry of the region have continued. For example, serious cement shortages occur in Romania due to organisational problems and to underinvestment (Romania Libera 1987a). Bulgaria too has problems with building material shortages (Radio Free Europe Research 1986b, Feiwel 1982, p. 222) as does the USSR (British Broadcasting Corporation 1988b) and Hungary (Osztlany 1985).

Table 1.7: Number of dwelling units built per 10,000 population

	1960	1965	1970	1975	1980	1985	1986	1986 ÷ 1980
Bulgaria	63	55	54	66	84	72	63	0.75
Hungary	58.4	53.9	77.8	94.5	71.9	68	65	0.90
GDR	46.7	40.1	44.6	83.6	101	128	130	1.29
Poland	48.0	54.1	59.7	77.6	61.0	51	49	0.80
Romania	137	101	78.6	77.9	89.1	46	47	0.53
USSR	121	97	93	88	75.0	72	75	1.00
Czechoslovakia	54.1	54.9	78.2	99.6	84.1	67	51	0.61

Source: Council for Mutual Economic Assistance 1979, p. 213;
Statisticheski Exhegodnik Stran-Chlenov Soveta
Ekonomicheskoy 1983, p. 170 and 1987, p. 161

Added to this are the effects of natural hazards - such as the 1977 earthquake which made 12,000 dwellings uninhabitable in Romania - and the continued migration from the countryside to the towns in all countries except Albania.

Table 1.8: Volume of capital investment in the national economy (1970 = 100)

	1960	1975	1980	1986	1986 ÷ 1980
Bulgaria	38	151	184	254	1.38
Hungary	44	140	158	136	.86
GDR	49	126	150	267	1.78
Poland	49	224	193	179	.93
Romania	34	172	259	260	1.00
USSR	51	140	166	213	1.28
Czechoslovakia	64	147	169	163	.96

Source: Statisticheski Ezhegodnik Stran-Chlenov Soveta Ekonomicheskoy 1983, p. 138; 1987, p. 144

In terms of East-West comparisons, the region under-performed relative to Western Europe until 1975, when it did slightly better. If we consider housing construction per thousand population, the average figure in the Comecon countries (the Eastern Europe region minus Yugoslavia and Albania) was 6 in 1960, rising to 8 in 1980, falling to 7 in 1983, while the average figure in 15 Western European nations was 7 in 1960, rising to 10.5 in 1974, falling to 5 in 1983 (Tosics 1987 Figure 1). After a relatively slow and even development, peak housing output was attained four or five years earlier in the western countries than in the eastern ones, and the decline following this peak was more severe in the former than in the latter. The trends are not related to political changes but to the economic recession

(delayed by six years in the eastern countries)
which took place as a result of the oil price
shock (Tosics 1987, p. 62). Hemström (1986, p.
3) shows by more detailed analysis that the greatest
housing decline took place in those countries most
dependent on oil.

THE EFFECTS OF THE HOUSING SHORTAGE

Homelessness is perhaps the most obvious
effect of the housing shortage. According to the
estimates in an underground monthly in Poland,
that country had 500,000 homeless in 1985
(Bivletyn Dolnoslaski 1985). However, homelessness is exceedingly hard to define, and waiting
times for a flat and overcrowding are perhaps
better indicators of the housing problem. Another
indicator is the legal restriction on living in
many large towns. Also, there are occasional
expressions of public opinion by means of surveys
in the more open societies such as those of
Hungary and Poland, also by means of letters of
complaint (e.g. in the USSR and Romania), and very
rarely by means of violent street demonstrations
(the last was in Brasov, Romania, in 1987).
Even reliable data on waiting times are
difficult to obtain. Sometimes waiting times are
calculated by means of estimates of building rates
and marriages, but this fails to take account of
families "solving" their problems by other means,
such as through the private sector. In Poland,
where a third of the housing is private (Tables
1.9 and 1.10) such methods have led to estimates
of 20 year average waits for a dwelling for the
country as a whole and a 50 year wait in Warsaw
(Radio Free Europe Research 1988a). The
waiting time in Warsaw has also been estimated
as 26 years (Zwiakowiec 1986). This kind of
unsatisfactory estimate must be supplemented by
other information. For example, what is known
for certain is that in Poland 46 per cent of all
young married couples have a place of their own,
38 per cent share with at least one room for their
exclusive use, while 15 per cent are forced to
wait in workers' hostels pending housing
allocations (Zycie Gospodarcze 1984). The
situation was considered serious enough in the
late 1970s for a special plan to be introduced in
1977 (Brus 1982 p. 126). Moreover more than one
third of the 64,000 complaints received by the

Table 1.9: New dwellings by type of investor

		Percent		'000 dwellings (per cent)			
		1970	1980	1980		1986	
Bulgaria	A	29.3	52.0	38.6	(52.0)	23.92	(42.7)
	B	70.7	48.0	35.7	(48.0)	32.04	(57.3)
	C	38.2	32.6				
	D	32.5	15.4				
Czech'ovakia	A	19.3	23.4	52.1	(38.8)	21.32	(24.8)
	B	17.0	16.7				
	C	38.2	35.1)				
	D	25.5	24.8)	82.1	(61.2)	64.82	(75.2)
GDR	A	75.8	49.6	59.6	(49.6)	69.57	(58.3)
	B	20.9	34.6)				
	C	3.3	15.8)	60.6	(50.4)	49.76	(41.7)
Hungary	A	41.1	34.1	30.4	(34.1)	7.62	(11.0)
	B	58.9	65.9	58.7	(65.9)	61.81	(89.0)
	C	43.7	60.2				
	D	15.2	5.7				

A = State and co-op (Bulg); state (Czech); state (GDR);
 state (Hungary)
B = Private (Bulg); enterprises (Czech); co-op (GDR);
 private (Hungary)
C = Aided (Bulg); co-op (Czech); private aided (GDR);
 aided (Hungary)
D = Not aided (Bulg); private (Czech); not aided (Hungary)

Notes: (1) 1971 data;   (2) 1979 data;   (3) 1985 data.

Table 1.9 (continued)

		Percent		'000 dwellings (per cent)		
		1970	1980	1980		1986
Poland	A	23.5	18.1	39.4 (18.1)		38.2 (20.6)
	B	48.9	56.2)			
			)	177.7 (81.9)		146.8 (79.4)
	C	27.6	25.7)			
Romania		(1)	(2)			(3)
	A	86.5	93.6	186.6 (94.3)		97.0 (91.9)
	B	13.5	6.4	11.2 (5.7)		8.6 (8.1)
USSR						(3)
	A	76.0	82.5	1667.0 (83.2)		1638.0 (82.3)
	B	10.9	8.4)			
			)	337.0 (16.8)		353.0 (17.7)
	C	13.1	9.1)			
Yugoslavia	A	34.5	38.1			
	B	65.5	61.9			

A = State (Poland); state and co-op with state aid
   (Romania); state and co-op (USSR); state (Yugoslavia)
B = Co-op (Poland); private, with and without aid
   (Romania); employees (USSR); private (Yugoslavia)
C = Private (Poland); collective farms and rural population
   (USSR)

Source: United Nations (1980, 1986, Table 5).

Table 1.10: Tenure of all dwellings

			'000 dwellings (percentages)		
		Total	Owned	Rented	Others
Bulgaria	(1975) Total	2698593	2086813 (77.3)	611780 (22.7)	–
	Urban	1601504	1044233 (65.2)	557271 (34.8)	–
	Rural	1097089	1042580 (95.0)	54509 (5.0)	–
Czechoslovakia	(1980) Total	5359530	2395812 (44.7)	2237317 (41.7)	709659 (13.6)
	Urban	3599509	1061087 (29.5)	1869670 (51.9)	656294 (18.6)
	Rural	1760021	1334725 (75.8)	367647 (20.9)	53365 (3.3)
Hungary	(1980) Total	3719349	2549656 (68.6)	1122634 (30.2)	47059 (1.2)
	Urban	2049854	1057862 (51.6)	955361 (46.6)	36631 (1.8)
	Rural	1669495	1491794 (89.4)	167273 (10.0)	10428 (0.6)
Poland	(1978) Total	10935768	3965050 (36.3)	5619655 (51.4)	1351063 (12.3)
	Urban	6782023	1114439 (16.4)	4914460 (72.5)	754124 (11.1)
	Rural	4152745	2850611 (68.6)	705195 (17.0)	596939 (14.4)

Source: United Nations 1985, p. 309

Table 1.11: Demographic trends

	Births per 1000 population		Deaths per 1000 population		Average life expectancy in 1984 (males and females)[2]
	1960	1986	1960	1986	
Bulgaria	17.8	13.4	8.1	11.6	68.4
Hungary	14.7	12.1	10.2	13.8	65.1
GDR	17.0	13.4	13.6	13.4	72.4
Poland	22.6	17.0	7.6	10.1	66.8
Romania	19.1	15.8 (1)	8.7	10.9 (1)	66.9
USSR	24.9	20.0	7.1	9.8	69.6
Czechoslovakia	15.9	14.2	9.2	11.9	66.9
Albania	24.5 (3)	25.2 (4)	-	-	68.0[5]

Source: Council for Mutual Economic Assistance 1979, p. 10
Statisticheski Ezhegodnik Stran-Chlenov Soveta
Ekonomicheskoy 1987, p. 11
Marmullaku 1975, p. 107, Appendix Table 4

Notes: (1) 1985 data

(2) Data are from World Health Organisation 1985
for all countries except GDR and USSR for which
1985 and 1986 data respectively are from
Statisticheski Ezhegodnik Stran-Chlenov Soveta
Ekonomicheskoy 1987, p. 12

(3) 1960 data

(4) 1971 data

(5) 1970 data; the 1960 figure was 64.9 years; the
1938 figure was 38.8 years

Housing Policy in E. Europe and the USSR

Polish Communist Party's Central Committee in 1986 referred to housing, and housing issues were the focus of critical comment in individual consultations with party members in the weeks preceding the PUWP 10th Congress in June 1986, (Radio Free Europe Research 1986b).

In the USSR, homelessness is now starting to be discussed and publicised and several recent detailed accounts have been written in Russian newspapers (Trehub 1988). The first explicit article was in Izvestia on 10 February 1988. However, no statistics exist on homelessness in the USSR. A well-informed journalist recently suggested a figure of "hundreds of thousands" (Moskovskie Novosti 1988, p. 11) although this may be an under-estimate, considering the fact that there are 4.5 million officially registered alcoholics (Pravda 1987), between 20 and 22 estimated alcoholics (Radio Liberty Research Bulletin 1987), over 5 million mentally ill people (Argumenty i fakty 1987, p. 5), and, according to a secret MVD report that was obtained by Agence France-Presse, at least half a million known "parasites" (Le Monde 1984, p. 6). To put these figures and the wide ranges in perspective, homelessness in the USA was recently estimated as anything from 250,000 to 3 million (Wickenden 1985, p. 20). Trehub comments on the current situation:

> Ironically, the problem of homelessness in the Soviet Union is finally being acknowledged at a time when the prospects for alleviating it are bleaker than ever. The reason for this is the shift of Soviet enterprises and factories to economic accountability, (Khozraschet). In theory, Soviet enterprises have long been obliged to hire vagrants sent to them by the police and the job placement offices. In practice... this obligation is usually ignored. In the past, the danger was that vagrants and 'parasites' would disrupt labour discipline and prevent the collective from fulfilling its plan and getting its bonus. Today in conditions of Khozraschet when enterprises must pay their way or face bankruptcy, the disincentives to hire the down-and-out are even stronger. Gorbachev's attempt to make Soviet enterprises more efficient appears to have complicated the task of rehabilitating

the homeless. (Trehub 1988 p. 5)

This view, however, ignores the major policy changes now taking place in the USSR (see below). The housing shortage in the USSR also expresses itself in widespread illegal subletting at exorbitant rates. One co-operative block of flats on the outskirts of Moscow had 10 per cent of its apartments illegally sublet, mostly to young people. The rate for a one room flat in this building was 50 rubles a month, for a two room flat, 70 rubles a month. In Moscow proper the respective rates are 80-100 and 150 rubles a month (Trud 1986). The average Soviet family spends a little less than 3 per cent of its income on rent and utilities, i.e. 12 rubles a month (Narkhoz 84, pp. 428 and 432). As the average Soviet worker earns about 190 rubles a month, it can be assumed that these black market rents are paid by those with second, black market incomes of their own (Trehub 1986b, p. 2). All this is because in Moscow housing, as elsewhere in the USSR, is in very short supply: some people who were on the housing list in 1979 were still waiting to be housed in 1988 (British Broadcasting Corporation 1988a). In Tallinn, the average wait is four and a half years (British Broadcasting Corporation 1987e).

Increasingly the frustrations engendered by the USSR housing shortage are coming out in the open. Unfair allocation practices are now investigated and even discussed at CPSU Central Committee level - for example, 78 dwellings misallocated in Frunze City and 60 in Kharkov (British Broadcasting Corporation 1987b) and the illegal distribution of housing in Tashkent (British Broadcasting Corporation 1987c). Also, those in rent arrears are being criticised (British Broadcasting Corporation 1987d).

Another effect of the housing shortage is that there are many unauthorised, poorly-built residential buildings. For example Sumgait, with a population of 250,000, has a shanty town of 20,000 people (British Broadcasting Corporation 1988b). Other examples abound, and are increasingly being investigated, publicized, and criticised in the Soviet press (for some samples see Voznesenskaya 1986).

In Hungary the Research Institute for Mass Communications is one of several major Hungarian research institutes that conduct regular surveys of public opinion. One major survey conducted by the Institute in 1986 found that when asked what their most pressing problem was, 62 per cent said inflation, followed by problems obtaining decent housing (Farkas and Pataki 1987).

In Romania the strong attraction of the large towns led to the authorities drawing up a list of 'overcrowded centres' (later referred to as 'large towns') to which migration was strictly controlled. This list was never made public but Bucharest was, according to many references in the newspapers, at the top of the list, and the list seemed to comprise all towns over 100,000 population, of which there were 21 in 1986. Sometimes local authorities issue regulations that are even more restrictive than the national law. The 1976 law on the right to reside in large towns states, for example, that aged and ill parents may live with their children. But in practice old people are banned from moving to Bucharest and other large towns, even if their children are willing to care for them (Ionescu 1988, Maier 1986). Although no evidence has yet come up of the implementation of the plan, there has even been an idea to "resettle" pensioners from the large cities to the countryside (Gafton 1985).

Because of the deteriorating living standards in Romania, more and more people there are writing to party and other officials to complain and to press for action. In 1984, for example, 1,647,000 letters of complaint were written, a 16 per cent rise on the previous year (Scinteia Tineretului 1985b). No statistical analysis of these complaints has been published. However, housing is probably only one of several topics of complaint, and probably it is not one of the most pressing problems, compared with food and energy shortages, price rises, and other problems.

THE EFFECTS OF THE HOUSING SHORTAGE UPON DEMOGRAPHY

Eastern Europe and the Soviet Union have seen their birthrates fall drastically since the second world war. This decline worries manpower planners in these countries, because their economies are so labour intensive. It also threatens ethnic identities in small countries such as Hungary, as well as in the Soviet Union where Muslim birthrates remain very high. These declining birthrates have been the result of four factors: (a) poor housing, (b) poor living conditions, (c) the increased employment of women, and (d) the encouragement of abortion as the major means of birth control (Kosinski 1977). Because of fears about falling birthrates the abortion regulations have been tightened somewhat. However, with the exception of Poland the existence of modern contraceptive methods as an alternative to abortions, together with reluctance by medical practitioners to refuse permission for abortions has meant that birthrates have continued to fall (see Table 1.11). This fall has continued despite national programmes to ease the financial burden of motherhood by providing progressive bonuses for second and third children, monthly income subsidies, and maternity leave with full pay as well as granting priority in housing allocation (Miskiewisz 1986a, Robinson 1984).

Table 1.11 also shows that death rates have risen everywhere except the GDR. The resulting life expectancy figures are alarming. Hungary has the worst life expectancy of 65.1 in 1984. This compares with 69.1 years for Portugal, which has the lowest average life expectancy in Western Europe. Although housing is not directly to blame, health officials in at least two of these countries, Poland and Hungary, have admitted that the health of their citizens has deteriorated because of the stress caused by the difficulties of daily life, poor nutrition, smoking, alcoholism, and long-term nervous strain. One of the stress factors is overcrowded and inferior housing. Other factors blamed are rapid industrialisation and inadequate environmental protection facilities leading to deaths from respiratory diseases, economic difficulties leading to shortages of medical personnel and facilities, and shortages of cleaning and disinfectant products (Miskiewicz 1986b).

HOUSING POLICY SINCE 1945

The Second World War had a devastating effect upon the housing stock in the region. Added to this was a large influx into the towns during and immediately after the war. In all countries in the region the level of investment was kept low and this led to a desperate situation, even in the high-priority areas of new industrial construction. The new economic and political structure after 1945 led to many payments in kind, partly to supplement a relatively egalitarian wage distribution, and partly to reward key workers such as managers and technological leaders. Of crucial significance in the system of payments in kind was the allocation of housing, which became a deeply entrenched institution.

Agencies of central or local government took over control of most of the housing stock in the towns. By 1950 house allocation was in the hands of housing administrators. Rents were kept at a low, sometimes even nominal, level, except in Yugoslavia. In Yugoslavia the bulk of housing plus business premises and building plots were only nationalised as late as 1958. Thereafter rents for housing rose from 16 to 40 dinars per square metre, on the argument that low rents were deemed inconsistent with distribution according to work (Sirc 1979, p. 63). New dwellings in the towns (except in Yugoslavia) were limited to state-built dwellings and were distributed almost free according to political and economic criteria. The share of privately constructed dwellings was limited to the countryside, where an acute shortage of building materials and in some cases the ideological tastes of officials meant that conditions were unfavourable to private house-building.

However, by the mid-1950s there was an increase in house-building (Kaser 1986, p. 66). Also, in most countries the contribution of the private sector was growing. In Czechoslovakia in 1956 privately built dwellings constituted almost half the total, compared with one seventh in 1953. In Poland the share of private construction rose from less than 5 per cent in 1953 to almost 40 per cent in 1956 (Kaser 1986, p. 65). Most of the private buildings were in rural areas and had little effect on urban shortages.

During the years 1957-65 housing policy underwent several institutional changes. Since

the end of the war industrialisation and
urbanisation had not been matched by an increase
in housing. Housing shortages were not as bad in
Eastern Europe as they were in the USSR, due to a
smaller migration to the towns and to less wartime
devastation. The problems of the USSR were made
worse by Stalin's refusal to invest properly in
housing in the period before the war. Since
investment in housing was generally not enough to
sustain the existing population, the extra
pressure put on the housing stock necessitated the
subdividing of apartments into increasingly
smaller units. This meant that several families
would be gathered together and squeezed into what
once had been an apartment for one family. The
prewar norm became one Soviet family per room,
with the toilets and kitchen shared. Whereas the
amount of living space in urban areas was 5.7
square metres in 1926, it fell to 4.5 square
metres in 1940. The target had been set at 9
square metres per person, a target which has only
recently been surpassed (Goldman 1983, p. 20).
However, housing shortages were severe enough
everywhere to cause significant social problems.
The average number of people per room was 1.8 in
Bulgaria (1956), 2.0 in Czechoslovakia (1961), 1.5
in Hungary (1963), 1.7 in Poland (1960), 1.4 in
Romania (1966), 2.4 in Yugoslavia (1961), and a
relatively luxurious 0.9 in 1961 in the GDR (Kaser
1986, p. 132). In the GDR wartime devastation had
been relatively less severe and the housing
situation after the war continued to improve
because of the declining population. By 1960 the
rate of house-building per head had picked up in
all countries (Kaser 1986, p. 133). After the
death of Stalin in 1953 an element of the "New
Course" which followed in the USSR and Eastern
Europe was a revival of private house
construction. This revival of private house-
building took place mainly in the countryside and
reflected greater peasant confidence and
prosperity, and a greater availability of building
materials. Private construction peaked in
1957-1960 in many countries after which it
declined simultaneously with a steep rise in state
and co-operative housing. In Yugoslavia the
private sector continued to expand after 1960
together with a static state sector, while in
Hungary after 1960 the private sector expanded
with government aid.

One problem afflicted all countries alike. The low investment in housing by the state, and the artificially low rents in private accommodation, meant that there was no money or no incentive to repair and maintain housing. Repairs went undone and maintenance work lapsed, and the physical quality of the housing stock everywhere declined (Morris 1984, p. 137).

This period also saw the ending of the hitherto uniform housing system (Kaser 1986, pp. 134-136). The traditional Soviet-style system was based on a state monopoly of the construction and distribution of new housing, financed from non-reimbursable budget grants. The full capital cost, the cost of land and maintenance costs, were all borne by the state. The development and implementation of housing allocation was in the hands of administrators and political and economic factors meant that important managers and technicians received the limited amount of housing available, despite their already higher incomes and the fact that such housing was allocated almost free.

In Yugoslavia rents were gradually moved towards economic levels, at least sufficient to cover maintenance costs and depreciation, and there was a trend towards owner occupation. Private house-building or private house-buying in publicly-built estates spread widely, with the help of cheap bank credits and personal loans provided by funds established by enterprises and local authorities. Interest rates and the availability of credit became more and more important methods of influencing house-building (Kaser 1986, p. 135).

Yugoslavia represents the most extreme mixing of market forces and administratively-allocated housing. Yet all countries saw some increase in the operation of the market, and of private sources of finance. The old system has survived longest in the USSR, Bulgaria, the GDR, and Czechoslovakia. In the GDR rents have remained stable since the late 1940s. In Bulgaria, there has been no change of rents since 1967. In Czechoslovakia, rents valid today were fixed in 1964. In the USSR the regulation of rents was set up in 1928. The influence of market forces and of private finance there have been deliberately restricted until recently. The GDR began at one stage in the late 1950s to increase the ratio of state to co-operative housing in favour of the latter but then during the 1960s this policy was reversed and the emphasis once more placed on state housing. In Czechoslovakia, Hungary, and Poland, abandonment of the Soviet-style system took place the earliest. The change was justified in various ways. From the social point of view the old system rewarded the better-off, and this was seen as unjust. The economic justification was that private house-building and house-buying using credit were efficient means of absorbing demand and lessening the burden on an overstretched state. These arguments are still topical in the USSR where an insufficiency of paid-for goods and services means that savings bank accounts have dramatically increased. Recently some expensive co-operative apartments have been allowed to be sold on the open market, and the size and number of these savings accounts are indicated by the fact that such dwellings sell fast at prices which are very high by Soviet standards (Goldman 1983, p. 99). In 1979 a Pravda editorial commented, "Savings bank deposits have almost trebled in eight years, and the cash incomes of the population have increased much faster than the supply of goods and services" (Pravda 1979). Other arguments included the more economic use of the housing stock, and the increasing of personal choice. In all three countries there came into being two separate systems - the old state-run system of almost free, administratively-allocated housing, and privately built or bought housing, the latter usually funded with credit from enterprises or banks. In Hungary an important extra type of housing has been that built by the state for sale. In Czechoslovakia in 1964 all rents were increased to cover maintenance costs, and from 1959 to 1964 there was a rapid expansion of the co-operative sector until it was equal in size to the state sector. Two main types of co-operative were formed: the tenants' co-operative, without property rights to the dwelling, and the owners' co-operative. In Poland the system developed in much the same way as in Czechoslovakia, with a rent increase in 1965. In Poland (as in Hungary) state-allocated dwellings required a lump sum payment or deposit. From 1956 to 1965 the co-operative sector grew rapidly until it provided a third of all new dwellings. As a result of these changes the contribution of private finance to the housing effort rose significantly, although by

Western standards it was still very low (Kaser 1986, p. 136). For example, almost 55 per cent of the construction costs of housing for the non-agricultural population in 1966-1970 were borne by the population in Poland, compared with 46 per cent in 1961-1965 (Kaser 1986, p. 158). Recently Poland has introduced rent reforms, the third phase of which was completed in 1985. The goal of the reform was to gain enough proceeds from rents to cover the cost of operating and maintaining housing. Rent differentials (depending upon the amount of equipment in the dwelling) have been increased. In the USSR by contrast, state housing is still allocated according to political and economic criteria. The party official and the factory manager will probably get preferential treatment in obtaining housing. However, given the general housing shortage, it is unlikely that this flat is larger than that of other workers.

Even in the countries where rents have remained stable, each country has its own individual approach to the setting of rents. In the GDR for dwellings constructed prior to 1945 the rents paid in 1944 have remained unchanged. The price regulations state that prices or fees for utilities and services above those valid on the day when the regulation came into force may not be demanded or paid without special appropriation of the price authorities. As a social measure the rent freeze of 1944 has been worthwhile. However, it has also meant that unjust differences of rents, e.g. between 0.5 marks per square metre per month and 1.3 marks per square metre per month have been maintained, often within the same building. For a post-1945 dwelling the regulations stipulate rents of between 0.9 and 1.05 marks per square metre per month, according to the type of equipment provided. In case of additional technical equipment, such as central heating or central hot water supply, the same surcharges are levied as in newly-constructed dwellings. Improvements may increase the monthly rent by up to 0.5 marks per square metre per month. But the basic rent may in no case be exceeded. Thus in the GDR rents are independent of family income. The price for the dwelling (i.e. rent) accounts for some 3 to 5 per cent of the gross family income. Families with three and more children receive rent allowances to the amount of that part of the rent that exceeds 3 per cent of the gross family income.

In Bulgaria, the rent system is to a great extent based upon criteria related to dwelling quality. By means of extra charges and price reductions the rent may be varied from +32 to -50 per cent. The range of influencing factors for calculating the actual rent is very wide. Such factors take into account the dwelling's location, the settlement size, and the availability of services in a town. The rent level depends upon the dwelling's location within the building - for example rents are 10 per cent less if above the fifth storey.

In Czechoslovakia, rent calculation relies on size, quality, and the type of equipment in the dwelling. Dwellings are divided into four quality classes. Generally speaking, class 1 corresponds to dwellings built since 1960, which are fully equipped with modern facilities including central heating. Rent calculation is done on the price per square metre. Dwelling space and ancillary space are distinguished, with two different prices. The price of dwelling space is applicable to rooms with a floor area of more than 8 square metres and with daylight illumination, natural ventilation, and heating. For kitchens, the dwelling space is applicable only to the portion above 12 square metres of floor surface. The cost of communal facilities is added to the rent. The cost of potable water, electric power, gas and heating is calculated according to actual consumption. These items do not form a part of the rent paid.

In Poland, rent is calculated according to living space per square metre. Differentiation is derived from the type of construction of residential buildings, the provision of technical equipment, and the location of dwellings within the building. The provision of technical utilities is the overriding factor. For dwellings without water supply indoors, the rent is reduced by 30 per cent. The following raise the rent in 30 per cent increments: inside toilet, bathroom, gas supply and central heating. Dwellings above the fifth storey have a 30 per cent lower rent. This rent calculation is interesting in that it results in a relatively wide range of rents, and also because it operates irrespective of the size of the settlement or its location in the settlement. Nor does it matter if the dwelling is located in a new or an old house.

Housing Policy in E. Europe and the USSR

In the USSR the rent system is based on a uniform and extremely low rent per square metre of dwelling space. Rents are calculated on the basis of "equal price for equal dwelling quality". The rent level is influenced by hygienic and technical aspects of dwellings, by the location of a dwelling within a building, and by criteria related to the dwelling's equipment. But the main factor in rent determination is the family structure, social aspects, and the social merit of the tenants.

The rent system in Hungary is interesting in that the rent is calculated on 1.8 times the building cost, with reductions for amenities lacking.

Therefore, while many countries have kept rents low and stable, criteria for calculation have varied between quality of dwelling (the majority of countries) and social merit of the tenant (USSR). Also, while some countries such as the GDR have explicitly kept rents below a certain proportion of family income others (such as Hungary) have started off with the initial premises of recouping a certain proportion of costs.

The relation of rents to costs has been one of the dimensions on which the countries of the region have begun to differ. Generally speaking, the difference between the revenues from rents, services, charges and fees, on the one hand, and dwelling costs on the other hand, is primarily financed from state funds. More than 50 per cent of the running expense for the housing stock has to be covered by the national budget. In some of the countries, rent reforms are currently under way to meet the cost of management and maintenance by rental revenues. In the GDR, one third of all costs are covered by rents, two thirds by the state. In the USSR the rent of 13.2 kopecks per square metre is extremely low, contributing only a very small amount towards housing costs. In Czechoslovakia rents amount to less than half of maintenance and management costs - the proportion would be very small indeed if construction costs were also brought into the calculation. In Poland rents cover 22 per cent of total housing costs. In Hungary, while reforms aim to shift the financial burden away from the state, rents do not contribute towards construction costs.

Another important aspect of state housing provision is the relation of rent to income.

Generally countries have attempted to keep a family's rent at between 2 and 5 per cent of average family earnings. The figure is between 3 and 5 per cent in the GDR. In Czechoslovakia, the level of rents has been 2 to 3.5 per cent of average family income. (The figures are 4 to 7 if utility services are included.) Families with children pay lower rents (5 per cent less for 1 child, 15 for 2, 30 for 3). In Poland, fluctuations in the ratio of rents to income have been due to macro-economic factors. In 1965 the rent for a dwelling of 50 square metres was 10 per cent of average individual income. This fell to 2 per cent by 1982 due to inflation. The first stage of rent reform in 1983 took the figure to 5 per cent. Subsequently later stages in the rent reform process have led to a higher rent-income ratio. Besides attempting to raise rents, the state wishes to reduce under-utilisation in Poland. The rents for dwellings where the occupancy standard of 10 square metres living space per person is not observed are due to receive higher than average rent rises. In the USSR the average rent as a percentage of individual income was 2.7 per cent in 1986, although this figure was due to rise to 5 per cent. These extremely low rents are reduced still further for those with social merit (war heroes etc.). In Hungary comparable figures were 3.2 per cent (1981), 4.7 per cent (1985), 5.4 per cent (1988).

Investment in the region has traditionally favoured industry over housing, although the balance has changed slightly in recent years (see above). This has meant that the creation of new jobs has always tended to outpace the construction of dwellings and has resulted in "underurbanisation" (Konrád and Szelényi 1971) especially in Romania (Ancuta 1971) and Hungary, where large numbers commute from the countryside to urban workplaces. However,

> although these peasant workers are doubly exploited - they spend long hours commuting and are deprived of the somewhat better, government-subsidised urban infrastructure, including government-built housing, better shops and schools - they often turn their disadvantages into advantages. They maintain semi-

43

> legal or illegal private enter-
> prises, and in the long run live
> better and may secure more autonomy
> for themselves, more highly skilled
> urban colleagues. (Szelenyi 1988 Preface)

Recently many important policy initiatives have begun in the USSR. One dimension of recent Soviet policy has been to attempt to increase efficiency in house-building and construction generally. In the past the output of the construction industry in the USSR was insufficiently linked to completions and too closely linked with the fulfilment of plans expressed in ruble value of work done. As many Soviet critics have pointed out, this meant that management was rewarded for spending more, using expensive materials, and was under no pressure to keep costs down. Also, there was insufficient incentive to complete dwellings, so that construction times were long, despite repeated campaigns and exhortations to avoid scattering of investment resources and to concentrate on a limited number of sites. Typically the effect of scattering (raspylenie) was to cause the construction time to be long, hence the process known as dolgostroi or "longbuild" (Nove 1982, pp. 25-26). As a reaction to these problems, new economic performance indicators were introduced in the late 1970s, relating to increases in labour productivity, profit, reduced use of raw materials, and net value of work completed (Nove 1982, p. 39). Similar criticisms have been levelled at the construction industries in other East European countries - for example, Hungary (Vajna 1982, pp. 200-204) and Bulgaria (Feiwel 1982, pp. 230, 236.)

Another dimension of recent Soviet thinking has been towards decentralisation. In a speech in 1986 Mr. Gorbachev complained of excessive centralisation which limited the ability of local soviets to deal with housing problems (Hill 1987, p. 46).

One aspect of the move towards decentralisation has been the encouragement of private house-building in the USSR, after years of neglect. In February 1988 the CPSU Politburo adopted a resolution aimed at roughly quadrupling the volume of private house-building by 1995. The resolution removed legal restrictions on private construction and raised the amount of money Soviet citizens can borrow to build or buy a dwelling (Pravda 1988a). In an interview recently Mr. Batalin, chairman of the USSR State Committee for Construction Affairs (Gosstroi) said that the Soviet Union had long underestimated the importance of private housing, and referred favourably to the high home-ownership levels in Hungary, Bulgaria, Czechoslovakia, and Poland (Pravda 1988b). He acknowledged that while Soviet citizens have long had the right to self-build, in practice they were hampered by building material shortages, bureaucratic and legal restrictions, and the envy and resentment of neighbours (Izvestia 1985). This led to a fall in the amount of private house-building, from 53.8 million square metres (49 per cent of total construction) in 1960 to 17.4 million square metres (15 per cent) in 1986 (Narkhoz SSR za 70 let 1987). One of the legal restrictions which may be lifted is the ban on private house-building in cities above 100,000 population (Sobranie postanovlenii pravitel 'stva SSSR 1962) which has always been ignored by the nomenklatura and which, with glasnost' and the emphasis on "social justice", is increasingly attacked in political speeches and in the newspapers. However, besides legal obstacles, money is a problem. A privately-built house costs 15-25,000 rubles, while the average Soviet worker earns 196 rubles a month and has a bank account of 1361 rubles (Narkhoz SSSR za 70 let 1987, pp. 431, 448). Even if these figures conceal wide variations it seems that large government credits would be necessary for the volume of private house-building to rise significantly. So the February 1988 resolution consequently makes available loans of between 3,000 and 20,000 rubles, with repayment terms of 50 years (rural areas) and 25 years (cities) compared with 10 years formerly. Such loans are now available for house-building and house improvement. An innovation designed to reduce the shortage of raw materials is the planned construction of a number of small, rural, brick factories. Also planned is the creation of building material co-operatives, such as the one which took over the failing brick factory in Sverdlovsk in October 1987 and has already started to make a profit. Also, economic incentives are to be applied to overcome the reluctance of contractors to take on private construction jobs. Private house-building is from now on to count towards fulfilling plan tasks. So the more that private housing contractors build, the more their bonus will be. Also, enterprises are allowed to build their own houses and sell them cheaply to their workers, a measure which in slightly different form has existed since 1980 (Sobranie postanovlenii pravitel 'stva SSSR 1980) but which until 1985 had only been taken advantage of by one factory (Trud 1985).

Although these innovations are important, the co-operative sector has not yet received as much encouragement. Building co-operatives still face strong resistance, from those bureaucrats who dislike any innovation, and from competition-shy state building enterprises (Izvestia 1988b).

Also, efforts are being made to increase productivity in the USSR state house-building industry. During 1987, for example, Mr. Melnikov, head of the construction department of the SPSU Central Committee, claimed that "scattering" (see above) had been reduced by concentrating investments on 12,000 fewer sites (British Broadcasting Corporation 1988c). However, according to several articles in Pravda in 1986 there is also much resistance to innovation at the design stage (Trehub 1986c).

There have been policy innovations in Bulgaria recently too. In July 1985 several measures to encourage the birth rate were introduced. Newlyweds are now entitled to loans of up to 15,000 leva for purchasing a dwelling, with 30 years payback time. If the family has a second child within four years after the birth of the first child, 3,000 leva of the debt is cancelled. Another 4,000 leva debt is cancelled after the third child (Radio Free Europe Research 1984, p. 3). Also, a decree has been issued (Darzhaven Vestnik 1986) which prohibits the construction of housing more than three storeys high in small towns and villages and encourages private construction of individual family houses in the suburbs. Also, the decree provides incentives to encourage private citizens to undertake the joint construction of condominiums to ease the housing burden borne by the state (Nikolaev 1986). The existing restrictions on the size of individual one-family houses and building plots in villages are to be revised so as to allow for increased agricultural production in private gardens. In Sofia and other cities the average height of new buildings is to be reduced by two or three storeys, and from January 1988 no one-room flats (even those with a separate kitchen in Sofia itself) will be built, in a drive to increase the average size of flats. Living in the outlying suburbs is being encouraged: people who build new houses in the suburbs and who vacate a city apartment will not lose the right of city residence should they wish to move back at a later date. This is an important provision given the very restrictive policy on issuing residence permits, especially in Sofia. However, earlier calls on those owning houses in outlying suburbs to use them as permanent homes and commute to work do not seem to have been successful in the past, largely because of the inadequate communications facilities - city dwellers continue to use their "villas" for weekends only (Nikolaev 1986 p. 8). The policy to reduce densities may come up against criticism from those who feel Bulgaria's arable land requires protection - such arguments often appear in the country's newspapers.

The movement away from state finance has occurred also in Poland, in the form of increased rents. The amount of rent increase has varied according to the type of housing. At the end of 1984 almost three quarters of urban housing was in the form of co-operatives, which requires occupiers to make substantial deposits and offers cheap credit over 60 years. In 1985 co-operative rents were increased by the government (to between 10 and 20 per cent of an average worker's income) as were rents on state-owned dwellings, and in 1986 workers' compulsory contributions to the welfare and housing funds of their enterprise were almost doubled by the government. The critics of government policy comprise economists and trade unions. A consensus exists among the government's critics within Poland on how policy should be changed. Credit should be decentralised and freed from government control. Housing provision should be self-financing. Private house-building should be encouraged. Small private firms should be freed of bureaucratic and legal restrictions. An encouraging sign is that so much statistical information and careful argument has been publicised for several years (Radio Free Europe Research 1986b).

CONCLUSION

In all the countries of the region the housing shortage has worsened. Between 1950 and 1975 the worsening was due to a fall in the proportion of all investment which went into housing. Between 1975 and 1986 the proportion devoted to housing actually rose in most countries, but general economic conditions were such that total investment fell or was stagnant. Because of this the housing shortage for 1986 (as estimated above in Table 1.2) stood at a little over 35 million dwellings, an awesome number and 8.3 per cent of the region's 1986 population of 415.7 million.

The result has been increased overcrowding, larger numbers in shared accommodation and dormitories, and longer waiting times for dwellings. One of the trends related to these problems has been the falling birth rate.

Since the war housing policy has gone through several phases but it would be true to say that the influence of socialist ideology has declined in the last five to ten years, with a decline in the importance of need and a rise in the importance of ability to pay as criteria of housing provision. Some countries have for an even longer period managed to mix political ideological objectives with pragmatism. One country above all others has continuously mixed state and private sources of finance and types of organisation - Yugoslavia now has more than 30 years experience of an elaborate system of local bank finance self-managed building co-operatives and central government controls. In the most recent years the major changes have been those in the USSR - towards reduced building costs, greater efficiency, an increase in the size of the private building sector, and a promised encouragement of co-operatives. The encouragement of the private sector has been popular with several governments since building rates peaked around 1975, for all governments in the region are seeking ways of reducing the burden on the state of house-building. A measure that many governments have used but which has obviously not been popular has been the attempt to make house occupiers bear a greater proportion of the building costs. This has proceeded farthest in Yugoslavia, but Poland and Hungary have also made significant strides in this direction.

The present time (1988) could well in retrospect be viewed as the turning point for housing policy in Eastern Europe and the Soviet Union. This above all else should be emphasised here, because it is not the purpose of this book merely to state that a housing problem exists in the region. It is hoped that a clear picture will emerge to readers of the details of this new housing policy. In some countries (Yugoslavia) the policy has already begun (since 1987). In some countries (the USSR) some small steps have been taken (since 1986) and much larger steps are being seriously debated. In other countries (Hungary) important new policies are about to be implemented (late 1988). Other countries (Czechoslovakia, Bulgaria, and Poland) have introduced some changes, while others (Rumania, Albania, the GDR) remain apparently uninterested in the new policies.

Moreover, the new housing policies, it will be suggested, arise from wider economic concerns, and also display a remarkable consistency across the region in their attempt to increase the role of the "market" versus administrative allocation of housing, and to increase the role of private funding.

Broadly speaking, the postwar period during which the governments of the region took housing seriously, that is 1960-1975, was one during which it was assumed that supply side measures would be adequate. This was because the problem was seen largely in quantitative terms - it was optimistically believed that the acute housing shortages were the result of past neglect due to prewar under-investment, war damage, or to early postwar concentration on heavy industrial investment. Despite the fact that housing investment accounted for a falling proportion of total capital investment between 1950 and 1975, the large increase in overall prosperity in the region during this period led to a rise in the annual numbers of dwellings constructed.

The supply side approach was based on a number of assumptions. One was that industrialised building methods and high rise housing would be cheaper and quicker than traditional brick-built, low-rise housing. The organisations to produce the components and assemble them on site needed to be large - employing large numbers of unskilled workers - with powerful political contacts in the customer housing departments. Such organisations often exerted (and still exert) a strong monopolistic influence over control of (and hence simplification of) design and price (amounting to protectionism of increasingly out-of-date technology and methods). The lack of participation of the eventual customers - the residents - or at least of any organisation (such as a local authority to represent their interests) was one factor in escalating construction costs and poor quality work. The result was a higher demolition rate and even higher costs to repair poorly constructed dwellings.

Another assumption was that the state should be the major supplier. However, because of poor quality work a demand arose (unmet by the state construction industry - itself geared up to construction more than to running repairs) for building services and materials which could only be supplied illegally.

Another assumption was that administrative allocation by large centralised bureaucracies was required. The agencies included workplaces, local authorities, and ministries. Resource management decisions for the supply of dwellings to these agencies by state construction companies were often influenced by political considerations and often need factors were not easily ascertainable or did not make themselves easily apparent. Administrative allocation, because of the persisting shortages, led inevitably to housing playing the role of wage supplements and political reward.

Criticising these assumptions with the benefit of hindsight is easy. Also, mistaken assumptions influenced Western policy makers in a similarly pathological way. Some criticisms emerged quite early on and led to reformulations of supply-side policies - more attention should be put into rehabilitation to reduce demolition rates, more emphasis on quality to reduce repair costs, to regularise illegal building workers and to reform the construction companies.

However, since 1975 there has been a major shift in policy thinking in the region (except in Albania, Romania, and the GDR). This shift can be characterised by the need to become more sensitive to the demand-side, to the needs of the customer. Whether or not this unusual interest in 'market-oriented' housing policy is a result of similar

shifts (though of course greatly offset to the
right in the political spectrum) in Western
political ideas is difficult to answer. What is
clear is that the delayed economic recession
affected the region in the mid- and late- 1970s
and the most noticeable effect from the point of
view of housing policy was a sharp fall in state
house-construction rates after 1975. This was
followed by a major reformulation of the role of
the state in housing supply, and consequently
ideas and policies to increase the contribution of
private funds and private effort in house-
building (sometimes via more ideologically
acceptable house-building co-operatives).

The swing to the private owner and private
builder has occurred several times in the past -
either at a time when state funds were unavailable
or at a time of political liberalisation. In the
past such swings have been short-lived. Is the
present change only a temporary one therefore?
This is really a part of a larger political
question about the likelihood of success of the
new policies that have emerged in the USSR since
1986 so that the question must remain unanswered
here. Besides, this economic liberalisation is
proceeding relatively independently in Yugoslavia
and Hungary - two of the countries where the new
housing policies are being developed. Despite
uncertainties about the future success of reforms
in the USSR (such as gloomy comparisons with the
short-lived reforms in nineteenth century Russia),
the changes there are not just pragmatic reactions
to current economic stagnation. The marketisation
concept in fact fits into a large web of ideas
("new" only in the sense of being spoken of by
leaders so frequently). Policies are being
suggested in the USSR which emphasise the
importance of paid-for services (otherwise why
work harder? - thus housing should be paid for
rather than being almost free), of openness (in
criticising, for example, abuses in the housing
allocation system), of social justice (better
flats should have higher rents), and of
democratisation (local councillors should be more
responsive to the housing needs of their
constituents).

What does this new housing policy consist of?
First, the individual owner or individual house-
builder should be encouraged to put his money or
effort into housing. Home-ownership is
traditionally high in the Balkans. But over and

above this, there has been increasing encouragement of the private sector in Yugoslavia, the USSR, Hungary, and Poland. In these countries also there has been greater encouragement of co-operatives, though often government policy is more guarded and ambiguous - in the past co-operatives have been the subject of investigations of corruption and profiteering, and have given rise to popular resentment due to their middle-class connotations.

Second, there have been moves to sell off state flats to sitting tenants - directly to private individuals in Czechoslovakia, Yugoslavia, and Hungary - indirectly (it is suggested) via local soviets or co-operatives in the USSR. At present these moves have not been spectacularly successful, mainly because tenants pay such low rents at present.

Third, there have been suggestions for raised rents and credit rates. In Yugoslavia it is intended that rents will reach "economic" levels by 1990. By the beginning of 1989 it is likely that mortgage rates for new borrowers in Hungary will rise from 6 to 18 per cent, and large rent rises are also probable. In Poland rent rises (relative to inflation) began in 1985, when co-operative rents were between 10 and 20 per cent of the average worker's income. Although low rents play such an important political role in the USSR, suggestions for raising rents are a central feature of current debate on housing policy there. A feature of historically different rent levels will be the abandonment of "universalistic" policies for "particularistic" ones and the acceptance of the "poor" - in Yugoslavia now there are special housing subsidies for the poor, rebates are being suggested in Hungary, and a similar policy may prove inescapable in the USSR. Such changes also represent a reduced role of planners and administrators - there is now much discussion about the increased "monetisation" in the USSR. Similar changes have already taken place, rapidly and recently, in Yugoslavia - new housing investment decisions are now made via banks based on enterprise (or individual) credit-worthiness.

REFERENCES

Ancuta, D. (1971) 'Clasa muncitoare in procesul favririj societatii socialiste multilateral dezvoltate', Lupta de Clasa, No. 9, pp. 10-17

Andrusz, G. (1988) Personal communication

Argumenty i fakty 1987, No. 10

Bivletyn Dolnoslaski, 1985, No. 6 (64), July

British Broadcasting Corporation (1987a) "Russian architects' Congress on Crisis in Soviet Architecture", Summary of World Broadcasts, SU/8517/B/4, 16th March 1987

_____ (1987b) "Central Committee discusses housing allocation violations", Summary of World Broadcasts, SU/8691/B/3, Part 1 USSR, 6 October

_____ (1987c) "Measures taken over "undesirable" housing practices in Tashkent", Summary of World Broadcasts, SU/0030/B/12, 19 December

_____ (1987d) "Millions owed by rent-debtors", Summary of World Broadcasts, SU/8686/B/7, Part 1 USSR, 30 September

_____ (1987e) "Problems of co-operative flats in Tallinn", Summary of World Broadcasts, SU/8664/B/6, 4 September

_____ (1988a) "Moscow City Soviet discusses housing problem", Summary of World Broadcasts, SU/0098 B/3, Part 1, USSR, 12 March 1988

_____ (1988b) "Tasks facing construction industry with introduction of self financing", Summary of World Broadcasts, SU/0040 B/6, Part 1, USSR, 5 January

_____ (1988c) "Failures of leadership in Nagorno-Karabakh Crisis", Summary of World Broadcasts, Part 1 USSR, SU/0113/B/1, 30 March

Brus, W. (1982) "Aims, methods and political determinants of the economic policy of Poland 1970-1980", Chapter 4 in The East European Economies in the 1970s (editors) A. Nove, H-H. Hohmann, G. Seidenstecher, Butterworths, London

Campbell, R.W. (1974) Soviet-type Economies, Macmillan, London

Council for Mutual Economic Assistance (1979) Statistical Yearbook, English Language Edition, Statistika, Moscow

Darzhaven Vestnik (1986) "On creating conditions for the large scale participation of the population in housing construction", No. 92, 28 November (Decree No. 65 dated 6 November)

Directia Generala de Statistica (1986) Anuarul Statistic al Republici Socialiste Romania 1985, Bucharest

Donnison, D. and C. Ungerson (1982) "Housing in Eastern
Europe" in Housing Policy, (editors) D. Donnison and
C. Ungerson, Penguin, Harmondsworth
Farkas, K. and Pataki J. (1987) "The 8th year of seven
lean years", Jel Kep, No. 2, pp.54-157
Feiwel, G.R. (1980) "The standard of living in centrally
planned economies of Eastern Europe" Osteuropa
Wirtschaft, Vol. 25, Part 2, P.73-96, (especially
p. 91-93)
_____ (1982) "Economic development and planning in
Bulgaria in the 1970s", Chapter 7 in The East
European Economies in the 1970s, (editors) A. Nove,
H-H. Hohmann, G. Seidenstecher, Butterworths, London
Gafton, P. (1985) "Plans to resettle pensioners in the
countryside", Radio Free Europe Research, Romanian
SR 85/14 Vol. 10, No. 41, Part 1, 1 October pp. 3-5
Goldman, M.I. (1983) USSR in Crisis, W.W. Norton, New York
Hanson, P. (1986) The Consumer in the Soviet Economy,
Macmillan, London
Hemström, E. (1986) "Housing construction in industrialised
countries 1945-85", Paper prepared for the IRCHP
Conference, Gävle Sweden, 10-13 June 1986
Hill, R.J. (1987) "State and ideology", Chapter 2 in
The Soviet Union under Gorbachev, M. McCauley
(editor), Macmillan, London
Ionescu, D. (1988) "The mirage of residing in Bucharest",
Radio Free Europe Research, Vol. 13, No. 17, Part II
Romanian SR 6/88, 29 April, pp. 13-15
Izvestia (1985) "Na peske...", November 21
_____ (1988a) "Ostatochnyi printsip 'esche v ostatke",
January 30
_____ (1988b) "Gosudarstvennyi zakas", February 17
Kaser, M.C. (editor) (1986) The Economic History of
Eastern Europe 1919-1975, Vol. III, Clarendon Press,
Oxford
Konrád, G. and Szelényi, I. (1971) "A Késleltetett
városféjlödés tarsadalmi konfliktusai", Váloság,
No. 12, pp. 19-35
Kosinski, L.A. (1977) Demographic Developments in Eastern
Europe, Praeger, New York
Lad, (1987) 9 August
Le Monde (1984) "Cinq cent mille 'parasites'", October 20
Lengyel, E. (1985) "Joint tenancies aggravate serious
housing problems", Radio Free Europe Research, Vol.
10, No. 33, Part 2, 16 August Situation Report
Hungary, 9/85, pp. 7-10
Maier, A. (1986) "Pensioners barred from moving to large
cities", Radio Free Europe Research, Romanian SR
5/86, 18 April, pp. 15-17

Marmullaku, R. (1975) Albania and the Albanians,
   C. Hurst and Co., London
Mieczkowski, B. (1975) Personal and Social Consumption in
   Eastern Europe, Praeger, New York
Miskiewisz, S. (1986a) "Demographic policies and abortion
   in Eastern Europe", Radio Free Europe Research, Vol.
   II, No. 51, Part 1, 19 December Background Report
   179/86
_____ (1986b) "Life is shorter in Eastern Europe",
   Radio Free Europe Research, Vol. II, No. 45, Part II,
   7 November, Background Report 152/86
Morris, L.P. (1984) Eastern Europe since 1945,
   Heinemann, London
Morton, H.W. (1979) 'Housing problems and policies
   of Eastern Europe and the Soviet Union',
   Studies in Comparative Communism, Vol. XII, No. 4,
   Winter, pp. 300-321
Moskovskie Novosti (1988) March 13
Narkhoz 84, (1985), Moscow
Narkhoz SSSR za 70 let (1987) Moscow
Nikolaev, R. (1986) "A new deal in urbanization and housing
   construction policy", Radio Free Europe Research,
   Bulgarian Situation Report, Vol. II, No. 53, Part 1,
   SR 12/86 30 December pp. 7-10
Nove, A. (1982) "USSR: economic policy and methods after
   1970" in The East European Economies in the 1970s,
   A. Nove, H-H. Höhmann, and G. Seidenstecher (editors)
   Butterworths, London
Organisation for Economic Co-operation and Development
   (1986) "Living Conditions in OECD Countries A
   Compendium of Social Indicators", Paris
Osztlany, Z. (1985) "The housing situation in Hungary
   today", Tarsadalmi Szemle, June, pp. 142-147
Polish Television (1986) 28 January
Pravda (1979) 27 July
_____ (1987) "Nastupat' na alkogol", November 15
_____ (1988a) "V Tsentral' nom Komitete KPSS i Sovete
   Ministrov SSSR", February 21
_____ (1988b) "Svoi dom" February 16
Radio Budapest (1987) November 7
Radio Free Europe Research (1984) Bulgarian Situation
   Report SR/7/84, Vol. 9, No. 21, 12 May
_____ (1986a) "GUS Household Census", Vol. 11, No. 13,
   Part 2, Situation Report Poland, 5/86, 19 March, item
   6
_____ (1986b) "Government struggles with the housing
   crisis", Vol. II, No. 15, Part 1, Polish Situation
   Report 6/86, 4 April 1986
_____ (1986c) "Economic plan not fulfilled", Vol. II, No. 9,
   Part 1, Bulgarian Situation Report 1/86, p. 17

_____ (1987a) "Housing prospects", Vol. 12, No. 37, Part III, Item 2, p. 5, 8 September

_____ (1988a) "Trade unions stage 'Congress of People without Housing'", Vol. 13, No. 6, Part 1, 12 February, pp. 29-31

Radio Liberty Research Bulletin (1987) "An interview with Vladimir Treml on alcoholism in the USSR", August 3, RL 317/87

Robinson, W.F. (1984) "Demographic trends in Eastern Europe", Radio Free Europe Research Vol. 9, No. 39, Part II, 28 September, Background Report 174/84

Romania Libera (1987a) 27 July

_____ (1987) 31 August, p. 3

Schrenk, M., Ardalan, C., and NA El Tatawy (1979) Yugoslavia: Report of a mission sent to Yugoslavia by the World Bank, Johns Hopkins University Press, Baltimore

Scinteia Tineretului (1985a) 25 December

_____ (1985b) 5 April

_____ (1986a) 13 February

_____ (1986b) 19 February

_____ (1986c) 27 February

_____ (1986d) 14 March

Shoup, P.S. (1981) The East European and Soviet Data Handbook - Political, Social, and Development Indicators 1945-75, Columbia University Press, New York

Singleton, F. (1982) "Objectives and methods of economic policies in Yugoslavia 1970-1980", Chapter 9 in The East European Economies in the 1970s (editors) A. Nove, H-H. Hohmann, G. Seidenstecher, Butterworths, London

Sirc, L. (1979) The Yugoslav Economy under Self-Management Macmillan, London

Slowo Powszechne (1986) "A few illusions, a few hopes", 25 February

Sobranie postanovlenii pravitel'stva SSSR (1962) No. 12, Article 93

_____ (1980) No. 3, Article 17

Statisticheski Ezhegodnik Stran - Chlenov Soveta Ekonomicheskoy (1983) Uzaimpo - Moschoi, Statistika, Moscow

_____ (1987) Uzaimpo - Moshchoi, Statistika, Moscow

Swidlicka, A. (1986) "Poisonous apartments", Radio Free Europe Research, Poland Situation Report 17/86, Vol. 11, No. 48, 'Part 1, 28 November, pp. 23-25

Szelényi, I. (1988) Socialist Entrepreneurs: Embourgeoisement in Rural Hungary, Polity Press, Cambridge

Tosics, I. (1987) "Privatisation in housing policy: the
    case of the Western countries and that of Hungary",
    International Journal of Urban and Regional Research,
    Vol. II, No. 1, March, pp. 61-78
Trehub, A. (1986a) 'How big is the housing deficit?',
    Radio Liberty Research Bulletin, No. 16 (3377),
    April 16, 1-5 RL 155/86
    _____ (1986b) "Housing and non-labour incomes in the USSR",
    Radio Liberty Research Bulletin, No. 25 (3386), June
    18, RL 155/86
    _____ (1986c) "Construction Committee under attack",
    Radio Liberty Research Bulletin, No. 37 (3398),
    September 10, RL 338/86
    _____ (1988) "Down and out in Moscow and Murmansk:
    Homelessness in the Soviet Union", Radio Liberty
    Research Bulletin, No. 12 (3477) March 23, Part 1 RL
    112/88
Trud (1985) "Dom dlya gorozhanina", September 17
    _____ (1986) "Medred' v Kooperative", February 2
Trybuna Ludu (1987) 6 August
United Nations (1980) Annual Bulletin of Housing and
    Building Statistics for Europe, Vol. XXIV, Economic
    Commission for Europe, Geneva
    _____ (1985) Compendium of Human Settlements Statistics
    1983, New York
    _____ (1986) Annual Bulletin of Housing and Building
    Statistics for Europe, Vol. XXX, Economic Commission
    for Europe, Geneva
Vajna, T. (1982) "Problems and trends in the development
    of the Hungarian new economic mechanism - a balance
    sheet of the 1970s", Chapter 6 in The East European
    Economies in the 1970s, (editors) A. Nove,
    H.H. Hohmann, G. Seidenstecher, Butterworths, London
Vienna Institute for Comparative Economic Studies, 1986,
    Comecon Data 1985, Macmillan, London
Voznesenskaya, Y. (1986) "Housing fit for workers",
    Radio Liberty Research Bulletin, No. 6 (3367),
    February 5, Part 1, RL 46/86
Warwick Statistics Service (1983) The USSR - A Statistical
    and Marketing Review, Occasional Paper no. 3
Wickenden, D. (1985) "Abandoned Americans", The New
    Republic, March 18
World Health Organisation (1986) World Health Statistics
    Annual, Geneva
    _____ (1985) World Health Statistics Annual, Geneva
Zwiakowiec (1986) 2 February
Zycie Gospodarcze (1984) 14 October
Zycie Warszawy (1984) 20 April

Chapter Two

HOUSING POLICY IN POLAND

Andrew Dawson

Introduction

Ordinary Polish people have been obliged to put up with very considerable frustrations and shortages since the Second World War, and one of the most severe of these has been housing. As in many countries, the housing situation in the aftermath of war was bad. About four fifths of Warsaw had been destroyed, and there had been extensive damage to other towns and cities in which fighting had been fierce, especially in the areas which had belonged to Germany until 1945. It is true that the population of Poland, and therefore the demand for housing, fell from 35,000,000 to 24,000,000 during the war, but natural increase was rapid thereafter, and population grew by an average of half-a-million every year during the 1950s. At the same time, the housing sector was obliged to compete for investment with others in the centrally-planned economy, and is generally acknowledged to have been starved of funds for several years, with the result that a huge backlog of demand remained unmet. The housing problem of the immediate post-war years became a persistent and pressing feature of everyday Polish life. Since that time much has been accomplished, and housing conditions in the mid-1980s were very different from those of thirty to forty years earlier. Nevertheless, problems remain, and the nature and severity of these vary widely between one part of the country and another. In particular, there are marked contrasts between the larger towns and the rural areas in the supply and quality of, and demand for, housing: a contrast that reflects in part the pattern of ownership. There are also differences within cities and between the east and centre of the country, on the one hand, and the Northern and Western Territories, on the other. Hitherto, the housing problem in Poland has not been considered widely in the English-language literature. Texts dealing with the recent economic history of the country have, of necessity, accorded housing rather brief mention (Kaser 1986, Landau and Tomaszewski 1985), and have dealt with it from an aspatial and economic point of view. Recent specialist writing about housing, in both English and Polish, in contrast, falls into several categories. Kramer (1980) discusses spatial variations in provision and quality for the period 1970-74, Kosinski et al. (1983) considers housing in the late 1970s, and Weclawowicz (1985) examines aspects of the Polish crisis of the 1980s as it affects housing. Other writers concentrate upon particular sectors of the housing market, areas of the country or the relationship between the availability and quality of accommodation and such social events as divorce, fertility and migration. Thus, co-operative housing and its management have been reviewed by Malicka (1979), rural housing by Dzun (1983) and Sikorski (1985), amongst others, and the housing situation in individual cities, and especially Warsaw, has been described by Ciechocinska (1987), Dangschat (1987), Dangschat and Blasius (1987), Kowalczyk (1986) and Wawrzynski (1986), examining in particular the social disparities within such cities. This account attempts to draw together some of these themes. It describes the changing demand for housing and then considers its provision and management over the post-war period in the country as a whole before assessing the extent to which the housing problem has been solved, both in general and between different parts of the country. The chapter concludes with an examination of the likely pattern of future developments.

THE DEMAND FOR HOUSING

The demand for housing has been driven by three engines since 1945: the increase in the number of households, structural change in the Polish economy and increases in the standard of living.

Housing Policy in Poland

The population of Poland rose from 24,000,000 in 1946 to 38,000,000 in 1987, when it was growing by about 300,000 each year (Rocznik Statystyczny 1987, p. 38). However, the post-war increase has not been steady, for there were substantially higher numbers of births between 1946 and the early 1960s and after 1975, and the population total also reflected a somewhat higher level of births between the mid-1920s and 1940. As a result, the number of those reaching the age of twenty, and thus approaching or entering the household-forming age group, was particularly high between 1945 and 1960 and between 1966 and 1981, while a further upsurge is to be expected after 1995 (Rocznik Statystyczny 1987, p. 48). Official statistics record a rise in the number of urban households from 3,246,000 in 1950 to 7,084,000 in 1984, but a slight fall of rural households from 3,989,00 in 1960 to 3,822,000, but do not give comparable figures for the earlier post-war period Rocznik Statystyczny 1966, p. 42; 1987, p. 49). Between 1960 and 1984 population grew by 24 per cent, whereas the number of households increased by 37 per cent, and the rate of household formation has almost certainly been faster than the growth of population during much, if not all, of the post-war era. Declining family sizes, a fall in the proportion of the population getting married and an increasing rate of divorce have made for a growing number of single-person and one-parent households. Between 1970 and 1984 the number of one-parent households rose by 34 per cent to 1,395,000 (Rocznik Statystyczny 1987, p. 49).

The demand for housing has also reflected the changing structure of the Polish economy. At the end of the war about half of the population depended upon agriculture for its livelihood, but that proportion had fallen to less than a quarter by the mid-1980s. Rapid expansion in mining and heavy manufacturing together with a more modest growth of the service sector, has meant that the entire increase in the workforce since 1945 has been absorbed by activities whose locations are chiefly in towns. Despite the fact that much of the natural increase in the population has occurred in rural areas, the rural population has declined absolutely as people have moved to new jobs in towns, though it must be noted that some rural areas and their populations have been absorbed into towns and reclassified as urban. In

1946 16,000,000, or double the urban population, lived in the countryside, the number in towns had trebled by the mid-1980s and was 50 per cent larger than that in rural Poland (Rocznik Statystyczny 1987, p. 38).

Third, the demand for housing has grown as standards of living have risen. Official statistics on this topic are difficult to interpret, but it would be widely accepted that family incomes have risen in real terms over the post-war period, and especially in the 1970s, even if they fell markedly at the start of the 1980s at the time of Solidarity and martial law. One of the difficulties in assessing the true extent of these changes in real incomes lies in the fact that, until the 1970s, inflation was not acknowledged to exist by the authorities, and was hidden behind the system of official prices, manifesting itself in shortages and advance payments for durable-use consumer goods, especially cars. However, it is clear that the more generous terms for private farmers, which were introduced by Edward Gierek after 1970, and the licensing of a wide range of small, private firms offering personal, professional, repair and retail services, have created a large number of relatively wealthy people in Poland. Many of these have chosen to invest their extra cash in housing, and, although the opportunities to build privately or to expand existing accommodation in the cities have been slight, rural dwellers have been able to collect building materials and subsequently construct better houses while urban people have been able to build second homes in the villages and major tourist areas. Whereas the typical Polish village of the 1960s was dominated by pre-war houses and contained only a few, modest post-war dwellings, that of the 1970s was characterised by piles of bricks and timber on plots within or around the built-up area, awaiting the moment when sufficient material had been accumulated to start building, and that of the late 1970s and 1980s by infill and peripheral areas of large detached villas, either in construction or complete: villas which are rarely in a style consonant with that of the older houses and contain much more floorspace than was the norm before 1970 or has been allowed in the towns.

Housing Policy in Poland

THE SUPPLY OF HOUSING

Just as the demand for housing has been rising, so has the supply (Table 2.1).

Table 2.1: Houses and flats in Poland 1950-1986

	1950	1960	1970	1980	1986
Total houses and flats (thousands)	5851	7026	8081	9794	10834
of which					
urban	2711	3560	4521	6133	6965
rural	3140	3466	3560	3661	3869

Sources: Rocznik Statystyczny 1966, p. 398; 1971, p. 435; 1987, pp. XLIV-XLV, 440.

In 1950 there were about 5,850,000 flats and houses in the country, of which the majority were in rural areas, but by 1986 the total exceeded 10,800,000, three fifths of which were in towns and cities (Rocznik Statystyczny 1987, pp. XLIV, 437). However, this increase of eighty per cent was not achieved smoothly. New building began again in 1947, but during the early 1950s fewer than 70,000 dwellings were constructed each year. Thereafter, numbers increased to reach 126,000 in 1957, 205,000 in 1972 and 283,000 in 1978, but by 1981 the figure had fallen back to 187,000 and remained below 200,000 up to 1986. About two thirds of the dwellings in any year were built in towns (Rocznik Statystyczny 1966, pp. 38-39; 1987, pp. XLIV-XLV, 437).

However, the supply of housing has been much more complex than these aggregate figures might suggest, and there have been marked variations in the sources of supply as government policy has changed. During the early 1950s - the Stalinist period of the Six Year Plan, when only about 65,000 flats and houses were built each year - about two thirds were provided by central and local government or by the industrial ministries and enterprises in the towns. Four fifths of the privately-owned houses were built in the villages.

After the return of Wladyslaw Gomulka in 1956, in contrast, construction by co-operative societies was encouraged, and they have made an increasing contribution (Rocznik Statystyczny 1966, p. 209; 1971, p. 440). By the mid-1970s about three quarters of all the dwellings built by the socialist sector were provided by co-operatives. During the 1980s, when the annual output of houses and flats has been just under 200,000, the contributions of the different sources of supply have varied somewhat from year to year, but in general have been as shown in Table 2.2. Slightly over half of the housing stock belonged to the socialist sector by the mid-1980s (Rocznik Statystyczny 1987, p. 440).

These changing levels of output and sources of supply reflect the shifting policies of Polish governments for the provision and management of the housing stock. During the late 1940s a Three Year Plan for Reconstruction was carried out, to repair the worst of the war damage, and little new building was undertaken during that time. It was intended to pave the way for new developments in the 1950s, but, following the establishment of a communist government in the Stalinist mould, the Six-Year Plan (1950-1955) gave the so-called 'non-productive' sector of the economy, which included housing, a very low priority. Some large apartment blocks were built by central and local government and the industrial ministries in the heavy, socialist-realist style in most of the major cities, and especially in Warsaw, but only about a tenth of all public-sector investment went into housing (Rocznik Statystyczny 1966, p. 91) and the private sector was severely hampered by a shortage of building material. About 45 per cent of public sector investment went into mining and manufacturing, in an attempt to widen the opportunities for peasants and their families to move into industrial employment, and many of the new jobs were in the towns, thus adding to the demand for housing there. However, urban housing increased only slowly, long-distance commuting from rural to urban areas became common, limits had to be placed on permission to settle in Warsaw and some other cities, and many households continued to live in congested, overcrowded conditions. The man who 'left' his wife for a mistress, but was obliged to bring the mistress to share the flat in which both he and his wife continued to live, illustrated the severity of the problem. What is more, a majority of the population continued to live in accommodation which had been built before the First World War, much of which was substandard, for it proved to be impossible to launch major programmes of either slum clearance or improvement under the investment priorities laid down by the government. Nor was the situation easier in the countryside. Very little of the meagre housing investment of the state was allocated to rural areas, and individual farmers were faced by a shortage of building materials and the problem of 'making ends meet' as a result of the imposition of compulsory, low-price deliveries of some of their products to the state and other forms of financial discrimination against them. At the same time, much of the rural housing in central and eastern areas was of wood, and of types which were usually rebuilt every fifty or sixty years. However, opportunities for rebuilding, let alone the installation of such amenities as running water, were slight between 1939 and 1955.

The chief tools for managing the limited housing stock were rationing and allocation by the state. Some dwellings had been seized by the authorities at the end of the war, strict public control, but not ownership, had been established over the rest of the stock by 1950, and rents were set or frozen far below economic levels. Brus (1986, p. 136) has calculated that only 2 per cent of the capital cost and 16 per cent of the maintenance expenditure on urban housing in the socialist sector were met directly by the residents in 1955-1956, the rest coming from government funds. Houseroom was rationed according to floor area, even in privately-owned dwellings, and little allowance was made for the numbers of rooms or configuration of existing properties in the setting of minimum numbers of residents by the authorities. Thus, many families in older properties in cities were obliged to make do with one or two large rooms in which four or five people might be required to live, and to take in lodgers if the floorspace was greater than the number of members in the family could justify. Even Mr. Gomulka was obliged to share a house in Warsaw for a short while after becoming First Secretary of the Polish Workers' party, or so it was believed. However, the story seems unlikely for the allocation of houseroom, and especially of new accommodation in the cities, was largely in

the hands of central and local government or of
the industrial ministries which had built it, and
much of it was given to high-ranking party, police
or government officials or to the more senior
managers of key economic enterprises. In short,
the more senior and better paid stood the best
chance of getting the new apartments.

The period since 1953 may be seen as one of
relaxation, at first only gradual, of the limited
and highly-managed role of housing in the Polish
economy described above. It may also be seen as a
belated attempt, also at first only gradual, to
remedy the acute shortage and low quality of
housing which had resulted from the inheritance of
many poor, pre-war dwellings, war-time damage and
the under-investment of the Six-Year Plan. The
major developments since the early 1950s have been
the growth of the co-operative sector and the
increasing role of private housebuilding.

Half the dwellings built in Poland in the
1970s and 1980s were provided by the co-operative
sector. In 1956 only about 35,000 people belonged
to 297 house-building co-operative societies, most
of which were building for rental, but by 1960
there were more than a thousand societies and
almost 3,000 by 1986. Membership exceeded
1,000,000 in 1970, 2,000,000 in 1975 and stood at
about 3,200,000 in 1986, though the contribution
of the societies to the housing stock has grown
more slowly. By 1956 co-operatives had provided
less than 20,000 dwellings, by 1970 600,00, by
1982 2,000,000 and by 1986 almost 2,500,00
(Rocznik Statystyczny 1966, p. 405; 1987, p. 447).
Co-operative housing is classified as part of the
'socialist' sector of the economy, and it is
subsidised by the state, though not to the same
extent as the housing provided directly by
government. Co-operatives are of two types:
those building for rental and those for sale to
members of the co-operative society. In both
cases members make substantial downpayments in
advance of the start of building, which restricts
membership to the better-off, and membership is
also limited to the number of flats which are
likely to be produced in the foreseeable future,
which is determined by the availability of funds
from the state, of land, of permits to acquire
building materials and of planning permission.
Members have little, if any, say in the operations
of the co-operatives. They simply pay and wait
while the state controls the rate, form and

location of construction. About a quarter of the members are either children, whose deposits have been paid by parents, or are single (Radio Free Europe, 1984, p. 24). Throughout the 1980s between 600,000 and 700,000 members of co-operatives have been waiting at any time for the completion of their flats, and more than half of these have been waiting for seven years or more. At the same time, the number of officially registered associate members, that is, those in the queue to join co-operatives when there is a foreseeable change that they may get a flat, stood at 2,000,000 or more during the early 1980s (Rocznik Statystyczny 1987, p. 447). "Rental" co-operatives were the dominant type in the 1950s and 1960s, many of which were organised through industrial enterprises or by local government, but 'building-for-sale' societies have become increasingly important, and were providing about three-quarters of the supply from the co-operative sector in the mid-1980s (Rocznik Statystyczny 1987, p. 446).

The second most important source of new housing in the 1980s has been the private sector. The number of new private houses has varied during the post-war period from about 30,000 per annum in the early and mid-1950s to 74,000 in 1978, but for most of the 1970s and 1980s the annual output has been of between 50,000 and 60,000 (Table 2.2). Most of these have been in the villages and smaller towns, and reference has been made above to the changes which have resulted in the appearance of those settlements as families have banded together to collect building materials and then to put up houses in their spare time and as resources allow with the help of friends, many of whom are rewarded with food and vodka. Everybody is entitled to a loan to help such operations, but only once building has begun. Two marked contrasts exist between the physical form of private and public (that is, co-operative and government) housing. First, whereas strict controls govern the floorspace allowed in public housing, private builders are much less constrained, and new private dwellings in the 1980s were almost twice as large, with an average of about 100 square metres, as against about 55 in public-sector dwellings (Rocznik Statystyczny 1987, p. 444). Second, whereas almost all public housing is in blocks of flats, private dwellings are almost entirely detached houses.

Opportunities for building such houses have been very limited in the larger towns and cities, where housing developments tend to be on a large scale and on publicly-acquired land. Thus, in the 1980s only about 6,000 private dwellings were built each year in cities of 100,000 population or more, in spite of the fact that 30 per cent of Poles live in such settlements, as against 60,000 public-sector dwellings (Rocznik Statystyczny 1985, p. 428).

Third in the provision of new housing in the 1970s and 1980s has been that provided by central and local government and by industrial enterprises for their employees. These sources have provided between 30,000 and 50,000 dwellings each year since the early 1950s, but their proportion of the total supply has fallen from about half to a fifth (Table 2.2). Most are built in towns in the form

Table 2.2: New houses and flats in Poland in 1986

	Total	Per cent
Total	185,000	100
of which		
urban	136,700	73.9
rural	48,300	26.1
Socialist Sector	127,600	69
of which built by		
co-operatives	89,400	48.3
enterprises	31,200	16.9
local government	5,700	3.1
of which		
urban	113,700	61.5
rural	13,900	7.5
Private Sector	57,400	31
of which		
urban	23,000	12.4
rural	34,400	18.6

Source: Rocznik Statystyczny 1987, p. 443.

of blocks of apartments. Enterprises continue to
use such accommodation to secure or retain senior
and skilled staff whereas, unlike the situation
before 1956, state housing has been reserved
increasingly for those on very low incomes - less
than 30 per cent of the average pay in the public
sector - who have been unable to afford co-
operative or privately-owned accommodation. Both
types, however, continue to be the most heavily
subsidised sources of housing in any economy in
which all housing enjoys financial support in one
form or another from the state.

Altogether as a result of these developments
investment in housing has increased greatly since
the early 1950s. Less than eight per cent of
public sector investment went into housing in
1950, and less than 12 per cent of all investment,
both public and private. However, by the early
1960s these proportions had risen to 14 and 19 per
cent, and moreover these were proportions of a
larger percentage allocation from the gross
national product to accumulation; and, by the
early 1980s about 20 and 27 per cent of the
investment in the public sector and the economy as
a whole went into housing (Rocznik Statystyczny
1966, p. 92; 1971, p. 188).

Nevertheless, it would seem that the supply
of housing has been very inadequate. Floorspace
in new dwellings has generally been less than that
in other developed economies, especially market
economies, and the supply of dwellings has always
been well below the number of marriages,
especially in rural areas. More than a fifth of
all new dwellings in the public sector were
allocated every year in the early 1980s to people
who had been in overcrowded conditions hitherto,
but in 1984 18 per cent of families were still
living in two-family households - a figure which
had changed little since 1960 - and more than one
per cent in households of three or more families
(Rocznik Statystyczny 1966, p. 403; 1987, p. 446).
Dangschat (1987, p. 39) claims that rentals of
private housing "on the black market are up to 80
percent of an average income" whereas approved
rents for state-owned dwellings are between five
and ten per cent and for co-operative housing
between 15 and 20 per cent. Furthermore, families
often wait many years for dwellings: Ciechocinska
(1987) and Dangschat (1987) refer to 15 and 20
years respectively as a not unusual period! Nor
is the quality of new housing in urban areas

improving. During the late 1950s and early 1960s most was in the form of small blocks of flats, usually of less than five or six storeys, often with balconies and in pleasantly-landscaped surroundings, in individual blocks containing small numbers of apartments. From the late 1960s onward, in contrast, larger scales of building became common, and huge, gaunt, standardized, slab-like blocks of ten or more storeys, in pre-fabricated concrete, each containing several hundred flats, were built. Few resources were made available to landscape these developments, which in any event cannot be disguised by shrubs and ornamental trees, and car parking was quite inadequate. Moreover, conditions inside such apartment blocks are far from satisfactory. Poky stairs in unfinished concrete and inadequate lifts from which people often steal the light bulbs reinforce the drab external impression, and those moving into flats in such blocks, who are usually young married couples with families, are obliged to spend months completing electrical and plaster work, installing plumbing and other fittings and decorating before they are suitable for habitation. Ciechocinska (1987, p. 11) reports that

> In the initial period housing co-operatives paid much more attention than the municipal authorities to the appearance of housing estates, various embellishments, verdure, the supply of services, the organisation of daily life of the residents, etc. However, as co-operatives became the dominant power on the housing scene, as big co-operatives grew at the expense of smaller ones, such concerns were by and large abandoned.

THE SITUATION IN THE MID-1980s

According to official statistics, the situation in the housing market in Poland is now much better than at any time since the Second World War. Not only has the housing stock been greatly increased, but the average floor area of dwellings had risen to 56 square metres, and to 16.6 metres per person by 1986. (Figures for these two measures are not available before 1970, when they were 50.7 and 12.9 respectively.) Moreover, the average number of people per room had fallen from 2.05 in 1946 to 1.75 in 1950, 1.66 in 1960 and 1.03 in 1986; household amenities have also been improved, and the number of rooms per dwelling has increased from 2.5 in 1960 to 3.3 in 1986 (Rocznik Statystyczny 1987, pp. 437, 439).

However, marked variations exist between different parts of the country. For instance, the average dwelling in rural areas had a floor area of 64 square metres in 1986, as against only 52 in urban areas, though the number of people in the average rural dwelling was also higher, with the result that the floor space per capita was almost the same. This contrast reflected that of ownership. Whereas public sector housing in the villages had an average floor area of 51 square metres in 1986, that for privately-owned houses was 67, and in the towns the averages were 47 and 65 square metres respectively. The picture regarding household amenities was even more extreme. Ninety-three per cent of urban dwellings in 1986 were connected to mains water, 82 per cent had a flush toilet and 78 per cent a bath, but in rural areas the figures were only 55, 37 and 42 per cent respectively (Rocznik Statystyczny 1987, pp. 439-440, 443).

These contrasts are even more marked at the level of the chief local administrative units - Wojewodztw or voivodships - of which there are 49. Figures 2.1 to 2.4 show the variations in the average size of public-sector and private dwellings in rural and urban areas, and Figures 2.5 and 2.6 the average number of persons per room. The patterns of these maps are remarkable. The average floor area of public-sector housing in towns (Figure 2.1) is almost uniform across Poland, varying between 42.8 and 49.7 square metres on average among the voivodships, and that in the villages (Figure 2.2), while being a little higher, does not vary much more. Private housing, in contrast, shows both a higher and a much wider range of averages almost everywhere, especially in the towns (Figure 2.3). Low values occur in the central and south-eastern parts of the country with much higher averages in the southwest (Lower Silesia) and northwest, and a similar, but not quite so marked, pattern occurs in the villages (Figure 2.4). Numbers of people per room present a mirror image of these patterns, with somewhat higher densities in the villages (Figure 2.6) in the south-eastern areas, and much higher densities in the towns (Figure 2.5), than elsewhere, and especially in the southwest.

Figure 2.1 Poland: Average
floor area of
public sector
dwellings in
towns in 1984

Figure 2.3 Poland:
Average
floor
area of
private
sector
dwellings
in towns
in 1984

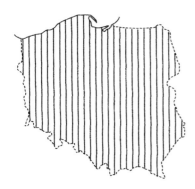

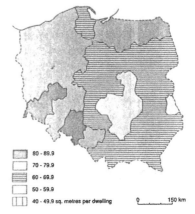

Figure 2.2 Poland: Average
floor area of
public sector
dwellings in
villages in
1984

Figure 2.4 Poland: Average floor
area of private
sector dwellings in
villages in 1984

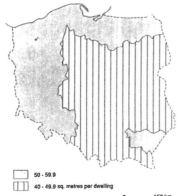

Source: <u>Rocznik Statystyczny 1985,</u> pp. 423-425

Figure 2.5 Poland: Average
        number of people
        per room in
        towns in 1984

Figure 2.7 Poland:
        Percentage
        of dwellings
        with mains
        water in towns
        in 1984

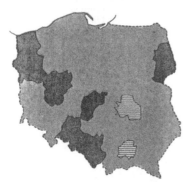

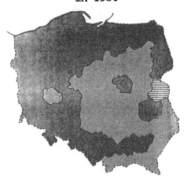

Figure 2.6 Poland: Average
        number of people
        per room in
        villages in 1984

Figure 2.8 Poland:
        Percentage of
        dwellings
        with mains
        water in
        villages in
        1984

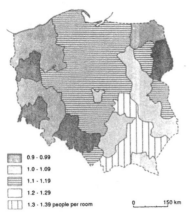

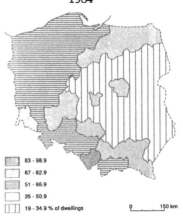

0.9 - 0.99	
1.0 - 1.09	
1.1 - 1.19	
1.2 - 1.29	
1.3 - 1.39 people per room	0 150 km

83 - 98.9	
67 - 82.9	
51 - 66.9	
35 - 50.9	
19 - 34.9 % of dwellings	0 150 km

Source: <u>Rocznik Statystyczny 1985</u>, pp. 423–424
        <u>Rocznik Statystyczny 1985</u>, p. 426

Figure 2.9 Poland: Percentage of dwellings with
flush toilets in towns in 1984

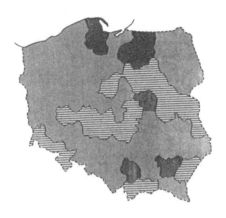

Figure 2.10 Poland:
Percentage of
dwellings with
flush toilets
in villages in
1984

Figure 2.11 Poland:
International
frontiers
crossing the
area of post-
war Poland
between
1815 and 1939

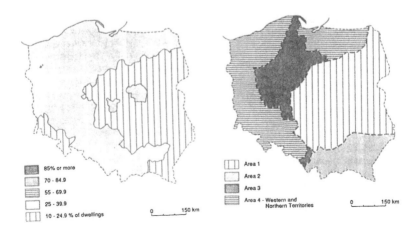

Source: <u>Rocznik Statystyczyn 1985</u>, p. 426

Housing Policy in Poland

The distributions of household amenities show similar patterns. For example, most dwellings in most towns were connected to mains water by the mid-1980s (Figure 2.7), but the level was generally higher in the north and west of the country than in the centre and east, and everywhere was much above that in the rural areas (Figure 2.8). Rural levels of connection remained very low - typically less than a third of dwellings - in the central and eastern areas, where wells were the chief means of supply, but in no voivodship were more than two-thirds of rural houses connected to water mains. Similarly only about three quarters of urban dwellings possessed a flush toilet (Figure 2.9), and levels of provision were lowest in central Poland, but in rural areas this amenity was still scarce in the mid-1980s (Figure 2.10), for in no voivodship did more than two fifths of houses possess it. The spatial distributions of other household amenities were similar.

These contrasts in the supply and quality of housing between one area of the country and another may be surprising in a state which, for forty years, has attempted to equalise the supply of necessities among the population, but they reflect the very great differences between the histories of development of the areas which go to make up post-war Poland. Figure 2.11 indicates the international boundaries which existed across present-day Poland between 1815 and 1939; and the four major areas into which they divide the country may be compared with those revealed in Figures 2.1-2.10. Area 1 in central and eastern Poland lay within the Russian Empire between 1915 and 1918 and in the independent Polish state between the World Wars, and suffered from little economic development. Polish attempts at economic development after 1815 were cut short by the uprising of 1830, and feudalism was not abolished until the 1860s. Industry developed in the second half of the nineteenth and early twentieth centuries in Warsaw, the textile town of Lodz and its satellites, and on the eastern edge of the Upper Silesian coalfield around Sosuowiec but much of the area was little affected by economic development, and there were few opportunities for rural improvement. Nor did the inter-war period, with its economic and political turmoil, enable significant changes in the structure of the economy or the level of wealth to occur. The result was that rural housing was largely of wood and contained few, if any, modern amenities, while many of the urban population were housed in congested and unsanitary tenements.

Area 2 suffered a similar fate, though it belonged to the Austrian-Hungarian Empire during the nineteenth century. Population pressure built up in rural areas during that period as a consequence of the paucity of alternative employment opportunities, farm sizes declined and rural poverty was acute and widespread. Housing standards were low.

Areas 3 and 4, in contrast, were part of the Prussian Empire and so were linked to one of the major economic powers of the late nineteenth century. Area 4 remained in German hands until 1945, but area 3 became part of the newly-independent Poland in 1918. Both, however, benefited from the commercialisation of agriculture in the last century, associated with the development of a dense network of railways and food processing industries, and the large-scale opportunities for rural workers to obtain employment in towns and cities. Standards of living were much higher than in Areas 1 or 2 and houses tended to be larger, to be built of brick or stone, and to possess more amenities.

The contrast between area 4 and the rest of the country also extends to ownership. Whereas few dwellings in rural areas in other parts are in the public sector - reflecting the low level of public investment in rural communities since the war - many properties in the Western and Northern Territories were taken into public ownership, having been abandoned by their German owners or seized from them at the end of the war. Whereas less than 15 per cent, and often less than 10 per cent, of houses are publicly owned in the rural central, eastern and southern areas, the figure ranged from a third to 55 per cent in the voivodships of area 4. Public ownership in urban areas also reflects location in Poland, with higher proportions in former German areas, partly because of the takeover of former German property and partly because many of the towns in area 4 were badly damaged during the German retreat. The extent of public ownership also reflects the size of towns, being higher in the larger settlements. Thus, in Warsaw, where most of the city has been rebuilt since the war, more than 80 per cent of the housing stock is in the public sector, and the same is true of Gdansk, Lodz, Szczecin and Wroclaw voivodships - all of which contain cities of more than 400,000 population and all of which, except Lodz, lie in area 4. However, Elblag, Gorzow, Jelenia Gora, Koszalin, Legnica, Olsztyn, Opole, Slupsk, Suwalki, Walbrzych and Zielona Gora - all of which have no towns of more than 150,000 but lie in area 4 - also have more than three-quarters of their urban housing in the public sector, that is, more than the average for the country as a whole. In contrast, less than half of the housing is usually in the public sector in the small towns in areas 1 and 2.

The extent of the pre-war housing legacy was indicated by the 1984 census. At that time, one third of all urban housing and 40 per cent of rural dwellings had been built before 1945, including 13 per cent which dated back before 1918 - a legacy which arises from the low level of investment in new housing in the 1940s and much of the 1950s, but also from the reluctance of the Polish authorities to demolish substandard accommodation. During the early 1960s only about 0.2 per cent of urban dwellings were 'lost' each year, including those destroyed by fire and flood, and the proportion was only half of that in the early 1980s (Rocznik Statystyczny 1966, p. 404; 1987, p. 440). Rehabilitation of tenement blocks, rather than demolition and redevelopment, has been the official policy, so that only about 3 per cent of new dwellings, at most, were replacements. The situation in the private sector, and especially in the villages has been very different for owners appear to have been only too willing to tear down traditional wooden and thatch, single-storey cottages when they have built modern villas for themselves. In the private sector as a whole as many as a fifth of new houses may have been replacements, and in rural areas the figure may have been a quarter, in the early 1980s.

Important distinctions also exist between the standards of housing, or at least perceptions of housing quality, within the cities. The historical legacies of construction and changing architectural style have created areas within towns which are characterised by pre-war villas, pre-1914 urban tenements, the smaller blocks of flats of the late 1950s and 1960s or the massive and unattractive 'slabs' and towers of tenements of the 1970s and 1980s. Moreover, housing type and location within cities is often closely

related, with the less attractive later apartment
blocks being located in peripheral sites at long
distances from city centre facilities or even such
everyday services as shops, schools, work places
or public transport in some cases. Weclawowicz
(1975, 1981) has described the spatial segregation
of social groups within the population and hinted
at the competition which takes place between them
for housing. Thus, party officials, managers of
enterprises, those in private employment in
service industries and people with access to hard
currency are allocated or can afford the largest
and best-equipped housing with most rooms, while
young families, those with only primary education
and poorer people are obliged to take what is
left. In other words, the allocation of housing
still leaves much to be desired, especially in
view of the fact that most Poles remain in the
same dwelling for much of their adult lives. This
occurs not only because of the shortage of
housing, but also because exchanges of flats
require official sanction and the payment of a
tax, and were prohibited for a few years during
the early 1980s. One consequence is that flats
and estates which were originally occupied by
young families gradually become under-occupied
areas of one-person households.

FUTURE DEVELOPMENTS

The economic crisis in Poland, which began in
1979 and was highlighted by the founding of
Solidarity and martial law, led to the abandonment
of ambitious long- and medium-term plans for the
further development of the country. Output fell
markedly, prices for co-operative flats rose
sharply, and the number of new dwellings during
the early and mid-1980s was only about two thirds
of the peak year of 1978. However, the demand for
housing is continuing to grow. It is estimated
that the population in the working age group will
increase by a million to 22,770,000 between 1985
and 1995, and thereafter the children of those
born in the post-war "baby boom" will be entering
the working, and therefore the household-forming,
age group (Rocznik Statystyczny 1987, p. 42).
Moreover, existing accommodation is decaying, and
it has been estimated that about 1,500,000 units,
or a sixth of the stock, is fit only for
demolition (Swidlicka, 1986, p. 25). By the mid-

Housing Policy in Poland

1980s the shortfall in housing had reached about 2,000,000 units, and an officially-backed Congress of People without Housing in 1988 was informed that construction would have to be doubled for the rest of the century if the shortages were to be met (Radio Free Europe, 1988, p. 29). However, the 1986-1990 5-year plan provides for no significant increase in output over the levels of the early 1980s.

At the same time, however, as the housing situation has been deteriorating, with the Polish authorities apparently incapable of increasing construction, new and ambitious plans have been announced for the economic and social development of Poland until the year 2025 which will require large-scale investment in new accommodation (Komisja 1985). It is envisaged that the population will have reached 45,000,000 by that year, with an increase of 8,000,000 in the towns between 1985 and 2025. Most urban settlements are expected to grow, but that growth is to be strictly limited in the major cities and conurbations of Warsaw, Upper Silesia and Lodz, and to be deflected to some of the scantily-populated northern parts of the country. In particular, Koszalin and Olsztyn, both of which had populations of less than 150,000 in the mid-1980s are expected to grow to between 500,000 and 3,000,000 by 2025, while the small towns of Chojnice and Gizycko are to have populations of 200,000 to 300,000. Notwithstanding the huge ranges of these targets it is clear that, as more people leave agriculture, the rural population will fall by at least 2,000,000, creating an increasing demand for jobs and housing in urban areas in the future. The plan indicates that some, but not all, of the fall will be accommodated in the towns, suggesting that pressure on urban housing and long-distance commuting to work in towns are both likely to continue, if not increase. What, however, is particularly striking about these proposals is that they seem to be blind to the current problems of increasing the housing stock and give no indication as to how the massive investment required to achieve these changes, in addition to that which the already-existing problems demand, will be possible in Poland in the foreseeable future.

In short, Poland's governments since 1945 have failed to solve the housing problem. Since that date a communist state, concerned with the equality of access amongst its citizens to basic goods and services, and in particular concerned to cherish those employed in the public sector, has contrived to coop up most of the urban/industrial population in gaunt blocks and tiny apartments, while those in the private sector, and especially in agriculture, are living increasingly in large, detached houses, comparable to those in many a western European country. Furthermore, it has failed to reduce the problems of overcrowding and has presided over a massive and growing housing shortage. Economic reform, if applied to the housing market, could rapidly lead to very substantial changes in the patterns which have been described in this chapter, but it does not appear that the authorities in Poland have any plans to extend their reforms to housing in any fundamental way, nor does it appear that the ambitious plans for Poland's future will either change the present uneven distribution of the nation's housing or solve its underlying and continuing housing problem.

Thus, the future of housing in Poland is uncertain, but the way forward for its study is much clearer. Almost without exception writings on the topic have depended heavily on official statistics, supplemented with anecdote. This chapter has employed similar material. But Ciechocinska (1987, p. 23) claims that "the unsatisfied demand for housing has persisted for decades, providing a fertile ground for profiteering, bribery and other abuses", and complementary studies have long been required into at least three major aspects of housing which are not caught in the published figures. Difficult though it would be to obtain information from a suspicious public about activities which lie on the fringes of, and perhaps beyond, what is legal, detailed investigations about who actually lives in different types of accommodation and the processes by which people arrange moves between dwellings are necessary if the extent to which the official system of management prevents the efficient use of the housing stock is for a full and accurate comparison between the prices of housing in the official and other markets. Much public housing is subsidised; some of the costs of private building are hidden, and there are substantial elements of monopoly in the market, especially in urban areas. Economic reform, if it

is to be meaningful, must move the allocation of
housing closer to that which would be produced by
an accurate assessment of costs and benefits, but,
until the extent of the present financial
distortions are recorded, the full impact of such
reform must remain a matter of guesswork.

REFERENCES

Brus, W. (1986) in The Economic History of Eastern Europe
    1919-1975, ed. M.C. Kaser, Oxford University Press
Ciechocinska, M. (1987) Government interventions to
    balance housing supply and urban population growth:
    the case of Warsaw, International Journal of Urban
    and Regional Research 11, 9-26
Dangschat, J. (1987) Sociospatial disparities in a
    'socialist' city: the case of Warsaw at the end of
    the 1970s, International Journal of Urban and
    Regional Research 11, 37-59
Dangschat, J. and J. Blasius (1987) Social and spatial
    disparities in Warsaw in 1978: an application of
    correspondence analysis to a 'socialist' city,
    Urban Studies 24, 173-191
Dzun, W. (1983) Regionalne zroznicowane warunkow
    mieszkaniowych na wsi, Wies Wspolczesna 27, 47-56
Kaser, M.C. (ed.) (1986) The Economic History of
    Eastern Europe 1919-1975, Oxford University Press
Komisja Planowanie Przy Radzie Ministrow, Zalozenie Planu
    Przestrzennego Zagospodarowania Kraju do 1995 roku,
    Warsaw 1985
Kosinski, J., E. Wysocka and T. Zarski (1983) Osadnictwo
    w Polsce w latach 1976-1980, Czlowiek i Srodowski 7,
    233-264
Kowalczyck, A. (1986) Second homes ownership in the urban
    environment. The case of Warsaw, Miscellanea
    Geographia, 215-221
Kramer, J. (1980) Przestrzenna Struktura Konsumpcji w Polsce
    PWN, Warsaw
Landau, Z. and J. Tomaszewski (1985) The Polish Economy in
    the Twentieth Century, Croom Helm, Beckenham
Malicka, W. (1979) Housing estates and town communities in
    postwar Poland, International Journal of Urban and
    Regional Research 3, 209-219
Radio Free Europe, Polish Special Report, 9/20, (1984) 13/6
    (1988)
Rocznik Statystczny 1966, 1971, 1985, 1987, Glowny Urzad
    Statystyczny, Warsaw
Sikorski, R. (1985) Uwagi o stanie i przemianach osadnictw
    a wiejskiego, Wies Wspolzesna, 29, 71-76

Swidlicka, A. (1986) Radio Free Europe, Polish Special
        Report 11/48, 23-25
Wawrzynski, J. (1986) Nowe Tychy: an assessment of a
        Polish new town, Planning Outlook 29, 34-38
Weclawowicz, A.M. (1985) Some aspects of crisis in housing
        in Poland, Geographica Polonica 51, 99-112
Weclawowicz, G. (1975) Struktura Przestrzeni
        Spoleczno-Gospodarczej Warszawy w latach 1931
        i 1970 w swietle analizy czynnikowej, Polska
        Akademia Nauk, Warsaw
_____ (1981) Towards a theory of intra-urban structures of
        Polish cities, Geographica Polonica 44, 179-200
_____ (1985) Some aspects of crisis in housing in Poland,
        Geographica Polonica 51, 99-112

Chapter Three

HOUSING POLICY IN CZECHOSLOVAKIA

David Short

INTRODUCTION

As reiterated by numerous authors (e.g.
Kansky, 1976, p. 67; Carter, 1979, p. 426;
Musgrave, 1984, p. 91; Musil, 1987, p. 30),
Czechoslovakia had one advantage over many
European countries in that it suffered relatively
little war damage, with the exception, pointed out
by Kansky, of Eastern Slovakia. One recent
Czechoslovak statement (Ončák, 1986, p. 28)
quantifies wartime destruction at 370,000 'flats',
but does not specify their location.

On the other hand, it is equally frequently
reiterated that Czechoslovakia, particularly the
western, Czech, half, had had a long capitalist
industrial history, which had produced not only a
huge stock of large and aging houses, a degree of
urban sprawl, but also, in the pre-war period, a
spatial segregation of classes with clearly
differentiated housing conditions (Musil, 1987, p.
29). The country also had its overcrowded slums
(de Felice, 1987, p. 112) and even pockets of the
crudest dwellings imaginable, especially on the
outskirts of Prague, documented photographically
in Raban (1986, pp. 9-12), and among gypsy
encampments in Slovakia, as can be inferred from
Slovak literature.

Studies such as Musil (1968), based on census
data for 1930, 1950 and 1961, reveal a general
trend towards the social homogenisation of urban
space in all areas of Prague, which is matched by
developments in other cities. Matějů et al.
(1979) confirm the trend as the outcome of (a) the
overall increase in house building, (b) the
increase in ownership forms with the launch of co-

operative schemes, (c) the start of the large peripheral estates and (d) consequent new types of intra-city migration. Thus the new post-war order had a major effect on existing patterns of housing and population distribution, but it did have problems with which to contend, particularly in the extent of disrepair of the housing stock and the stagnation of the construction industry in the post-war period (Ončák, 1986, p. 28). Overcrowding was a serious problem inherited after the war from the inter-war period. De Felice quotes 1930 statistics for occupancy of one-room flats (p. 112): 55% of workers, 31% of minor bureaucrats and even 10% of higher level managers and civil servants 'enjoyed' such limited accommodation, and 50% of working-class families were overcrowded. The problem was in part overcome by the municipalisation of rented buildings, the effective prohibition of further private renting or selling, freezing of rents at below maintenance and replacement cost (which still applies) and redistribution according to need (de Felice, 1987, p. 113). Later house building programmes were instituted, but only in second place to the demands of construction for industry, and more relief came with the institution of co-operatives.

TRENDS IN HOUSING CONDITIONS AND CONSTRUCTION

By 1961 there were 3,820,000 dwellings or 278 dwellings per 1000 population in the country as a whole (CSSR), with a lower density in the Czech Lands (Bohemia and Moravia, CŠR) (296 per 1000) than in more rural Slovakia (SSR) (236 per 1000). By 1970 the figure had improved to 296 per 1000 nationally (315 per 1000 in CŠR, 253 per 1000 in SSR). These figures are based on a total stock of dwelling units of 4,239,000. In 1980 the ratio of dwellings to population had risen again to 321 per 1000 (340 per 1000 in CŠR, 283 per 1000 in SSR). The total number of dwellings in 1985 was 4,909,000 and current estimates suggest a national ratio of dwellings to population of 336 per 1000.

After 1960 there was an overall predominance of two-room dwellings, since this size represented the bulk of new building, but the numbers with three or more began to rise sharply. The official percentage statistics are given in Table 3.1, but

Table 3.1: Czechoslovakia: Relative size of new dwelling in a) the state, co-operative and enterprise sector, and b) the private sector, expressed as a percentage of all new stock in each sector

	1956	1960	1965	1970	1975	1980	1981	1982	1983	1984	1985
CSR											
bed-sits											
a)	3.0	2.8	7.6	6.1	8.5	5.0	3.2	3.6	3.7	3.5	5.1
b)					0.1	0.1	0.1	0.1	0.1	0.1	0.1
1 room + kitchen											
a)	7.6	3.4	6.0	8.9	16.4	12.9	12.7	12.3	12.3	12.0	12.7
b)					4.3	2.1	1.8	1.6	1.4	1.7	1.5
2 rooms + kitchen											
a)	70.3	74.9	41.2	28.1	32.7	18.5	20.0	17.2	18.4	15.0	17.2
b)					22.7	12.6	10.6	9.2	8.6	8.3	8.3
3 rooms + kitchen											
a)	18.0	17.9	41.1	52.7	38.7	51.3	49.8	50.6	49.4	51.6	48.0
b)					38.0	28.0	27.0	25.2	23.7	22.9	22.2
4+ rooms + kitchen											
a)	0.4	1.0	4.1	4.2	3.7	12.3	14.3	16.3	16.2	16.1	14.3
b)					34.8	57.2	60.5	63.9	66.2	67.0	68.0
SSR											
bed-sits											
a)	1.8	4.2	1.5	2.8	3.8	3.0	1.4	2.1	1.6	0.7	0.5
b)					0.0	0.0	0.0	0.0	0.0	0.0	0.0
1 room + kitchen											
a)	7.4	8.2	9.8	1.6	20.0	13.4	13.5	8.0	6.6	6.8	7.3
b)					1.1	0.6	0.5	0.8	0.7	0.6	0.6
2 rooms + kitchen											
a)	72.4	64.7	38.6	23.6	25.5	19.5	16.9	18.1	14.7	14.3	14.5
b)					14.3	8.3	6.7	5.6	4.8	5.2	5.6
3 rooms + kitchen											
a)	17.5	21.6	45.5	52.2	41.8	46.6	48.9	56.8	60.9	64.9	68.0
b)					38.0	24.9	21.5	18.0	16.5	16.2	15.7

Table 3.1 (continued)

	1956	1960	1965	1970	1975	1980	1981	1982	1983	1984	1985
SSR											
4+ rooms + kitchen											
a)	1.0	1.3	4.6	9.8	8.9	17.5	19.3	15.0	16.2	13.2	9.7
b)					46.6	66.2	71.3	75.6	78.0	78.0	78.1
CSSR											
bed-sits											
a)	2.8	3.2	5.7	5.3	7.0	4.3	2.6	3.0	2.8	2.4	3.4
b)		0.9	0.1	0.1	0.1	0.1	0.1	0.1	0.1	0.1	0.1
1 room + kitchen											
a)	7.6	4.6	7.2	9.6	17.5	13.1	13.0	10.6	9.9	10.0	10.7
b)		11.6	4.6	5.5	3.1	1.5	1.3	1.2	1.2	1.2	1.1
2 rooms + kitchen											
a)	70.8	72.3	40.4	26.9	30.4	18.8	19.0	17.6	16.9	14.8	16.2
b)		57.6	46.7	28.1	19.4	10.7	8.8	7.7	7.0	7.0	7.1
3 rooms + kitchen											
a)	18.2	18.8	42.5	52.6	39.7	49.6	49.5	53.0	54.2	56.7	55.4
b)		21.1	35.2	41.0	38.0	26.7	24.6	22.2	20.8	20.1	19.5
4+ rooms + kitchen											
a)	0.7	1.1	4.2	5.6	5.4	14.2	15.9	15.8	16.2	16.1	14.3
b)		8.8	13.4	25.3	39.4	61.0	65.2	68.8	70.9	71.6	72.2

Source: Statistická ročenka, 1958, 1964, 1967, 1973, 1982 and 1986; in cases of discrepancy the later figures are given

there is inevitably a problem of interpretation when looking at statistics from different sources. We can see this if we compare the figure for average rooms per dwelling given in the UN Yearbook for 1967 (quoted in Schöpflin, 1970, p. 28) - 2.7 in 1961 - with that quoted by Musgrave (1984, p. 93) - 2.5 average for 1964 to 1973 (for new housing only).

The tradition of having to make do with a relatively small number of rooms, not to mention the not uncommon phenomenon of more than one household (generation) having to share accommodation, possibly explains why the figures most commonly cited related not to persons per room, but to average floorspace (Table 3.2).

The 1984 figure conflicts with the 46.3 square metres quoted in a recent economic report on Czechoslovakia (Lloyds Bank, 1986, p. 20), which probably excludes the private sector. Per capita living space, countrywide and irrespective of category of apartment or type of occupancy is currently quoted, for 1980, at 14.1 square metres (Anon, 1983, p. 106; Musgrave, 1984, p. 93). It is stated elsewhere, however (Lloyds Bank, 1986, p. 20), that 40% of families still live in apartments with less than 8 square metres per capita. This presumably applies to older state properties, which were smaller than the current average and in which the original occupants have stayed while their families have increased.

Between 1975 and 1984 there has been an increase in all the indices for average floorspace, but this has to be weighed against a drop in overall rates of construction (Table 3.3).

These figures confirm what every Czech 'knows', namely that many Slovaks are able to afford surprisingly large private dwellings. What truth there is in this is connected with the relatively more agrarian nature of Slovakia and the relative wealth in recent years of agriculturalists. Collective farmers also pay no income tax.

When it comes to amenities, we have to mention the official categorisation of dwellings. There are four basic categories, which apply nationwide and are unaffected by such considerations as land values or location. Category I has all amenities plus central heating; category II all amenities but no central heating; category III apartments are partially equipped with amenities, and category IV lacks all

Table 3.2: Czechoslovakia: Average size of dwelling in given years (square metres)

Country or Region	1948	1950	1953	1955	1960	1965	1970	1975	1977	1978	1979	1980	1981	1982	1983	1984	1985
ČSR	58.0	53.2	40.0	40.2	38.5	40.3	43.7	44.8	46.7	47.7	48.6	48.3	51.1	52.3	52.4	53.5	51.9
SSR	NA	50.0	40.9	38.8	39.8	42.0	47.5	46.3	49.0	49.8	50.7	51.6	53.8	53.4	53.6	54.4	53.9
CSSR	NA	52.0	40.3	39.7	38.9	41.0	45.0	45.3	47.5	48.4	49.4	49.6	52.0	52.7	52.9	53.8	52.6

Source: Statistická ročenka for various years with some reference to other statistical handbooks*

Note: *This particular table seeks to give the real averages but it is sometimes difficult to determine precisely what the figure should be. The sources, even different volumes or editions of entirely comparable works, give (apparently) contradictory data, due probably to the fact that some calculations are made on the entire housing stock for a given year, some on the newly completed stock only, or the latter with or without the private sector, or indeed the state sector only. Not all tables in sources are properly labelled.

NA = Not Available
CSSR = Country as a whole
ČSR = Czech part
SSR = Slovak part

Housing Policy in Czechoslovakia

Table 3.3: Czechoslovakia average living area in square
metres by construction category

	Nationwide		ČSR		SSR	
	1975	1984	1975	1984	1975	1984
Average overall	45.3	53.8	44.8	53.5	46.3	54.4
of which public	39.6	46.5	39.5	46.2	39.8	47.1
co-operative	43.8	46.3	44.3	46.8	42.5	45.6
enterprise	40.8	48.6	41.3	48.8	40.0	45.9
private	55.7	69.7	53.0	67.9	61.2	72.6

Source: Federální statistický úřad, 1986, p. 92

Note: ČSR = Czech part
      SSR = Slovak part

amenities. In 1984 about 70% (1978 - 60%) of all
the housing stock came under category I (Ončák,
1986, p. 29), while category IV is rapidly
becoming an anachronism, through urban
redevelopment and modernisation of rural
properties, often as second homes. In the same
year, 75.4% of all dwellings were connected to
water mains (ibid, p. 28). On the other hand, as
recently as 1961, about three quarters of all
dwellings lacked a bathroom or shower cubicle.
Since then, however, the position has changed
radically (Table 3.4).

Table 3.4: Czechoslovakia dwellings with bathroom or shower
cubicle (percentages)

1961	1980	1986
24.8	79.7	85.0*

Source: Raban, 1986, p. 76
Note: * estimate

88

In 1984, 61.6% of the population lived in accommodation connected to municipal sewers (Ončák, 1986, p. 28), and practically all have a lavatory. In cases where main sewer connections are absent, lavatories (chiefly in rural areas) may be of the flushing type, emptying, along with other household effluent to a cess-pit, with or without a set of overspill tanks, or rarely of the chemical type. Emptying of cess-pits is usually by courtesy of the local farm. Exterior toilets are restricted to the remaining unmodernised rural properties, often inhabited by the last of a family or by a new town-dweller, including descendants of the original occupant, for whom it is a second home and will therefore be modernised at leisure and subject to finance.

Mains gas is by no means universal outside towns and cities (40-45% of all properties in 1983, from 19.7% in 1961), but its use continues to increase (about 45% of all properties in 1986 [Raban, 1986, p. 76] and almost all new [Bauerová, 1986, p. 41]), while electricity is to all intents and purposes universal (Nový, 1983, p. 40). Quite a problem for tenants in some areas (e.g. Prague Vinohrady) has been the fact that the current was only 110 volts. Over a decade or so the many buildings involved have been rewired and upgraded to 220 volts, but until that happened, electricity was only a partial amenity, given the shortage or total absence of purchasable consumer durables to run on 110 volts. There are still pockets awaiting the change.

Central heating has increased from 8.6 percent of dwellings in 1961 to 56.8% in 1980 and an estimated 65% in 1986. Modern blocks are built to work from common district heating plants - a major element in the infrastructure, and one over which increasing care is taken in forms of design and siting. Systems serving just one flat or dwelling are increasingly available, and many are installed by self-help. As alternatives to central heating, coal-burning stoves (or more rarely open fires) have practically disappeared from towns to be replaced by gas and electric fires, but stoves of all types are found in rural areas.

Current requirements for amenities are that all flats, regardless of size, must have bathrooms or shower cubicles, separate toilet, kitchen or kitchenette with either gas or electric cooker, ventilation, kitchen sink and hot and cold running water. In new flats of over 71 square metres in size, there must be a second sanitation facility, (Raban, 1986, p. 76).

The proportion of actually substandard housing is difficult to assess; it depends how 'standard' is interpreted. The slums and shanties have gone, so too have the 17,000 'provisional structures' mentioned by Raban as still existing in 1950. Lloyds Bank Economic Report (1986, p. 20) refers to 25% of dwellings as seriously substandard. In part, from the context, this will have to do with overcrowding, as confirmed by Raban (1986, p. 40). However, there is official recognition of what else is considered unsanitary (by the decision of the public health inspectorate or housing authority), and this covers such obvious health hazards as damp or poor lighting (ibid, p. 39). Sanitary regulations also apply to accommodation which is not itself unsanitary, but does not suit the state of health of the occupants (e.g. upper flats occupied by heart-sufferers).

TENURE TYPES

There are four main types of tenure: rental from the state (administered by local councils at various levels); rental from an enterprise or through an enterprise co-operative; housing co-operatives and private ownership. Private rental has practically disappeared, quite rapidly, under socialism, though in 1950 it still represented 20.8% of all tenures (23.6% in ČSR, 14% in SSR). New regulations made it financially unviable, with rents being fixed by the state, restrictions on the amount of property one individual could own, mass expropriation etc. As the law stands, and it still affects 1.4% of tenants, the owner forfeits the right to the use of his property, income accruing going into a special account at the State Bank, released only for repairs and maintenance, and it is subject to a 25 to 51% housing tax (Raban, 1986, pp. 14, 64).

Rental in the public sector was basically the norm in the early post-war period, the state through its local agencies managing all previously public, expropriated or abandoned properties. It is still widespread, but the overall proportion is declining as co-operatives and private building expand. For the proportions in terms of occupation in 1980 (a census year) see Figure 3.1.

Figure 3.1: Structure of permanently occupied homes
in Czechoslovakia (1980)

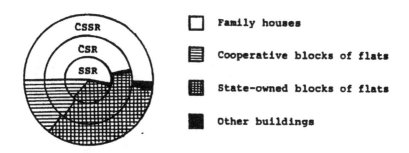

☐ Family houses

▤ Cooperative blocks of flats

▦ State-owned blocks of flats

■ Other buildings

Source: Raban, 1986, p. 69

Up to the 1960s enterprises (or hospitals, mines etc.) could provide accommodation to attract workers under advantageous terms, but later these developed with co-operative-like schemes to ensure proper financing and prevent job-switching once a flat had been obtained. Subsequently, in exchange for a guarantee to stay in the job, the employing body could write off up to 90% of the down payment it had placed with the co-operative. Other tenants could stay in their flat even if they left their job with the enterprise, provided they bought out the loan. Some housing continues to be built by enterprises, but it is of decreasing importance, most conspicuously in Slovakia (Table 3.5).

Co-operative housing was introduced in 1959 and is very widespread, especially in the country and in small towns. Over the last two decades it has accounted for about one third of all new building - just over 35% in 1980, but with the tail-off in state construction it has gone up to 48.6% in 1985 (45.2% ČSR; 54.3% SSR) (Table 3.4).

There has always been some degree of private ownership and it survived the early post-war years especially in the country (de Felice, 1987, p. 113). It is also in the country areas that 7 percent of Czechoslovakia's land surface is privately owned - mostly associated with housing, only marginally with farming. Schöpflin (1970, p. 448) shows that in 1960-1961 there was 50%

91

Table 3.5: Czechoslovakia: Dwellings completed according to construction sources

Source	Year						
	1970	1980	1982	1983	1984	1985	
CSSR							
Total	112135	128876	101829	95701	91863	102536	
of which:							
Public housing	18850	28362	21173	17263	15782	18179	
	16.8%	22.0%	20.8%	18.0%	17.2%	17.7%	
Co-operative	44240	45974	42040	45016	43520	49852	
	39.5%	35.7%	41.3%	47.0%	47.4%	48.6%	
Enterprise	19639	22149	8526	4395	3580	4897	
	17.5%	17.2%	8.4%	4.6%	3.9%	4.8%	
Private	29406	32391	30090	29027	28981	29608	
	26.2%	25.1%	29.5%	30.3%	31.5%	28.9%	
ČSR							
Total	73445	80661	61400	56897	57278	64798	
of which:							
Public housing	13062	17066	11979	10000	10554	12479	
	17.8%	21.2%	19.5%	17.8%	18.4%	19.3%	
Co-operative	34774	27447	23520	24428	25353	29257	
	47.3%	34.0%	38.3%	42.9%	44.3%	45.2%	

Table 3.5 (continued)

Source	Year					
	1970	1980	1982	1983	1984	1985
ČSR						
Enterprise	13566	16371	7239	3919	3362	4663
	18.5%	20.3%	11.8%	6.9%	5.9%	7.2%
Private	12043	19777	18662	18550	18029	18390
	16.4%	24.5%	30.4%	32.6%	31.5%	28.4%
SSR						
Total	38690	48215	40429	38804	34565	37747
of which:						
Public housing	5788	11296	9194	7263	5228	7700
	15.0%	23.4%	22.7%	18.7%	15.3%	20.4%
Co-operative	9466	18527	18520	20588	18167	20495
	24.5%	38.4%	45.8%	53.1%	52.6%	54.3%
Enterprise	6073	5778	1287	476	218	234
	15.7%	12.0%	3.2%	1.2%	0.6%	0.6%
Private	17363	12614	11428	10477	10952	11218
	44.9%	26.2%	28.3%	27.0%	31.7%	29.7%

Source: Federální statistický úřad, 1986, p. 92

owner occupation, a figure which had only risen to 55% of all permanently occupied homes in 1983. Figures for private building, as opposed to occupancy, are given for the entire post-war period in Table 3.6. Together with co-operative ownership about two-thirds of all dwellings are currently outside the state sector (Figure 3.1).

What constitutes a private house is regulated by law: it must be in use all the year round, it may not have more than five rooms (excluding kitchen, hall, bathroom, basement and loft), and if it has more than five rooms, the habitable area must not exceed 120 square metres (Raban, 1986, p. 65).

HOUSING TYPES AND OCCUPANCY

Detached houses are mostly to be found in the country, in small towns, or in what were (and in many cases still are) the more prosperous suburbs of larger towns. In the latter case the houses (villas) are often larger than would be tolerated under the post-war regime and have been divided into flats, managed by the local council. New construction of detached houses is highly restricted in towns, chiefly for lack of plots not thought better suited to other purposes, though it is widespread in the country. An innovation, still at the planning stage, is a return to the type of large detached house, divided into flats, as projected for the new Písnice estate in Prague; these are actually termed <u>vila-domy</u>, 'villa-houses' (Novotný, 1988).

The flight to the towns, typical of early industrialisation, has largely ceased, partly because of the more even distribution of new industry, with adequate commuting facilities from surrounding centres, and partly because of the new wealth in agriculture, which has led to a burgeoning of construction of often quite flamboyant new homes, first in Slovakia (early 1970s) and later more generally. Approximately 50% of all accommodation is in 'family houses'.

Walk-up housing is typically in the form of urban apartment blocks, pre-First World War or from the interwar period. All are administered by the local housing department as flats, individual flats varying from two to four rooms plus bathroom. In recent years there have been instances where two such have been knocked into one, one of the negative factors operating on the total dwelling stock. These urban blocks may rise to three or four storeys, often with un-numbered mezzanines, put in to by-pass pre-war regulations on maximum height. Being of mixed age, such buildings have different levels of amenities, though all are gradually being modernised. Many more recent public sector and co-operative types are also of the walk-up type, especially, but not solely in the smaller towns and villages, where anything taller would not be countenanced by the planners. The most striking developments are the five- to twelve-storey high-rise blocks that go to make up the modern estates, chiefly on the outskirts of the larger towns and cities. They will have all amenities including lifts (rarely out of order!). Such blocks are the mainstay of state housing and form entire townlets, self-sufficient (or intended to be) for shopping, education, leisure. In appearance these estates bear more than a passing resemblance to their cousins throughout Central and Eastern Europe.

Mobile homes as understood in the West are non-existent (except for increasing numbers of holiday trailers). Gypsies have travelled in this manner in the past, but this has been banned since 1958, when their caravans' wheels were forcibly removed (Puxton, 1980, p. 12). One type of caravan is used as temporary accommodation, namely on building sites serviced by contractors from far away. The function of such caravans is similar to that of many of the workers' hostels, and while the latter, like student accommodation are frequently included in statistics, the caravans are probably not.

Czechoslovakia's territory consists of 90% countryside, which is permanent home to 50% of the population. One quarter of the total population who are gainfully employed work in these areas, while another quarter live in the villages but commute to neighbouring towns and cities (Nový, 1983, p. 36). Commuting out of the smallest communities (up to 500 pop.) now affects 56% of their economically active population, while in communities of 500-2000 it is as high as 61%. All this means that there is a large population for whom existing village dwellings suffice (not adding to the pressure on larger towns), but at the same time there is a measure of new construction, of additional bigger detached houses and corporate housing. In part this is in

Housing Policy in Czechoslovakia

imitation of the kinds of housing already existing in towns; and in part it is due to the lower level of public control over existing stock, which can lead to neglect and destruction of traditional buildings and their replacement by newer housing. Not only is part of the rural architectural heritage lost, but unsightly modern low-rise blocks or stylistically incongruous urban-style detached residences can ruin the village or small townscape centred on parish church, town-hall and manor house (Nový, 1983, p. 72). In Slovakia family houses are favoured overwhelmingly, while they account for only about 50% of new construction in, for example, south Moravia. About 420,000 rural or semi-rural properties have become second homes for town-dwellers (over 80% of these in the Czech Lands), who, as elsewhere in Europe, dilute the rural population, but on the other hand this weekend population has generally no architectural ambition and is happy to re-use redundant farm buildings and cottages. Only in extreme cases have they gone for 'completely unsubstantiated snobbish stylizations' (Nový, 1983, p. 71) to out-folk folk architecture.

The suburban fringe is where the most striking developments have occurred, for it is here that the new tower block estates have arisen on green-field sites. In many cases they constitute entirely new self-contained suburbs of 5,000 to 12,000 population, and at their (aesthetic) worst, as to the north of Bratislava, they offer the impression of monstrous concrete barriers defending the town's approaches. South of Bratislava, across the Danube at Petržalka is a vast overspill area, home to about 100,000 people. Bratislava local planners have authorised annual construction of about 3,000 new apartments, mostly in multi-storey blocks (Carter, 1987, p. 821). Similarly in Prague the new 'Northern Town', 'Southern Town' and 'South Western Town' (not yet all completed) represent massive estates or agglomerations of estates of between 80,000 and 120,000 population.

Musil (1987, pp. 34-35) describes the extent of private housing or low-rise complexes that also appear in suburbs; Prague is the case in point. They are, he points out, usually situated beyond the big housing estates, cover relatively small areas in environments that have traditionally attracted 'better' housing. He also refers to the homogeneity of the occupants here, achieved not by class, but by access to status (top specialists, artists), funds (high incomes) or the ability to reduce costs (by having access to useful information on available sites, means of construction, knowledge of procedures etc.). The proportion of homes in this category is small and they are in areas not registered as separate statistical territorial units (Musil, 1987, p. 35).

HOUSING FOR SPECIAL POPULATION GROUPS

Housing for special population groups does not appear to have been widely described, but certain categories are well-defined in the literature.

Squatting as it is known in the West is non-existent, and would probably be precluded under the laws on parasitism. That there is, or has been, a small mobile, young, often criminal, drop-out population is attested, but chiefly in émigré literature (e.g. Pelc, 1985), but the characters who might elsewhere become squatters tend to put up with friends who have regular accommodation of one sort or another.

Visible (officially recognised) 'homelessness as one of the extremes in the housing system was eliminated after the socialist transformation of society' (Szelenyi, 1987, p. 2) or, to put it in the country's own terms: 'In view of the care of the state for the satisfaction of citizens' housing needs, there is virtually no likelihood of there being persons without shelter in Czechoslovakia' (Raban, 1986, p. 59).

People, especially families, on low incomes benefit from being given priority on waiting-lists, either in the local authority rented sector or for enterprise-run housing schemes with very low rents (2-4% of the average budget) (de Felice, 1987, p. 115 and note 13). Many also benefit from inclusion in co-operative schemes financed by interest-free loans granted by the employer as the member's contribution to the housing co-operative. Such loans may lapse after a certain number of years. Similarly, single-parent households may benefit from priority in waiting lists. Reference is made to the increase of such households due to the divorce rate, though single-motherhood is also by no means unknown. The divorce rate in particular is sufficiently high to have a major impact on housing need, otherwise Ončák (1986, p. 28) would scarcely need to mention it. In the absence of an immediate opportunity for rehousing, one or other divorcing partner may well return to the parental home, thus aggravating a situation (largish number of two-generation households) that the state has long sought to overcome. Only about one third of young couples have their own flat when they marry, and it can be five or more years before they acquire their own flat (Havelka and Špačková, 1983, p. 34) and if they have children before then, three generations may be living together. While once this would not have been unusual, the trend now is strongly against this pattern, and the grandparental generation increasingly lives alone. About one third of the over-65s continue to live with their children, one third are couples with their own household, and one third are solitary, the latter predominating in towns.

In 1980 there were 2.4 million Czechoslovaks over the age of 60. With improved heealth care the numbers of those over 80 are inceasing (270,000 in 1970, an estimated 350,000 in 1990, or 13.5% of all over-sixties) (Havelka and Špačková, 1983, pp. 9-10). While the policy is to encourage the middle generation to feel responsible for their parents, whether or not they live together, there is still a need for special accommodation for the most elderly and, usually, infirm. The non-infirm, as elsewhere, prefer to stay in their own home if possible. About 4% of all the aged are in the category for which a home is the only solution (Havelka and Špačková, 1983, p. 59).

Residential homes are either purpose-built (about 25%) or in conversions of former grand houses (about 75%); the latter have obvious drawbacks in the lack of lifts, less adequate kitchen facilities etc. In 1970 the number of flatlets available was 329 in 17 homes, while in 1980 this had increased to 7,314 in 331 homes (Havelka and Špačková, 1983, pp. 22, 43) and to about 9,900 in 1984 (Ončák, 1986, p. 27). The average age of beneficiaries of sheltered accommodation is 78-80 (Havelka and Špačková, 1983, p. 22). While homes average 10-15 flatlets, some of the most modern are much larger, such as that at Bratislava-Dúbravka, which houses 240.

Table 3.6: Czechoslovakia: Private house-building as a percentage of all new dwellings

Country or Region	1948	1950	1953	1955	1960	1965	1970	1975	1977	1978	1979	1980	1981	1982	1983	1984	1985
CSSR	45.2	23.8	23.9	30.7	24.3	24.7	26.2	26.7	29.7	29.4	29.1	25.1	30.8	29.5	30.3	31.5	28.9
ČSR	63.6	8.1	8.9	21.6	17.8	13.2	16.4	26.5	29.8	28.9	29.0	24.5	28.5	30.4	32.6	31.5	27.6
SSR	26.2	46.7	46.9	45.9	38.9	43.3	44.9	27.0	29.7	30.0	29.1	26.2	35.2	28.3	27.0	31.7	29.6

Source: Statistická ročenka for various years

Housing Policy in Czechoslovakia

Some flatlets are furnished; increasingly, however, the occupants bring in their own belongings as psychologically preferential. The premises are reckoned as part of the local authority housing stock and the occupants pay their own bills, on top of the 690 crowns (1983) 'rental'.

As with all categories of non-private housing there are waiting lists of varying lengths and there are some problems of distribution. Prague in particular is short of sheltered accommodation in the city and operates a home at Heřmanův Městec - over 60 miles away. The disadvantages of this situation for enlisting the co-operation of residents' families back in Prague, and their separation from their former environment are acknowledged shortcomings (Havelka and Špačková, 1983, pp. 54-55).

Ethnic groups as a special demand on housing are only a minor issue and this relates chiefly to Czechoslovak gypsies. The accommodation of Cuban or Vietnamese <u>Gastarbeiter</u> is outside the scope of this paper.

The Romany population of Czechoslovakia was claimed by the World Romany Congress in 1980 to be about 367,000 and rising fast. They used to be concentrated in Slovakia but many have moved, legally or otherwise, to industrial centres in the Czech Lands, where, 'housed in hostels and barracks, they later brought their families and relations, creating crowded conditions as bad as those prevailing in the old settlements in Slovakia' (Puxton, 1980, p. 12). Some arrived in the Czech Lands under a twinning scheme between Czech and Slovak villages and were allocated flats by district councils or enterprises. Again the arrival of relations led to overcrowding and new gypsy ghettoes (ibid.) Part of the problem is alleged to be that despite all the planning and funding put into resettlement, the Roma themselves have not been involved in the planning. Current policy (1980) is to improve conditions at the Slovak settlements, rather than break them up (ibid.). In the cities, whether Košice or Prague, where buildings on certain estates have been set aside for Romany occupation, the same horror stories are told, by Czechs and Slovaks, as I recall hearing in the 1950s, about gypsies on estates at Speke in Liverpool; similar tales were recently (1988) reported in a British television programme on Hungarian gypsies, but were strenuously denied. Musil (1987, p. 35) points out that gypsies are increasingly a new element in the population of run-down inner-city districts. Newlyweds as a social category enjoy the financial benefit of loans available to the under-30s at 1% interest if for flat purchase. The maximum loan is 30,000 crowns and 2,000 crowns is written off after the birth of the first child and 4,000 crowns for each additional child.

FINANCE

The overall share of housing out of total investment was at its peak in 1955 at 24%, dropping by 1966 to 12.7% to peak again in 1970 at 17.9% (calculated from Statistická ročenka at 1967 prices). Since then there has been a gradual decline to 12.5% in 1985 (at 1977 prices). Actual building finance comes from three main sources. a) The state pays the full cost of all housing in its care, and it contributes through grants about 55.5% of co-operative housing costs and 12.3% of private building costs. All told, the state's financial involvement in all construction is 53.7% b) Personal finance from savings accounts for 64.1% of the costs of private building, and 17.9% (the membership fee) of co-operative housing. Both these figures contain a loan element from sources other than: c) state loans for private building, which amount to 23.6% of the cost, and bank credits for co-operative building which amount to 28.6% of the costs. Overall the financial involvement of private individuals in total housing construction costs amounts to 29.6%, while the credit element accounts for 16.7% (Figure 3.2).

Out of the total costs from whatever source, about 34.8% of all construction is on state-owned flats, about 26.5% on co-operative buildings, and about 38.9% on private housing.

Private building is encouraged on multiple sites as the most economical and other sites may be made available at the discretion of the local council. User rights are granted quite cheaply or even gratis. Loans are available from the State Bank - up to 250,000 crowns repayable over 30 years at 2.7% (never called mortgages!) - or from the employer - 25,000-50,000 or higher for priority professions (e.g. miners) or those working particularly difficult terrain. Employer loans may not even be claimed back if the employee 'meets his obligations' to the employer. Pensioners, students etc. may get similar non-returnable loans from the local council, though the numbers of individuals involved are not quoted (Raban, 1986, p. 69). The total average cost of a privately built family house in 1980 was 205,101 Kčs (Ončák, 1986, p. 29), while the average state subsidy in 1984 was 38,000 crowns (ibid.). Clearly not everyone has the means to contemplate private building or a co-operative venture, and the cost of state or municipal accommodation in an average (60 square metres) flat is lower at about 10% of national average earnings than the 17% average repayments under co-operative schemes (de Felice, 1987, p. 114). State flats benefit from considerable invisible subsidies in that rental is not influenced by building costs, prices of materials, transport etc. (Musil, 1987, p. 28).

Co-operative housing is cheaper on larger residential developments than on small sites cleared for redevelopment, and the cost of the membership share (3,000 crowns fixed deposit, or half to two-thirds monthly income per household, plus 25,000-31,000 crowns according to size of apartment and overall costs of building) can be brought down by self-help in the construction process. The State Bank gives single-purpose investment credits to meet any shortfall in co-operative resources and other state contributions, but there is a ceiling to discourage wastage. Interest is charged, but cannot exceed 3% of the whole sum (Raban, 1986, p. 58).

The state subsidy amounts to 11,000 crowns per flat regardless of floor area, plus 1,200 crowns per square metre of the usable area of all the flats in one building. Extra subsidies can be made available for unusual conditions, such as the higher costs arising from a particularly large building, or for buildings in areas where plant movement costs exceed the average, or where the existing technology cannot be used with the same cost-effectiveness as elsewhere. The state will also meet all costs connected with electricity, water and gas supply, and, if the co-operative undertakes to supply the labour, also roads, septic tanks, wells etc.

Local enterprises may also subsidise co-operatives to the extent of 3,000 crowns per flat, provided a contracted number of flats is reserved

Figure 3.2: Structure of financial coverage in
housing construction costs in
Czechoslovakia in percentages (1985)

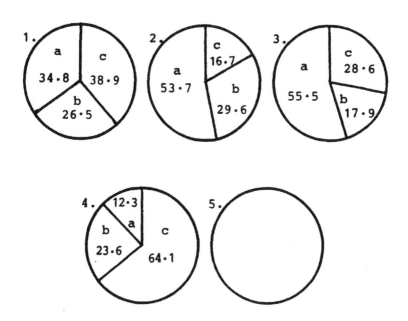

1. Total construction

   a communal
   b co-operative
   c private

2. Sources of finance

   a society as a whole
   b private resources
   c credits from the
     state

3. Financing of co-
   operative
   construction schemes

   a state contribution
   b membership fee
   c bank credit

4. Financing of private
   construction

   a state contribution
   b state loans
   c private resources/
     loans

5. 100% state coverage of communal housing construction

Source: Raban, 1986, p. 68

for the current employees. The company then pays the employees' membership share, which is written off after an agreed number of years. This pattern of involvement of enterprises in housing co-operatives has practically replaced other forms of enterprise sponsorship (Raban, 1986, table p. 58).

Loans to co-operative members may be up to 100% of costs and are available through the trade union organisation at 0%. I have seen no statistics for the number of applicants and recipients of such loans.

Rents in the state sector have remained basically unchanged since the war and have been frozen since 1964. By now this amounts to a 50% reduction in real terms, given increases in incomes, currently 2,700-3,000 crowns per month (Musgrave, 1984, p. 92). Most households will have two incomes.

The basic rent for a category I flat is 26 crowns per square metre per month, and for category III 10 crowns per square metre per month, but if the available living space exceeds 12 square metres per person, the rent will be higher. The rent rises according to the precise nature of amenities (gas stove, storage, basement, gas and electricity supply, balconies etc.), but to a maximum additional charge of 400 crowns per annum.

On the other hand, rent decreases with the number of dependants of the tenant: 30% with three children, 50% less with four. There are also rebates for flats shared by more than one family, for invalids etc., and reductions for such negative factors as basement location or exposure to noise (Raban, 1986, p. 41), but there is no rent differential between town and country (Carter, 1979, p. 429).

Rents are independent of means or any increase in income. The 1980 average monthly declared income of a working-class family is 4,556 crowns per month; the average size of such a family is 3.5; and the average rent amounted to 5.6% of the combined income of the tenants. Of the tenants in flats administered by the local authority for the state, 15.6% pay up to 100 crowns per month. For a summary of rents in the state sector and percentages of tenants paying them, see Figure 3.3.

Figure 3.3: Czechoslovakia: Monthly rents in the state
sector and percentage of tenants paying them
(1985)

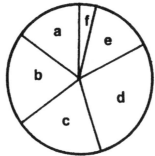

a   15.6% paying up to 100 Crowns p.m.

b   20.0% paying 100-149 Crowns p.m.

c   19.5% paying 150-199 Crowns p.m.

d   27.5% paying 200-299 Crowns p.m.

e   15.0% paying 300-399 Crowns p.m.

f   2.4% paying 400-699 Crowns p.m.

Source: Raban, 1986, p. 42.

RATES OF CONSTRUCTION

        The annual rate of construction over recent
years has been of the order of 8-10 per 1,000
population.   This annual rate multiplied
approximately six-fold between the war and the
mid-1970s, since when it has been trailing off
(Figure 3.4).
        Between 1945 and 1984 a total of 3,115,000
dwellings in all categories were built (Ončák,
1986, p. 28) of which approximately half date from
the period since 1970. Against this has to be
weighed the loss, through demolition and decay,
(45% of the increment in new housing in 1970-1980
[ibid.] or 20,000 dwellings per annum [Lloyds Bank
Economic Report, 1986]). Figure 3.4 shows how the
totals built since the war have generally risen,
with the recent tail-off, related to other
economic factors, while Table 3.5 additionally

shows for selected years, the changes in proportions in the different forms of development. It follows from this that the decline in total building has affected the state, and through it, the enterprise sectors, while co-operative private effort together has generally risen in absolute as well as percentage terms. Musil (1987, p. 34) notes that from the 1970s private building even began to increase in Prague. Previously there had been very little scope for it. Co-operative building has also increased in the big cities, reaching about 33% in Prague and Brno. The latest figures, from Statistické přehledy (1988/3, p. 75), reveal that the trend in the early 1980s has been reversed, with state-financed and co-operative construction taking an increase and decrease respectively in both absolute and proportional terms.

In 1950 there were some 3,613,000 dwellings, of which 2 million were over 70 years old. By 1985 a total of 3,217,000 new dwellings had been built, which means that now about 65% of the population is housed in post-war stock. Of the old stock much has been pulled down, and some has been lost statistically by knocking together of existing units or, in the country, by houses passing into use as second homes. Actual proportions of housing stock by age, for 1961 and 1980, are given in Figure 3.5. The total number of homes in 1985 was 4,909,000, of which 3,495,000 were in the Czech Lands and 1,414,000 in Slovakia. The peak of housing construction came in 1975 and the sixth five-year plan (1976-80) (Figure 3.4).

Over the whole country about 51% of all dwellings are family houses, 14.5% are in co-operative blocks of flats, 33% are in state-owned blocks of flats and 1.5% in 'other buildings'. For the Czech Lands the figures are about 46%, 16%, 37% and 1%, and for Slovakia 53%, 14%, 32% and 1% (calculated from Raban, 1986, p. 69; figures are for 1980). For a more detailed picture, including comparison with the situation ten years previously, see Table 3.7.

CONSTRUCTION DESIGN AND MANAGEMENT

Much of the larger-scale housing is to a handful of standard designs, the earliest ones imported, along with the production plant, from the Soviet Union (Carter, 1979, pp. 429-430).

Figure 3.4: Czechoslovakia: Dwelling construction 1948-1985

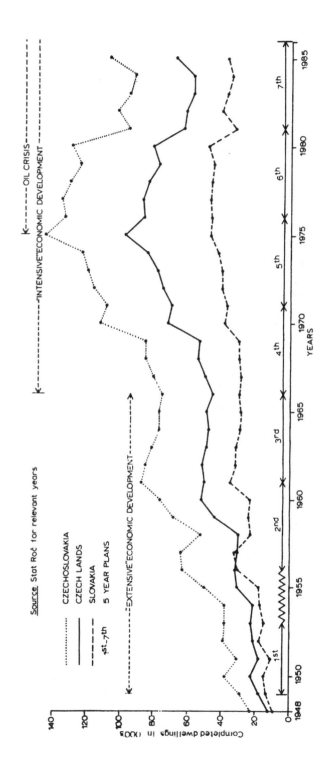

Source: **Statistická ročenka** for relevant years; adapted from Carter, 1987

Housing Policy in Czechoslovakia

Figure 3.5: Czechoslovakia: Age-proportions of housing
stock, 1961 and 1980

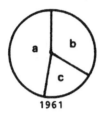

a   Housing completed by 1920

b   Housing completed by 1945

c   Housing completed by 1961

d   Housing completed 1961 - 1980

Source: Raban, 1986, p. 38

Up to 1960 they consisted of brick walls with
horizontal prefabricated concrete skeletons, but
more recent technology relies on reinforced
concrete building blocks and panels within steel
frames.   Production is serial, with erection on
site.

Self-help housing is based largely on
traditional technologies and is entirely dependent
on the availability of bricks or breeze blocks and
other raw materials.   Roofs are most commonly
covered in diagonally placed asbestos-cement
squares, frequently edged at the eaves with
galvanised iron sheets or squares.   An
alternative, especially in the country is tarred
paper over timber.   Individual areas evince a
dominance of cement or terracotta tiles.   Exterior
chimneys are often of fire-brick, subject, again,
to availability.   The private builder can also
avail himself of an assortment of pre-cast units,
doors, window frames etc., but the supply from
regular sources cannot match demand.   Pilfering
from other building sites is by no means uncommon
and there is plenty of work for moonlighting
carpenters and other craftsmen.   It is said to
take 4-5 years to build the average family house.

In the post-war period there have been
nationwide standards for design, overseen by the
State Design Organisation, Stavoprojekt.

Housing Policy in Czechoslovakia

This has subsidiary offices for regional planning
and for special engineering projects. There are
also a handful of minor co-operative design
studies scattered throughout the country. The
industry employs about 40,000 architects,
structural engineers, technicians, cost surveyors
etc. in a network of design organisations. One
function of these bodies is to ease the tension
between the demand for housing and aesthetic
considerations, but in the early period at least
quantitative considerations outweighed the
qualitative. Novotný (1988), commenting on a
recent exhibition to mark 40 years of the
'creation of the urban dwelling environment',
notes a trend away from 'economising' towards
'humanising' in the design of all dwelling types,
including tower-blocks of prefabricated panels.
State and co-operative design offices are
responsible for about two-thirds of all blueprints, especially in towns and cities. Other
designs, especially private ones, can be passed,
along with the issue of building permits, by local
authority bureaucrats. This has disadvantages in
that the latter are not necessarily so well
qualified as architects or planners, and the
system is also open to abuse. On the other hand,
private builders have recently been encouraged to
use state-approved designs as a condition of the
release of land (Musgrave, 1984, p. 92) and
current policy favours designs for two-generation
occupation (ibid.), which echoes the stated policy
of encouraging renewed awareness in the younger
generation of its responsibilities towards their
elders who are no longer fully self-reliant
(Havelka and Špačková, 1983, p. 91).

Apart from self-help in private and co-
operative ventures, all housing construction has
been in state hands since the big contractors were
nationalised in 1948. Nationalisation did not
solve all the problems in the construction
industry, which did not become fully stabilised
until the late 1950s. It was in 1958 that the
local national committees became the main
organisers of government housing policy,
regulating the distribution, use and building of
homes. The 1958 (11th) party congress outlined
the main directives for solving the national
housing problem, encouraging, among other things,
the development of projects using private
resources, i.e. the start of co-operative schemes.

Table 3.7: Czechoslovakia: Total housing stock (based on 1980 census)

CSSR

Type of building		Houses				Flats			
		total	permanently occupied	occasionally occupied	un-occupied	total	permanently occupied	occasionally occupied	un-occupied
Family houses	1970(1)	2295882	2180072	38817	76993	2579449	2461973	34121	83355
	1980	2423036	2181703	119223	112110	2699278	2431369	122576	145333
Houses consisting of flats	1970	204860	203102	70	1688	1736687	1699486	2144	35057
	1980	286265	282753	137	3375	2510837	2424614	4192	82031
Other buildings	1970	129857	60385	4228	65244	90279	77530	1556	11193
	1980	52229	39687	4747	7795	67668	52795	5136	9737
Total	1970	2630599	2443559	43115	143925	4406415	4238989	37821	129605
	1980	2761530	2504143	124107	133280	5277783	4908778	131904	237101

Table 3.7 (continued)

Type of building		Houses				CSR	Flats			
		total	permanently occupied	occasionally occupied	un-occupied	total	permanently occupied	occasionally occupied	un-occupied
Family houses	1970(1)	1494604	1408079	36402	50123	1740945	1652786	32245	55914
	1980	1567863	1384080	108863	74920	1813288	1604843	112045	96400
Houses consisting of flats	1970	172824	171396	44	1384	1406332	1376080	1710	28542
	1980	220542	218063	89	2390	1913556	1847659	2631	63266
Other buildings	1970	97660	48188	2698	46774	69354	59975	1200	8179
	1980	42486	32161	3975	6350	54567	42344	4298	7925
Total	1970	1765088	1627663	39144	98281	3216631	3088841	35155	2635
	1980	1830891	1634304	112927	83660	3781411	3494846	118974	167591

Table 3.7: (continued)

Type of building		Houses				Flats			
		total	permanently occupied	occasionally occupied	un-occupied	total	permanently occupied	occasionally occupied	un-occupied
					SSR				
Family houses	1970(1)	801278	771993	2415	26870	838504	809187	1876	27441
	1980	855173	797623	10360	47190	885990	826526	10531	48933
Houses consist-ing of flats	1970	32036	31706	26	304	330355	323406	434	6515
	1980	65723	64690	48	985	597281	576955	1561	18765
Other buildings	1970	32197	12197	1530	18470	20925	17555	356	3014
	1980	9743	7526	772	1445	13101	10451	838	1812
Total	1970	865511	815896	3971	45644	1189784	1150148	2666	36970
	1980	930639	869839	11180	49620	1496372	1413932	12930	69510

Source: Statistická ročenka, 1982, p. 105
Note: (1) for 1970 includes farmhouses

Each national committee or its subsidiaries is responsible for roads, water supply, heating plant, sewerage, transformers, and all other amenities - health centres, schools, shopping facilities and public transport. They also have sub-committees for investment and construction which contact design organisations on behalf of co-operatives, guide projects through the blueprint stage and maintain links with supply organisations to oversee actual contracts for materials and labour when required. It also acts as mediator for the co-operative in financial matters, acts with the co-operative on the takeover of the final structure and may assist a co-operative in making claims against builders. General management of the housing in its care is through the council's maintenance agency (OPBH), against which complaints are frequently to be heard (slow, inefficient) and in 'competition' with which other solutions are often found more successful (teams or individuals from various types of workshop with access to tools and materials, or unauthorised 'private' workmen, often moonlighting or 'daylighting' on their employers' time; this is a major area in the country's black economy).

Co-operative schemes are managed during construction and gradual completion by membership meetings (to determine, for example, priority of occupation), but the day-to-day management of an existing co-operative development is in the hands of an elected secretary in each high-rise block (Musgrave, 1984, p. 92) who all liaise with relevant departments of the National Committee.

RENEWAL AND REHABILITATION

Within towns there has been a considerable amount of redevelopment. Over 50% of the population live in what are towns by the official definition (the parameters are: population of at least 5,000; population density at least 100 per hectare; presence of administrative functions; workers in agriculture fewer than in the industrial or non-production sector; good standard of amenities).

In the first three post-war decades large-scale demolition was applied to housing deemed unfit for habitation. Partly this was to do with age, with 60% of the 1945 housing stock being over

70 years old. Eventually government measures were introduced to halt widespread continuing demolition, which is now the second resort after refurbishment and modernisation. Without knowing what guidelines or interests prevail in individual cases, one could not fail to be pained, in the early 1980s to see an entire street of modest Renaissance houses falling under the demolition contractor's iron ball in central Bratislava. Demolition has accounted for up to 18% of annual building programmes (cf. Donnison, 1967, p. 126). The rate of loss of housing through demolition or deletion from stock for a number of other reasons is currently (1981-1985) about 16%, i.e. the net annual national increment of dwellings is 84% of the total new units built. The net figure is significantly lower in the ČSR as opposed to the SSR (83% and 89% respectively in 1981, 81% and 90% in 1985), and has been lowest of all in North Bohemia (71% in 1981, but 83% in 1985), Southern Bohemia (73% in 1981, but 80% in 1985), Western Bohemia (77% in 1981 and 76% in 1985) and Central Bohemia 87% in 1981, but down to 72% in 1985). Prague itself shows a slowing down of the loss (77% in 1981, 87% in 1985). In Slovakia the net annual increments are currently fairly constant: 87%-88% in West and East Slovakia, 92% in Central Slovakia and Bratislava. If we look at some earlier yearly figures, it is clear that, for example, 1970 was a particularly poor year (net increment nationwide 72%, ČSR 71%, SSR 73%), for despite the relative leap in new building (see Figure 3.4) official figures reveal record losses in all categories, most of them in the Czech Lands. The net figures for the 1950s and early 1960s suggest a fairly steady rate of 80%-85%, but the true rates for the 1960s in particular are less easy to tell with accuracy, since those for 1966, showing a drop to a mere 53% in ČSR, 74% in SSR and a nationwide aggregate of 61%, are an officially admitted distortion, incorporating an unspecified number of units not duly recorded in 1965 and earlier (Statistická ročenka, 1967, p. 172).

The larger cities have had to contend with the problem of older properties (apartment blocks) gradually vacated by the death of older inhabitants or the transfer of the infirm to old peoples' homes. Prague's one-time working class area of Žižkov is a case in point where a great deal of reconstruction has been carried out, releasing modernised smaller flats for young families (Havelka and Špačková, 1983, p. 70). Medium-sized towns have been stagnating since the 19th century and have not seen much in the way of renewal or conservation. The popular preference is in any case for more modern accommodation with higher standards of comfort and hygiene. It is now seen as a matter of urgency to arrest decay. The listing of buildings for protection and conservation is now a standard practice, and the list is growing as 19th century buildings are added. Only 40 towns enjoy total protection. In the less historic centres new construction is generally limited to public buildings or commercial premises, new housing being left to the fringes. The 15th party congress, seeking to speed up construction to match the needs of a stabilised workforce, took measures to reduce building costs, including a programme of modernisation of older housing (up to 50,000 units by 1985). However, between 1961 and 1980, 989,000 homes were phased out of permanent use, so that 47.8% of new homes were to replace those that were outdated. This has had the effect that by 1980 65.1% of all citizens (57.8% in ČSR, 80.2% in SSR) were housed in properties not older than thirty-five years. Urban refurbishment can raise flats from Category 4 to Category 1, with up to 10-fold rent increases. This in turn can displace remaining tenants, some of whom will become priority cases for new housing, while some will go into homes; here there is a typical waiting-list problem (Musgrave, 1984, p. 93). One aspect of rehabilitation quoted by Carter (1979, p. 442) is the case of the centre of Prague. Here pedestrianisation has been accompanied by the restoration of historic buildings, some as dwellings, behind the original façades, to maintain their appearance, specifically to make it psychologically easier for the citizens to adjust from the old to the new.

Redevelopment of towns is fairly strictly controlled in particular to prevent the emergence of new 'false dominants' on the sky-line, i.e. within towns and cities high-rise blocks are a rarity. Also in these larger urban units the authorities tend to be more competent and do consult such bodies as ancient monuments commissions. This same virtue can, unhappily, be viewed by the private individual wishing to refurbish as one more irritant. Wishing, for

example, to install central heating in a dwelling
within an urban conservation area, he needs the
approval not only of the housing authority for his
ward, but also of a structural engineer and of the
conservationists, who must approve the siting of
the flue to avoid defacing of an historic facade.

A number of towns (31 Czech, 13 Slovak) have
been earmarked for no further development, while
some towns have come into being only during the
post-war period: Kladno, Havířov, Prievidza. Only
Havířov was born on a completely green-field site,
as were many of the agglomerations of estates on
the fringes of, say, Prague or Bratislava.

Rural renewal of buildings relates to
farmsteads and other farm buildings that have
become redundant since collectivisation and the
construction of new buildings for man, machinery
and livestock. Many are now being re-adapted as
dwellings and it is a matter of course that they
are provided with bathrooms, toilets and central
heating where they are for permanent habitation.
Cottages coming into use as second homes (chalupy)
will certainly be made structurally secure, but
not every new town-dwelling owner will want to
invest in total modernisation. Indeed as second
(weekend, recreational) homes they must not by law
be better than an owner's first home (Herold,
1981, p. 52). In the period since the war the
stock of buildings available for recreational use
has been enhanced by the expulsion of Germans, in
particular at the rural fringes, and by the in-
migration to the towns of parts of the Czech rural
population (ibid., p. 58). The huge spread of
recreational houses since the 1960s, is due to a
number of factors, including more leisure time
(non-working Saturdays since 1965, long dreamed of
before as anglická sobota 'English Saturday'),
increases in disposable income, and vastly
increased car-ownership (ibid., p. 59), though
urban overcrowding and the general crisis in
housing is probably rightly seen as a contributory
factor (ibid., p. 60), not least of all because
the average floor space of recreational homes (in
Central Bohemia) was 25% more than the average
floor space of Prague flats (ibid., p. 60). To
supplement the stock of cottages for second use,
there is a long tradition of building chalet homes
for the purpose. Previously controls in their
siting and construction were fairly lax, but the
blight they came to represent in some popular
picturesque areas and their somewhat random

intrusion on forest or agricultural land has led, since 1972, to much stricter inventorisation and control, with limits set (maximum 75,000 new units to be permitted from 1975-2000) and locations (for details of these and related provisions see Herold, 1981, p. 61).

HOUSING NEEDS AND MECHANISMS OF DISTRIBUTION

After the war the housing shortage was reckoned at about one quarter of a million, a large part of the problem being to do with unsuitability through age. By 1961, 50% of the housing stock was pre-1919 (Herold, 1981, p. 60), though 19th-century stock and older was down to 11% by 1983 (Nový, 1983, p. 60). By the beginning of the 1960s post-Second World War stock was beginning to deteriorate seriously, blame for which lies with poor maintenance (ibid.), though poor building quality, a common complaint of inhabitants in buildings of all ages, is probably just as much to blame.

In the 1970s it was estimated that 12,000 to 15,000 old units were required in Prague alone to satisfy general requirements for residential occupation (Carter, 1979, p. 439) if only to alleviate the high population density in some blocks. At the same time, failures of the transport system were a disincentive to any move to the suburbs. Since then, however, the huge northern and southern (essentially self-contained) suburbs have been built and are nearing completion, as is the underground railway and an improved bus network. Renovation of historic cores may as much as halve the number of dwelling units and, owing to regulations on space per capita, sanitary improvements etc., this net loss has to be compensated for in addition to new building. The cost of renovation may also be up to ten times the cost of new building (cf. ibid., p. 441).

A more recent estimate of current housing need is based on the fact that 5.7 million households are accommodated in 5.2 million permanently occupied units. While 92% of households live in separate accommodation and while some are 'known to prefer living together' (i.e. younger and older generations of the same family) there is still a shortfall calculated at 310,000 of which 150,000 are in Slovakia (Ončák,

Housing Policy in Czechoslovakia

1986, p. 29). This was chiefly resolved by the state planners and their agencies within the national committees. Private house-building is limited by the absence of a land market, though local committees may assign otherwise unwanted parcels for that purpose. Private initiatives are extremely costly, compared with state rental arrangements, though 'private house-building has been an instrument for improving the housing situation in smaller communities, small towns and in the outskirts of cities and it has been implicitly supposed that private builders are willing to spend more money and manual work in order to get better housing' (Musil, 1987, p. 28).

The major contributor to solving the housing problem in recent times has been the system of co-operatives. The share of co-operatives in overall investment in housing has increased steadily ever since they began in the early 1960s. They were initially encouraged as a means to serve a 'combination of collective and individual interest' (Musil, 1987, p. 28) connected with the ability to participate in house-building.

Involvement in a co-operative begins with the membership deposit. Thereafter the member continues to be involved financially, but may also be involved by provision of labour (of his own or additionally paid for to a third party) during construction. Management of a co-operative is in the hands of elected bodies who oversee assignment of the flats (studios, garages, etc.) built, who see, in theory, to the proper and timely maintenance of the co-operative's properties so that users enjoy the unrestricted benefits of use, and who ensure the provision of related services and organise any relevant additional projects (cf, Raban, 1986, p. 47). Founder-members approve statutes and elect appropriate committees, while the status of a co-operative as such is approved by the respective Czech or Slovak Union of Housing co-operatives. The supreme body is the membership meeting, which endorses or alters statutes, determines the order of proceedings, directs work and appoints delegates to conferences etc. The executive agent is the presidium, a steering committee which manages the property, admits new members, determines the size of deposits and deadlines for payment, allocates flats and sets rents. An auditory commission monitors all activities while the inspectorate is supposed to see that any shortcomings are removed. In law co-operative membership is open to all without regard to professional or political affiliations (even foreigners may join a co-operative), and is individual, i.e. no part can be taken by, say, an enterprise as block member. Since only one flat is allowed to an individual, it is not possible to become a member of two co-operatives and there should be no room for speculative involvement. The minimum age for membership is 18. Transfer of rights is possible, i.e. one member resigns to be replaced by another, and membership can be inherited in the direct line or otherwise disposed of by testament (cf. Raban, 1986, pp. 49-50).

The allocation of dwellings in all categories is fairly rigidly controlled. The free market as understood in non-socialist societies cannot exist as such, but there are certain areas in which the initiative of the individual can come into play, whatever the legal constraints on manipulation with property. It is entirely possible for an individual to apply for a new flat even before his existing accommodation has been declared unsuitable (for whatever reason). The extent to which this possibility is exploited in reality and the rate of successful applications is not recorded, but there is reason to doubt whether it has any major role to play. On the other hand the sense of proprietorship of even rented accommodation is very strong and flat-swapping is a widespread form of barter, as if with one's own disposable property, though any exchange is subject to administrative approval, which will also cover any financial make-weight to be considered. What will not show in the transaction is any monies that change hands as private inducements to the party presumed to benefit most from the exchange.

Land deemed suitable for building and owned by a party not intending to build cannot be disposed of through private sale. It must be offered to the state which then sees to the allocation. Thus even if the buyer knew of an intending purchaser, it may not go to him in the end, though clearly there is scope for abuse to ensure that the parties' wishes are met. While 'there is no market in land, and land values do not affect differences in the costs of housing as in the market economies, and practically do not influence the costs of ... housing' (Musil, 1987, p. 28) there is a fixed system of evaluation of real estate for transfer to the state and of sites intended for private occupancy. The average value of a site for the construction of a family house in the countryside (and no private housing site may exceed 800 square metres) is 3,000 crowns, in a medium-sized town 6,000 crowns and in Prague 10,000 crowns, with variation governed by ground plan and local conditions. How much less than these figures a seller receives or how much more a buyer (if allowed to purchase) pays - and some administrative charges must be involved - is not stated.

In principle there is very little scope for the exercise of private needs or aspirations in housing type or choice of where to live. The greatest scope is in private building or purchase from an existing private owner. A great deal of private building now goes on, and private purchase is by no means unknown. However, it is inevitably subject to local authority control, preeminently to set the price and administer the conveyance. But here too additional monies may change hands, allegedly two or more times over the 'agreed' price, and even gazumping can occur (verbal second-hand account of a case in E. Bohemia).

When it comes to private building, the builder is subject to a variety of legal constraints, chiefly implemented by the local national committee. The underlying legal provisions refer to planned growth and take into account the sites and materials available, owners' ability to complete a structure, and general ecological, town-planning and architectural considerations. The local committee has delegated to it powers to designate sites for private building (where these are not already in private hands), sets specifications for a given site (whether the proposed premises may be detached or semi-detached, the number of floors etc.) and deadlines for completion. The proposal, including a description of amenities and transport facilities is then open to public scrutiny. The committee can then sign a contract with a prospective private builder-owner, which guarantees freedom of use of the site though the site itself remains in law 'social property'. Subject to the use of standardised designs such plots are on a 99-year lease (Musgrave, 1984, p. 92). Similar general functions apply in the allocation of sites to co-operatives. In all cases the local National Committees are first and foremost direct instruments of the state, but not necessarily in the interests of the local community to whom they are theoretically accountable. In the summer of 1988 I met an instance where, despite the general accountability of local committees to the state before the interests of the community, it was an issue of great local importance that motivated one particular committee's actions: a number of adjoining plots of under-used or unused low-grade privately-owned meadow were bought up at Kčs 4.00 per square metres by the committee, with little or no resistance from the owners, to be re-parcelled and sold at a mere Kčs 2.00 per square metres to young couples, provided an undertaking was given to build a private house within three years - less than the average time such an enterprise takes, but presumably the parties concerned would continue to receive every encouragement and help needed. The remaining Kčs 2.00 was to be recouped from the state in due course. The underlying motive was to encourage residence by people likely to have children in the foreseeable future, since the village school was under threat of closure because of falling rolls.

Local committees not only issue building permits, but are also in charge of territorial and area planning and the distribution of flats and non-residential premises (Raban, 1986, p. 33). They are involved in all aspects of investment in housing, including quality control, safety, aesthetics and environmental considerations, and in granting occupants permits to convert to living space areas not so used previously.

For the distribution of state flats, applicants are placed on a waiting list. The system seeks to be egalitarian, and probably is more so than in the earliest post-war period, when many larger properties, having been vacated by or confiscated from the wealthier property-owning classes were re-allocated to deserving members of the new society or the new intelligentsia, top party bureaucrats, high-ranking officials and any personnel etc. (cf. Szelenyi, 1987, p. 2). Currently waiting lists are based on a mixture of social need and social involvement. A points system is operated and lists are open to public inspection (Musgrave, 1984, p. 92). The primary categories affecting position on a waiting list are: low income families, applicants with no home in the constituency, married couples living apart,

persons who have flats of their own, but in other
communities than their place of work, applicants
whose present flats are considered unsanitary or
otherwise unsuitable, people evicted from existing
accommodation by court order, applicants whose
claim is based on their situation in the community
or the importance of their work (which is judged
by the standing of the profession generally and an
applicant's performance in it; in another context
Raban [1986, p. 59] points out that being a worker
or a teacher is a bonus), and victims rendered
homeless by natural disaster. The last category
and those evicted by court order (resulting, for
example, from a requisition order in the national
interest or redevelopment plans) are those most
likely to short-circuit housing lists. Just as
the Local National Committee oversees incoming
tenancies, it also takes charge where tenancy is
voided by death, renunciation or exchange of
flats, court order or its own decision. Grounds
for eviction are many and varied and also include
cases where an employer has been involved in the
tenancy arrangement and needs the accommodation
back from a former employee to make way for a new
one, cases of failure to fulfil duties connected
with the use of a given flat, the holding of two
flats, and leaving a flat unused or almost so. By
law an occupant being evicted must be found
suitable alternative accommodation or
accommodation adequate to his genuine needs (which
are, of course, determined by the authorities).
If space in a flat exceeds 18 square metres
tenants can be moved to smaller accommodation, a
rule said to be applied rigidly in the case of
younger people, but more leniently in the case of
the elderly (Musgrave, 1984, p. 92).

Allocation by co-operatives is governed by
generally similar considerations (getting people
out of unsanitary accommodation as soon as
possible, key-worker settlement etc.), but here
the waiting lists (which may be related as much to
the order in which units are built as to the order
in which they are allocated to members) are in
principle approved by membership meetings. The
members themselves divide into those with urgent
and those with less urgent demands on the new
housing; a points system is operated. Urgency
theoretically includes non-possession of accommodation of one's own at all, and needs which vary
from the facilities afforded by the current
accommodation. This covers young families that

have outgrown existing accommodation, families already divided against their will and being accommodated with both sets of parents or otherwise separately, and any members who are currently having to share accommodation. In practice, however, an unmarried adult, aspiring to the independence his/her flat may give him/her, and in practice 'sharing' the family accommodation, and therefore entitled to be viewed as fairly urgent, may well be treated as non-urgent if the original family home is not actually overcrowded, whatever tensions may otherwise exist there. The second parameter governing co-operative waiting lists is duration of membership; membership will here be treated merely as an investment and if no case of urgency applies it can take years before a flat is allocated. Despite all the provisions, the allocation system is wide open to abuse by management, as is evinced by a notorious case in Karlovy Vary, first reported in Rudé právo on 28 February 1988. The full account of the chairman's machinations, dilatoriness and gross abuse of the points system was subsequently reported on 30 August 1988. The entire management and auditing committee had to be removed by decision of the Czech Union of Housing Co-operatives and the culprits were to be dealt with by the party. Curiously, the whole scandal had gone on since at least 1985; equally curiously it seems to be the District Committee of the Party, and not the courts, which has had the main say over the legal aspects.

CURRENT PROBLEMS AND POLICIES

Having eliminated the bulk of the 19th-century and inter-war stock, so that currently over 50% of dwellings are of post-1945 construction, the country is nevertheless faced with a 'severe and persistent absolute housing shortage', although the gap between households and dwellings has dropped since 1970 (671,000) and 1980 (278,000) (Musgrave, 1984, p. 92; Musgrave advises caution in the handling of such statistics, but as with many here quoted, there is little beyond official statistics with which to operate). In the two decades before that overcrowding had been particularly acute, and between 1950 and 1961 Czechoslovakia had been the only country in Europe where housing conditions had actually deteriorated (Herold, 1981, p. 60; also cf. Table 3.2 above). Many of the current problems are a throw-back to the immediate post-war period, when urbanisation was slow to get off the ground, investment in housing generally being given low priority as against the massive concentration in industry.

Early shortages of building materials and skilled labour were a further impedance and this too has not been adequately overcome (cf. the difficulties known to be faced currently by the would-be private builder). Musgrave also notes the excess over need in 1970 of 500,000 small dwellings, and the general conflict between household sizes and the size distribution of dwellings. As mentioned above, many (though progressively less) new young households have to live for different lengths of time with one set of in-laws, or indeed each spouse with his own parents. Building completion rates are at first sight fairly satisfactory and rising: in 1987, 79,310 dwellings of all types were completed (19,200 state, 33,500 co-operative, 1,800 company and 24,800 private), but actual completions are very unevenly spread through the year, a recurrent complaint in issues of Statistické přehledy (cf. 1988/3, p. 75), and produce such distortions as in December 1987, when 29.8% of the annual capacity was handed over. This is a clear indication of haste in meeting deadlines and catching up with the state plan. Statistické přehledy (ibid.) speaks in general terms of many instances of unsatisfactory quality of flats as handed over and of shortcomings in the infrastructure on new estates (common complaints of the population at large), and lays the blame in part at the door of the planners. Even with the haste to meet the 'building objectives of the state plan' targets are not always met and housing is among the areas to suffer most. In addition to the plan to construct 480,000 new homes between 1986 and 1990, which may well not be met, the aim is also to modernise 30,000 in the same period. In 1987, 4,474 dwellings were modernised ('1500 more than in 1986', Statistické přehledy, 1988/1), which at the same annual increment would reach that target, but the current rate is deemed 'unsatisfactory', suggesting that no such increment is envisaged. Increased rates of upgrading are planned (Raban, 1986, p. 89), but a threefold rate of increase in 1990 over 1986 seems practically inconceivable.

The long-term aim is to modernise and convert all third and fourth category flats into first and second category by the year 2,000. Special provisions in the 1986-90 plan envisage the construction of 11,000 smaller flats in retirement homes, with all relevant services (assisted shopping, cleaning, washing, laundering, meals on wheels), and single-persons' hostels with a total capacity of 10,000 to serve in the main employees in the health service (ibid.). Other policies designed to speed up the solution to the housing problem are additional support for co-operatives, whose share in the overall output is to grow, a ten-year extension to credit-repayment deadlines, which should make co-operative enterprises even more attractive by reducing the monthly repayment and similarly, additional support for private builders through increased subsidies, higher bank loan limits and a ten-year extension of repayment terms (ibid.). With regard to infrastructure, areas designated for improvement in the current period are health-care facilities, provision of shops, sewage disposal and water supply.

The relative costs of different types of housing have long weighed heavily in favour of council-managed properties, which, as noted above, are intended in the first instance for lower-income families with a number of children. Rents were frozen between 1939 and 1964 and even now represent only a small proportion of the family budget. There appears to be no intention at the present for rents in the state sector to be revised. Rumour has been (mid-1988) much concerned with whatever other increases in retail prices or service charges might be round the corner, given that pensions are all to be adjusted upwards, but nothing is known for certain. The television phone-in on the pensions increases dealt almost exclusively with details of administration of the change and how various anomalies were to be removed. Rumour is also somewhat concerned with the suggestion, genuinely being discussed in high places, that the currency be put on a different footing to eliminate profiteering on black-market currency deals once and for all. If such a change were to be brought in, it seems likely that wages would receive a massive boost, prices would have to be restructured and in all probability then, if not before, there could be some adjustment to rents. At the time of writing, however, this can be no more than conjecture. One other factor that could influence things in the near future is that Czechoslovakia, like the rest of Eastern Europe, is undergoing a version of general economic restructuring. The introduction of <u>khozraschot</u> (financial self-sufficiency of enterprises) could lead to certain price increases in, for example, the building materials industry. This would in turn increase the unit cost of public sector housing and with it the proportion of state subsidy. One way to reduce any additional burden in this area, which could be brought on by other factors as well in a period of change, would be the imposition of a general increase in rents. However, this too is conjecture and time alone will tell.

Until the recent increases in subsidies, the incentive to build privately was low, though surprising numbers have made the sacrifices in money, time and effort required. Despite forty years of socialism, a strong sense of proprietorship survives (extending of course beyond housing) and the increasing rates of second-home ownership (over 25% of the inhabitants of the largest urban centre, Prague, now own one) and of co-operative building are indicative of this. It helps to explain why many people join co-operative schemes while having no actual urgent housing demand. The attractions are not only affordability, but the ultimate acquisition of a property to be disposed of at will, with the certainty of a return of investment, or vacations. The opportunity to comment on design is also more likely to be taken up on co-operative designs than on state-sponsored projects, although there too the chance exists in theory, through the public exhibitions of new designs. Novotný's article (1988) contains a searing indictment of the system and suggests that no more than lip-service is paid to public discussion. The contrasts between the findings of a 1954 survey conducted by the Construction and Architecture Research Institute and the Central Council of Trade Unions into public aspirations regarding housing and the housing actually built in the sequel could not be greater; not one of the public's main preferences was put into practice. He is sceptical whether even now the situation would be any different. More official sources claim that public involvement at the blue-print stage has led in some instances to alterations before large-scale building was undertaken. Clearly, this is not enough, and, while Musgrave found 'no evidence to show that the residents of Czech high-rise flats feel strong antipathy to the building type as such' (1984, p. 92), the official sources at least admit that there has been criticism of excessive uniformity and stereotypes. A common complaint about the earlier types of post-war blocks is the lack of noise-insulation on the flimsy party-walls, which is so bad that the 'house regulations', hanging in the vestibule of every block, contain an injunction against 'disturbance of the peace of the night' from 10.00 p.m. Even the operation of a (usually noisy) washing-machine after that hour can constitute an infringement and is therefore a source of secondary dissatisfaction. An early, graphic reflection of noise-penetration between flats in the literature of the 'thaw' is to be found in Páral (1965, passim). Currently, some of the residential dissatisfaction might be overcome by a greater role played in design, of estates if not of buildings, by sociologists, who have complained that they are not involved enough (Musgrave, 1984, p. 93).

Raban notes that the inhabitants of housing estates spend on average more time at home than those living in detached houses or older urban housing. This is used as an argument against any allegations of 'monotony and the nondescript nature of life on housing estates' (1986, p. 78). In part, however, it will be due to the predominance of young families on the estates, where the average age is 15 years below that of the rest of the population; this sector of the population will have less time or need to be out and about. Another determining factor is probably distance from town/city centres, at least in the largest cities. At the same time Raban repeats that certain other categories on the estates (workers - presumably not those with young families - and pensioners) visit friends more often, go to the cinema more often, and, in the case of pensioners, do more gardening (where, is not stated). They also pursue entertainment more than their co-evals not on estates. None of this speaks of strong dissatisfaction, but is symptomatic of at least something partially lacking from estate life.

Other sources of dissatisfaction are the inability of the state to match supply with demand (the newlyweds' problems referred to earlier) and

127

flat size and condition. Provisions exist for larger families to be moved to larger accommodation (over-crowding being one of the grounds for the public health inspectorate to declare accommodation unsuitable and/or unsanitary), or for, say, tenants with a heart condition to be moved from a higher to a lower-floor flat, but there are inevitable delays before any particular situation is resolved.

Thus there are still outstanding problems in the overall housing situation although the main policy aims of the post-war period are deemed to have been met. These were a) 'to liquidate the inherited differences of a material, social and cultural character, to give women equal rights in society and to level the living standards of all working people', b) 'to remove the existing differences between industrialised and economically less developed regions, especially between the Czech Lands and Slovakia', and c) 'to remove the basic differences between towns and villages and to protect their environments' (Nový, 1983, p. 24.) This meant the provision of adequate housing and public amenities in all towns and villages. It also meant a reduction in the maximum-minimum wage differential, and the mixing of the population in accommodation and by marriage. There was also to be differentiation in capital investment, economic incentives, preferences and subsidies, industrialisation was to be (more) evenly spread, ahead of urbanisation. The labour pool was to be gradually relocated to towns, while villages were kept in existence, with improvements in housing for a largely commuting labour force.

RELATIONSHIP OF HOUSING TO INDUSTRY

Changes in the pattern of industry have meant that between 1937 and 1980 employment in industry doubled, to 20% of the population (40% of the gainfully employed). The spread of industry means that where previously there were just three or four large concentrations of industry there are now some 1,240 works and 15% of the secondary sector in communities with as few as 1,000 population, 3,000 in communities of 1,000-5,000 (31% of the industrial population). Forty-three per cent of the industrial population work in factories in towns in the 5,000-20,000 population bracket, while only 26% are in towns and cities of over 100,000 (cf. Nový, 1983, p. 30). New industries have absorbed the agricultural labour force rendered redundant after the collectivisation and modernisation of agriculture without giving rise - in official eyes - to sprawl. Between 1950 and 1975 the number of workers in agriculture dropped by 50% (to under one million) but they retained their housing and either work in new local industrial enterprises or commute to nearby towns and their new works. One of the keys to the priority of industry over housing was the need to build up that sector which produced housing construction materials (Nový, 1983, p. 38; Musgrave, 1984, pp. 91-92). Industry has shown the greatest increments in manpower, rising by 1.8 million since the war, partially from the loss in the agricultural workforce, only very partially from an increase in the overall population (which grew from 12.5 million in 1950 to 15.3 million in 1980) and partly through the entry of women into the labour market. New and expanded industry and the accompanying back-up services are the main 'town-constituting factors', since settlement in the towns has always been controlled, reliance being placed on commuting in from nearby smaller centres, and no one centre suffering any major influx of population. Slovakia shows the most change, as a product of its industrial 'catching-up'. There the population increase was also proportionately higher, from 3.4 million in 1945 to 5 million in 1980. Urban population growth, with a proportionately much larger pool of rural workers to move in, has been more striking: 15% of the total 2 million increase in urban population were absorbed by Košice and Bratislava, 48% by towns of 5,000-20,000 inhabitants and 37% by towns over 20,000 (Nový, 1983, p. 53).

Related to the post-war industrialisation was the policy of using new state housing in the newly industrialised areas (Musil, 1987, p. 28). As Donnison put it (1967, p. 126): 'The large number of small towns, closely spaced, throughout Bohemia gives workers a considerable range of jobs to choose from, and in a country whose government has been reluctant to use wage rates or any other price mechanism to distribute resources, housing privileges have provided an alternative means of attracting and retaining labour for the industries to which highest priority is given.' To a lesser

extent this will still apply, but private
buildinghas become much more widespread and is
being encouraged, and it no longer applies that
successful applicants for plots and permits have
to work in priority industries (cf. Donnison,
1967, p. 127). Moreover, it is currently
encouraged, despite a measure of land loss, partly
because of the constitutional matter of family
cohesion and concern for child welfare and the
higher housing standards in which individuals are
prepared to invest. One incentive for the
government to see housing increased and improved
is the need to reduce the high divorce rate and a
desire to see an increase in family size.

The ratio of state, private and co-operative
construction is still determined by government
policy, but various official documents suggest a
belief that the right balance has now been
achieved. The workforce is now more or less where
it needs to be, though key industries (mining) may
still send out recruiting agents all over the
country. The main solution to housing new young
single employees is hostel accommodation attached
to the enterprise, or, for the lucky ones, a
proper company-owned flat.

The housing situation in rural areas and
small towns is felt to have been stabilised, with
few, if any, serious shortages, and the aim now is
to create a surplus in the larger towns. The
cardinal point in future territorial plans is to
harmonise immediate benefits and the long-term
responsibility to future generations (Nový, 1983,
p. 49).

FUTURE TRENDS

Factors that affect future planning are the
continuing rapid increase in car-ownership (i.e.
need for increased garage space); noise and other
pollution; increased space inside flats to
accommodate more domestic machinery - washing
machines, driers, freezers etc., representing in
part an admission of the inevitable desire of the
population to keep abreast with households in W.
Europe, and in part a recognition that some
communal services (such as osvobozená domácnost
['liberated household'], which offers, amongst
other things, laundering, garment-repair, curtain-
hanging etc.) can be slow, incompetent and
inefficient; strict control on conversion of

additional agricultural land to housing; continued and accelerated modernisation of existing stock, including a finer balance between renewal, demolition and new construction; and the more effective use of existing sites in urban areas. The problems are, as elsewhere, to integrate human demand for improvements in the size and quality of housing with all other priorities (conservation, recreation, transport). The official line continues to refer to socialist society as striving 'for the maximum satisfaction of the material and spiritual needs of citizens, but it does not consider property a yardstick of the social status of the individual' (Raban, 1986, p. 84). Some of the same citizens would of course insist that it is, however, a mark of the political status of some individuals. For example, the outsider is occasionally treated to tales of second, or even third, flats held by the powerful in the land for the use of mistresses or whatever, the offending accommodation being registered in the name of sundry maiden aunts etc. actually living elsewhere.

While on the one hand some satisfaction of general civic aspirations may come from the widening choice of types of accommodation, even in the state sector, it is doubtful whether the state will succeed, as it hopes, in eliminating 'such unhealthy trends as the prestige-motivated hunt for outsize luxury apartments' or so much property that the owner cannot use it rationally (ibid.); much still depends on who the owner is.

Despite the main official line of preventing the creation or acquisition of outsize or over-luxurious properties, one recent (since early 1987), though as yet not wide-scale trend, is the return to the original owners of some properties confiscated after 1948; the state and its local government agents simply do not always have either the means or the manpower to ensure proper maintenance, and if a former owner is interested his property can, apparently, be returned by a simple act of conveyance (private communication). It is conceivable that such transfers of stock out of the state sector may increase, given the deepening economic crisis facing the country.

Housing Policy in Czechoslovakia

CONCLUSION

Unless something intervenes to upset the current policies and policy orientation, the goal currently pursued by Czechoslovakia is for every household to have its own flat/house by the year 2000. It is not clear to what extent planning has taken into account the recent upsurge in the birthrate, any failures or indeed successes in the reduction of the divorce rate, any effects on the general labour market should Cuban and Vietnamese Gastarbeiter return home, any frustrations occasioned by thwarted aspirations to a second home, once the existing stock has been taken up, or any worsening of the situation with the supply of materials. Outside official documents for foreign consumption, which tend to underplay such difficulties as they do concede, there is clear official dissatisfaction (in, for instance, the commentaries in Statistické přehledy) with both the rate of completion and the quality of housing currently going into service. There is certainly no complacency here, in view of the costs of rectifying shortcomings, but there is little evidence of hope for improved performance in other than purely numerical terms. It remains to be seen how far economic restructuring will go and what effects this will have on public and private building costs, rent levels, and any consequent reformulation of the targets for the turn of the century, not to mention any further developments in transfers from the public to the private sector.

ACKNOWLEDGEMENTS

I wish to thank for their assistance with this chapter my colleagues Dr F. Carter and Dr A. Smith, who gave readily of their advice. Particular thanks are due to the former for having made available to me some of his unpublished materials (Carter, 1987). I am also grateful to Mr J. Pavlíček, Press Attaché at the Czechoslovak Embassy, for making available to me many of the official publications quoted herein.

REFERENCES

Anon: Socialist Czechoslovakia. Prague: Orbis, 1983
Bauerová, Jaroslava: Czechoslovakia in Facts and Figures,
    Prague, Orbis, 1986
Carter, Francis W.: 'Prague and Sofia: an analysis of
    their changing internal city structure'.
    In R.A. French & F.E. Ian Hamilton (eds):
    The Socialist City: Spatial Structure and Urban
    Policy. Chichester, New York, Brisbane,Toronto:
    John Wiley & Sons, 1979, pp. 425-459.
____ Unpublished MS originally intended as part of
    'Czechoslovakia' in A.H. Dawson (ed.):
    Planning in Eastern Europe. London: Croom Helm,
    1987, 103-138; page nos. quoted are from the MS.
Donnison, D.V.: 'Housing in Eastern Europe -
    Czechoslovakia'. In The Government of Housing,
    Penguin, 1967, pp. 125-129.
Federální statistický úřad, Český statistický úřad,
    Slovenský statistický úřad: Čísla pro každého.
    Prague: SNTL; Bratislava: Alfa, 1986.
de Felice, Micheline: 'Housing', in Pierre Kende and
    Zdeněk Strmiska (eds): Equality and Inequality in
    Eastern Europe. Leamington Spa, Hamburg, New
    York: Berg, 1987, pp. 111-127.
Havelka, Jaroslav and Špačková, Dana: A Contented
    Old-Age in Czechoslovakia. Prague: Orbis, 1983.
Herold, Laurance C.: 'Chata and chalupa: recreational
    homes in the Czech Socialist Republic'. Social
    Science Journal, vol. 18/1, Jan. 1981, pp. 51-68
Kansky, Karl Joseph: Urbanization under Socialism:
    the Case of Czechoslovakia. New York, Washington,
    London: Praeger, 1976 (esp. Chap. 4 pp. 66-125).
Lloyds Bank: Economic Report, Czechoslovakia. London:
    Lloyds Bank, 1986
Matějů, P. et al.: 'Social structure, spatial structure
    and problems of urban research: the example of
    Prague'. International Journal of Urban and
    Regional Research, vol. 3, 1979/2, pp. 181-202.
Musgrave, Stephen: 'Housing in Czechoslovakia'.
    Housing Review, vol. 33, 1984/3, pp. 91-93
Musil, Jiří: 'The redevelopment of Prague's ecological
    structure'. In R.E. Pahl (ed.): Readings in
    Urban Sociology. Oxford: Pergamon, 1968, pp. 232-259
____ Urbanization in Socialist Countries. White Plains,
    NY: M.E. Sharpe; London: Croom Helm, 1980 (esp.
    'Introduction', pp. 3-16, and 'Chap. 1 -
    Czechoslovakia', pp. 17-44)

Housing Policy in Czechoslovakia

——— 'Housing policy and the socio-spatial structure of cities in a socialist country: the example of Prague'. International Journal of Urban and Regional Research, vol. 11, 1987/1, pp. 27-36

Novotný, Jan: 'Podle mého názoru', Tvorba, 23 November 1988, p. 20

Nový, Otakar: Town and Country Development in Czechoslovakia. Prague: Orbis, 1983

Ončák, Oto: Living Standards and Social Certainties in Czechoslovakia. Prague: Orbis, 1986

Páral, Vladimír: Veletrh splněných přání. Prague: Československý spisovatel, 1965

Pelc, Jan: ... a bude hůř. Cologne: Index; Paris: Svědectví, 1985

Puxton, Gratton: Roma: Europe's Gypsies. 3rd edn. London: Minority Rights Group, 1980

Raban, Přemysl: Housing Policy in Czechoslovakia. Prague: Orbis, 1986

Schöpflin, George (ed.): The Soviet Union and Eastern Europe - A Handbook. London: Anthony Blond, 1970

Statistická Ročenka, Prague, SNTL, 1967

Statistické Přehledy, Prague: SNTL, 1988

Szelenyi, Ivan: 'Housing inequalities and occupational segregation in state socialist societies'. International Journal of Urban and Regional Research vol. 11, 1987/1, pp. 1-8

Chapter Four

HOUSING POLICY IN ROMANIA

David Turnock

INTRODUCTION

As in other socialist states the central
government in Romania has become deeply involved
in housing. Some 4.4 million apartments and
houses have been built by the state so that a
large part of the population can enjoy modern
living conditions. And the government believes
that a solution to the housing problem should be
to hand by the end of the decade by which time a
further 2.4 million homes will be available (ICCE
1983 pp. 285-286). This effort has been made
possible by the progressive mechanisation of the
building industry. Earth movement has seen an
increase in the mechanisation level from 56.7
percent in 1960 to 87.7 in 1970 and 95.2 in 1980
while concrete work and the handling of components
has improved from a level of 78.9 percent in 1970
to 91.8 in 1980. However these achievements have
not been scored without complications and
controversies. The Communist Party leadership has
pressed hard not only for the careful siting of
buildings, with soil conditions and servicing
costs in mind, but also for high densities and for
economy in the use of land and building space. In
1966 it seemed laudable that the party leader
should attack 'the tendencies towards
grandiosity' evident in 'a series of ministries
and in building design organisations' which
'unjustifiably raised the value of
constructions' (Ceausescu 1969-1980 I p. 244);
also that he should reject 'architectural
monotony, lack of variety in building heights and
in the use of materials' resulting in 'the
extension throughout the country of certain
architectural elements... without taking into

account the diverse relief forms and architectural
traditions of the various regions' (Ibid p. 245).
However the enlightened thinking at the end of the
Six Year Plan (1960-1965) came after a decade of
distinctly leisurely progress in housing (only 9.6
thousand homes in 1950, 13.8 in 1955 and 36.8 in
1960). Equally it came before the formulation of
principles for conservation of agricultural land
and for the strictest economy in building. The
former has limited the scope for the private
sector (especially in the towns) and the latter
has inhibited architectural innovation generally,
creating a distinct impression that housing
economics are a vehicle for ideologically-
motivated social engineering to limit the
independence of the individual. So despite the
transformation of the housing stock post-war
policies have been distinctly variable and the
state's current ambivalence towards the
aspirations of the individual is a matter of deep
concern.

SOCIALIST ROMANIA: HOUSING PROGRAMMES

During the post-war period there has been a
substantial increase in population from 15.87
million at the census of 1948 to an estimated
22.72 million in 1985. Although rates of natural
increase have fallen the government is committed
to a policy of stimulating the birth rate and this
should ensure a marked increase in numbers to the
end of the century. The demand for housing has
also been increased by the allocation of more
resources to improve the living environment for
the proletariat previously obliged to accept slum
conditions in many instances. And there had also
been an unprecedentedly large redistribution of
population as the industrialisation programme
(Montias 1967, Turnock 1986) has resulted in
sustained rural-urban migration.
Table 4.1 shows the trends in house building
in Romania as a whole from 1951 to 1985. When
building is related to the population level a very
rapid improvement can be seen in the late 1950s
because the rate increased from 3.12 houses per
thousand of the population in 1950 and 3.20 in
1955 to 7.27 in 1960. The rate fluctuated over
the next 20 years between 5.76 in 1968 and 8.91 in
1980, but it is noticeable that since 1980 there
has been a steady reduction in effort in both town

and country down to the level of 4.65 in 1985, the lowest rate since the 1950s. Major differences occur in the contributions by the state and by private investors. Private building was much more important than the state programmes in the 1950s but the state sector became the larger of the two in 1969 and the gap has widened since, with the state contribution at least ten times larger than the private sector during the last seven years. The reasons lie primarily in the party's decision to become more involved in housing so that apartments have become available for renting or purchase on very attractive terms. This has been especially true in the larger towns. At the same time urban planning has focused more and more narrowly on the apartment block and so it has become increasingly difficult for plots to be found on which new private houses can be built. It should however be stressed that the figures relate only to greenfield development. They do not include major reconstruction of old houses (which may amount to almost total rebuilding) which is a major preoccupation among private builders in the rural areas.

Table 4.1 also deals with the area of living space (suprafata locuibila) in new houses. It is clear that the trends are upwards from 26.5 square metres per house in 1951-1955 to 34.9 in 1981-1985 but the increase has been much more rapid in the rural areas (26.6 to 44.6) than in the towns (26.3 to 34.0); also in the private sector (26.5 to 52.4) compared with the state sector (26.0 to 33.6). Evidently the norm of 12 square metres per person is being increasingly exceeded, although many large privately-owned houses accommodate extended families. Information on the size of new houses and apartments (number of rooms) is published each year on a national basis and Table 4.2 has been drawn up to give a summary picture over five-year periods, differentiating between accommodation of one room, two rooms and three or more. It is evident that the importance of the smallest and largest categories has increased progressively since 1956-65 while the share of two-roomed homes has fallen from 67.5 per cent in this first period to 57.8 in 1971-1975 and 45.1 in 1981-1985. Evidently family accommodation is becoming slightly more generous while more emphasis is also being placed on flats for single people.

Table 4.1: Romania: House building in urban and rural areas
          1951-1985

Milieu	Period	State Sector		
		A	B	C
Romania	1951-1955	9.3	26.0	0.53
	1956-1960	20.8	29.7	1.13
	1961-1965	44.0	30.7	2.31
	1966-1970	66.6	29.1	3.49
	1971-1975	102.5	29.8	4.82
	1976-1980	151.1	32.6	6.80
	1980-1985	131.1	33.6	5.79
Urban	1951-1955	7.0	26.8	1.27
	1956-1960	18.6	30.3	3.14
	1961-1965	41.8	30.9	6.51
	1966-1970	65.1	29.0	8.91
	1971-1975	100.4	29.8	10.92
	1976-1980	146.9	32.6	13.33
	1981-1985	126.0	33.6	10.65
Rural	1951-1955	2.3	23.7	0.19
	1956-1960	2.2	25.4	0.18
	1961-1965	2.3	27.4	0.18
	1966-1970	1.5	31.6	0.13
	1971-1975	2.1	29.5	0.18
	1976-1980	4.2	33.7	0.37
	1981-1985	5.1	33.0	0.47

Source: Anuarul statistic

Note: A = Houses built per annum ('000s)
         B = Average area (square metres)
         C = Houses per thousand of the population

Table 4.1 (continued)

Milieu	Period	Private Sector			Total		
		A	B	C	A	B	C
Romania	1951-1955	77.3	26.5	4.42	86.6	26.5	4.95
	1956-1960	151.4	26.7	8.23	172.1	27.0	9.35
	1961-1965	137.1	25.5	7.21	181.1	26.8	9.52
	1966-1970	62.9	34.9	3.29	129.5	31.9	6.78
	1971-1975	47.9	40.1	2.25	150.4	33.0	7.08
	1976-1980	17.1	49.5	0.77	168.1	34.3	7.57
	1980-1985	10.2	52.4	0.45	141.3	34.9	6.25
Urban	1951-1955	11.8	25.9	2.15	18.7	26.3	3.42
	1956-1960	35.3	26.8	5.98	53.9	28.0	9.71
	1961-1965	28.0	26.2	4.37	69.8	29.0	10.88
	1966-1970	12.3	35.1	1.68	77.4	30.0	10.59
	1971-1975	12.1	42.2	1.32	112.5	31.1	12.25
	1976-1980	4.2	53.2	0.39	151.2	33.1	13.72
	1981-1985	2.4	54.6	0.20	128.4	34.0	10.85
Rural	1951-1955	65.6	26.7	5.46	67.9	26.6	5.65
	1956-1960	116.0	26.6	9.29	118.2	26.6	9.47
	1961-1965	109.1	25.3	8.65	111.3	25.3	8.83
	1966-1970	50.6	34.8	4.29	52.1	34.7	4.42
	1971-1975	35.8	39.3	2.96	37.9	38.8	3.14
	1976-1980	12.8	48.3	1.15	17.0	44.7	1.52
	1981-1985	7.9	52.0	0.73	12.9	44.6	1.20

Source: Anuarul statistic

Note: A = Houses built per annum ('000s)
       B = Average area (square metres)
       C = Houses per thousand of the population

Housing Policy in Romania

Table 4.2: Romania: Housing units 1956-1984 by size

Period	State Sector			Private Sector			Total		
	A	B	C	A	B	C	A	B	C
1956-1965	na	na	na	na	na	na	8.8	67.5	23.7
1966-1970	15.2	51.8	33.0	14.2	44.1	41.7	14.7	48.3	37.0
1971-1975	8.0	57.8	34.2	13.7	39.2	47.1	9.8	51.9	38.3
1976-1980	12.8	47.6	39.6	13.5	26.6	59.9	12.8	45.5	41.7
1981-1985	12.6	45.1	42.3	13.3	21.9	64.8	12.6	43.5	43.9

Source: Anuarul statistic

Note: A = Houses with one room (per cent)
        B = Houses with two rooms (per cent)
        C = Houses with three or more rooms (per cent)

The differences in building rates between town and country are particularly striking (Table 4.3). Whereas the urban rate was lower than the rural rate in 1951 (2.49 houses per thousand of the population as against 3.32), the urban rate overtook the rural in 1955 and it has stayed ahead ever since, apart from 1960 and 1961. The ratio has widened in recent years, 3.0:1 in 1970, 5.7:1 in 1975, 13.5:1 in 1980 and 19.2:1 in 1985. In the urban areas the growth of the state sector has been particularly impressive through the 1960s and 1970s while the private sector has registered an absolute decline since 1970. In 1985 45 houses were built by the state for each one built privately compared with 15 in 1975, 7 in 1965 and 1.5 in 1955. In the countryside private building is still prominent although the state sector is beginning to catch up, reaching two-thirds of the private sector level in 1985 compared with one-seventh in 1975 and one-twentieth in 1964. While private building on new sites is falling off, because it is administratively simpler to reconstruct existing properties, the state is acquiring land in many of the key villages for the construction of small apartment blocks.

Table 4.3: Romania: Annual housing record 1951-1985

	Urban Sector			Rural Sector			Total		
	A	B	C	A	B	C	A	B	C
1951	1.34	1.15	2.49	0.07	3.25	3.32	0.36	2.76	3.12
1955	2.39	1.20	3.59	0.21	2.81	3.02	0.89	2.31	3.20
1960	4.63	1.72	6.35	0.22	7.49	7.71	1.63	5.64	7.27
1961	6.32	1.44	7.76	0.22	7.56	7.78	2.19	5.58	7.77
1962	6.41	1.45	7.86	0.32	6.88	7.10	2.25	5.10	7.35
1963	6.32	1.08	7.40	0.23	5.83	6.06	2.24	4.26	6.50
1964	7.26	0.98	8.24	0.27	5.43	5.70	2.61	3.94	6.55
1965	7.41	1.07	8.48	0.27	5.01	5.28	2.68	3.68	6.36
1966	7.34	1.02	7.36	0.28	5.11	5.39	2.59	3.55	6.14
1967	na	na	1.67	na	na	4.78	2.76	3.57	6.33
1968	na	na	na	na	na	na	2.86	2.90	5.76
1969	na	na	na	na	na	na	4.29	2.89	7.18
1970	11.50	1.98	13.48	0.12	3.87	3.99	4.76	3.10	7.86
1971	10.47	1.91	12.28	0.09	3.53	3.62	4.32	2.86	7.18
1972	9.85	1.53	11.38	0.10	3.06	3.16	4.16	2.42	6.58
1973	11.27	1.35	12.62	0.16	3.04	3.20	4.83	2.83	7.16
1974	11.63	1.2	12.87	0.23	2.99	3.22	5.10	2.24	7.34
1975	13.79	0.92	14.71	0.33	2.19	2.52	6.15	1.64	7.79
1976	11.56	0.69	12.25	0.31	1.70	2.01	5.24	1.26	6.50
1977	11.73	0.48	12.21	0.34	1.30	1.64	5.79	0.91	6.70
1978	13.98	0.40	14.33	0.33	0.96	1.29	6.94	0.69	7.63
1979	16.28	0.27	16.55	0.39	0.84	1.23	8.13	0.56	8.69
1980	18.48	0.24	16.72	0.46	0.76	1.22	8.40	0.51	8.91
1981	12.98	0.21	13.19	0.47	0.76	1.23	6.73	0.49	7.22
1982	12.54	0.22	12.76	0.46	0.78	1.24	6.68	0.49	7.17
1983	11.01	0.21	11.22	0.55	0.82	1.37	6.00	0.50	6.50
1984	9.94	0.2	10.14	0.44	0.66	1.10	5.41	0.42	5.83
1985	7.69	0.17	7.86	0.40	0.61	1.01	4.27	0.38	4.65

Source: Anuarul statistic

Note: A = Total number of new houses in the state sector
per thousand of the population;
B = ditto private sector;
C = Total

Table 4.4: Romania: Housing building by regions 1970-1985

Region	Period	A	B	C	D	E
Central	1970-1985	269.4	7.65	32.34	12.0	88.0
	1970-1974	79.2	7.74	17.52	22.1	77.9
	1975-1979	93.4	8.49	8.98	9.6	90.4
	1980-1984	87.4	7.42	5.05	5.8	94.2
	1985	9.5	3.96	0.79	8.3	81.7
North	1970-1985	315.9	7.31	73.01	23.1	76.9
	1970-1974	84.9	6.43	39.24	46.4	53.6
	1975-1979	116.6	8.64	21.56	18.5	81.5
	1980-1984	102.5	7.31	10.62	10.4	89.6
	1985	12.0	4.24	1.59	13.2	86.8
Northeast	1970-1985	365.5	6.58	80.88	22.1	77.9
	1970-1974	104.3	5.94	50.66	48.6	51.4
	1975-1979	121.1	6.98	19.85	16.4	83.6
	1980-1984	123.1	6.87	8.84	7.2	92.8
	1985	17.0	4.66	1.53	9.0	91.0
Southeast	1970-1985	1013.8	7.64	137.75	13.6	86.4
	1970-1974	330.6	8.45	93.92	28.4	71.6
	1975-1979	313.7	7.56	26.34	8.4	91.6
	1980-1984	326.3	7.25	15.18	4.7	95.3
	1985	43.2	4.95	2.31	5.3	94.7
Southwest	1970-1985	229.3	6.07	63.57	27.7	72.3
	1970-1974	76.7	6.66	37.01	48.3	51.7
	1975-1979	69.2	5.86	17.63	25.5	74.5
	1980-1984	71.1	5.91	7.54	10.6	89.4
	1985	12.3	5.05	1.39	11.3	88.7
West	1970-1985	273.8	6.51	50.95	18.6	81.4
	1970-1974	79.4	6.32	28.97	36.5	63.5
	1975-1979	94.3	7.17	14.33	13.7	86.3
	1980-1984	88.5	6.62	6.66	7.5	92.5
	1985	11.6	4.31	1.00	8.6	91.4

Note: A = Total houses completed ('000s)
B = Completions per annum per thousand of the
population
C = Completions in the private sector ('000s)
D = Private sector completions as a percentage
of the total
E = Public sector completions as a percentage
of the total

Table 4.4 (continued)

Region	Period	F	G	H	J
Central	1970-1985	8817	32.7	60.23	1.83
	1970-1974	2470	31.2		
	1975-1979	3042	32.6		
	1980-1984	2977	34.1		
	1985	328	34.6		
North	1970-1985	11335	35.9	71.10	1.76
	1970-1974	2978	35.1		
	1975-1979	4198	36.0		
	1980-1984	3674	35.8		
	1985	485	40.3		
Northeast	1970-1985	12455	34.1	65.58	1.26
	1970-1974	3502	33.6		
	1975-1979	4146	34.2		
	1980-1984	4226	34.3		
	1985	581	34.1		
Southeast	1970-1985	34144	33.7	289.44	2.18
	1970-1974	10558	31.9		
	1975-1979	10974	35.0		
	1980-1984	11110	34.1		
	1985	1502	34.7		
Southwest	1970-1985	7688	33.5	33.09	0.93
	1970-1974	2512	32.8		
	1975-1979	2383	34.4		
	1980-1984	2395	33.7		
	1985	398	32.4		
West	1970-1985	9684	35.4	90.40	2.06
	1970-1974	2745	34.5		
	1975-1979	3232	34.2		
	1980-1984	3283	37.1		
	1985 .	424	36.6		

Source: Anuarul statistic

Note: F = Area of land developed by new housing ('000 ha)
      G = Area of land developed per house (square metres internal space)
      H = House sales by the state ('000s)
      J = House sales by the state per annum per thousand of the population

Regional variations are investigated in Table 4.4. The statistical yearbooks give figures for new housing in each county (number of units and area of living space) for the state and private sectors, but unfortunately without any division into the urban and rural areas: such contrasts can only be investigated at the national level. The counties are grouped into six macroregions for calculations covering 1970-1985 as a whole, three separate five-year periods and finally the year 1985 alone. Housebuilding rates are slightly higher in the economically dynamic Centre and Southeast but the period of greatest effort was 1970-1974 in the Southeast and Southwest, whereas in the other four regions, the highest building rates were recorded in 1975-1979. The decline in building in 1985 is evident in all regions but in the Southwest the 1985 rate is 83.2 per cent of the average for 1970-1985, while in the other regions it is much lower: only 64.8 per cent in the Southeast and 51.8 in the Centre. As is the case nationally the six regions all show progressive increases in living space per house during the period under review although there are slight anomalies: reductions in average size in the North, Southeast and Southwest in 1980-1984 compared with 1975-1979. However the greatest variations emerge in the relative importance of the state and private sectors in house building. The importance of the state increases in all regions (apart from 1985 when the contraction in building was particularly marked in the state sector) but the contribution by the state over the entire period 1970-1985 varies from 88.0 per cent in the Centre and 86.4 in the Southeast through 81.4 percent in the West and 77.9 in the Centre. Unfortunately these variations cannot be easily explained but they raise interesting questions about the regional allocation of capital and the process of phasing in mechanised building systems.

A finer focus can be obtained by looking at the situation in individual counties (judete) (Figure 4.1). In this illustration building rates, state participation and average house size are examined in terms of deviations from the national average picture for 1970-1985 as a whole. In the case of the first two criteria information is also given for 1970-1974, 1975-1979, 1980-1984 and 1985. What comes out very clearly is the above-average building effort in those counties where the largest cities are found.

Figure 4.1: Romania: Building rates, state participation, and average house size as deviations from national average, 1970–1985

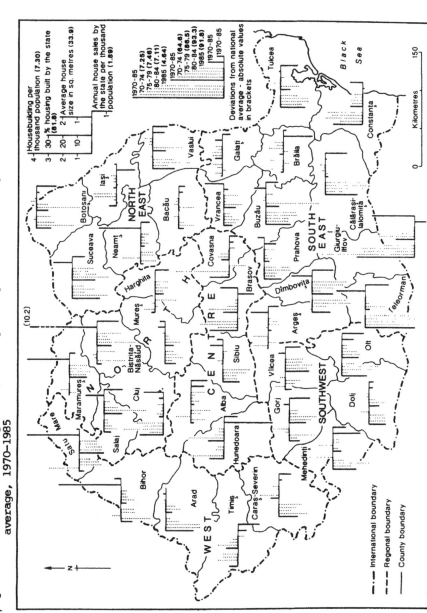

These cities will have attracted migrants from well beyond the limits of the relevant county and therefore building rates will be high when related to the population of the county alone: Brasov +1.91, Cluj +1.61, Constanta +2.02 and Giurgiu-Ilfov (covering Bucharest) +3.98. It is likely that there is a link here with above-average levels of state participation in house building since a large predominantly youthful immigrant population lacking any local family ties will require substantial assistance from the state building programme. However such a simple explanation cannot fit all cases and when the counties with the lowest levels of state participation are identified they form a block in the north of the country where the share of housing constructed by the state is below the national average by a substantial margin: Bistrita-Nasaud -9.8, Maramures -12.0, Salaj -12.6, Satu Mare -17.0 and Suceava -15.8. There is another grouping in the Subcarpathians in the Southeast and Southwest where the state contribution is relatively small: Buzau -9.6, Dimbovita -14.5, Gorj -11.8, Olt -11.4 and Vilcea -16.6. The northern counties are well known for a particularly cohesive community life in the villages where the population continues to increase (thanks to the availability of seasonal work in agriculture and forestry in other parts of the country). This expansion obviously creates a demand for additional house plots, whereas, in the rest of the country, rural population is falling and demand for new housing can be satisfied more easily by the redevelopment of existing sites. Thus there is an indication that rural population psychology plays a part in these variations. The situation in the Subcarpathians may be similar.

HOUSING ORGANISATION AND ALLOCATION MECHANISMS

Almost invariably state housing in Romania now takes the form of apartment blocks made largely from prefabricated concrete units (Damian 1973). The programme is determined nationally by central planning and then disaggregated through planning in each county where there are institutes concerned with siting and design. There is however some grouping and not every county has its own organisation. Building work is carried out by the local <u>trust de constructii</u> which complements the local government planning offices. Over the years not only has the average apartment increased in size but the facilities have improved as well. It is now unusual for any new accommodation to be built with a level of comfort (<u>confort</u>) inferior to the highest grade. This means that each apartment will have a kitchen, bathroom and balcony; also there will be no transit rooms. By comparison the second level of comfort would mean only a small kitchen and bathroom and no balcony at all and the third level (now encountered only rarely) would mean no bathroom and a water supply restricted to the kitchen. On the other hand, the standard of building may leave much to be desired. The prominence of brick is decreasing while pre-fabricated concrete components are increasing. Problems may be encountered on the top floor of the block through leaking roofs and also on the first floor above ground (where the defective apartments are often allocated to gypsy families). Standardisation is excessive and some fittings are of poor quality. Electricity is universally installed (though sometimes to the exclusion of gas which is usually preferred for cooking), but there is some concern over safety through the poor quality of the plastic components and the substitution of aluminium for copper. Many tenants find themselves replacing doors or fitting double glazing, to overcome defects in building as well as to bring about improvements. As car ownership increases the lack of garages becomes a problem. These are not normally provided in state housing projects and there appears to be no particular desire at the moment to tackle this problem since private/co-operative planning applications for garage construction in Bucharest have not been approved. For the present vehicles stay out in the open, usually covered by a waterproof sheet when the car is not being used regularly, but this is unsightly as well as unsatisfactory for security and maintenance.

State housing is offered to families who have lost their private house through requisitioning and received insufficient compensation to contemplate purchase of a new property. It is also made available to young families through the union (<u>sindicat</u>) which provides a link between each enterprise or institution and the local housing authority. Annual requests and allocations are made to ensure that priority needs

are met. Generally speaking the availability of
apartments is satisfactory but when there are
shortages preference is given to young married
couples (below the age of 30) and families with
children. To encourage skilled workers some
advantage also accrues to would-be tenants with
higher education. Priority cases may wait only a
matter of weeks before an apartment is allocated
but since some neighbourhoods are much more
desirable than others the first offer usually
involves one of the less attractive districts.
Since 1970 the state has encouraged private
ownership by selling apartments on attractive
terms. It would appear that there has been an
element of subsidy (though prices have become more
realistic with increases in wages) and also state
loans have been available, to be repaid over ten
to twenty years through monthly deductions from
salary. Usually the would-be buyer can make a
choice between different areas of a town, with
costs varying more according to the size of the
apartment than to location. There is certainly a
great interest in private ownership and there is
something of a national housing market in the
sense that vendors often advertise in the
newspaper <u>Romania Libera</u> (especially for
properties in Bucharest and the spa towns) as well
as the local papers. After a price has been
agreed the legal profession is involved in drawing
up the necessary papers and a tax is payable to
the state.

However sales of apartments by the state
have been sharply curtailed since 1986, largely it
would seem, because of government concern over
growing speculation and inequality in housing.
Apartments have been purchased in areas perceived
as desirable and the value of these properties has
then increased by two or three times over little
more than a decade as families unable to buy what
they want from the state have competed for
apartments previously sold to the occupants.
Hence the decision to curtail house sales not only
by the state but also, it would appear, by private
owners as well. Coupled with curbs on inheritance
this could eventually result in nationalisation of
the housing stock, although it is too early to say
that this extreme scenario is seriously envisaged
by the party. It is still possible to rent from
private owners and here competition is again rife
because families with ample resources are able to

offer rents well above the level set by the state for good accommodation in a desirable area. Income from letting has in fact become important enough for people with interests in several properties (registered in names of different members of the family to get round the regulation that an individual may own only one house) to gain a considerable augmentation to their income. Equally, families with ample good quality accommodation will often find it rewarding to let a single room and to this extent the need for state regulation on living space is eliminated. However in the past when there were severe shortages the state imposed the standard strictly and families with excess space had tenants and rent levels imposed on them. But as the supply of state housing has increased faster than the growth of the urban population there has been greater scope for private bargaining over housing matters.

SOCIALIST ROMANIA: THE URBAN DIMENSION

The major housing effort has been made in the towns which accommodated only 3.05 million people in 1930 and 3.71 in 1948 compared with 6.63 in 1966 and 10.49 in 1980 (respectively 21.4, 23.4, 34.7 and 47.3 per cent of the total population). As already noted, this increase has been sustained largely by migration although rates of natural increase are now higher in the towns than in the countryside (Miftode 1978a). There has been an increase in the number of towns, currently 237 compared with 176 in 1966 and 141 in 1930, but indications are that there has also been an increase in average size, with the greatest increases accruing to the capital city and the major provincial towns. The distribution of towns remains uneven however and it is very evident that the central planning responsible for urban development process in general (Ronnas 1982) has encouraged growth in the areas offering greatest economic potential, where urban agglomerations have emerged (Karteva 1966). Less developed areas have a relatively sparse urban network and one comprising relatively small towns which are just beginning to benefit from a great decentralisation of economic activities away from the regional administrative centres (Ianos 1982).

After 1948 the state placed a very low priority on new house building in order to concentrate investment in industry. The resulting growth of population in the capital city and the main provincial cities (especially those which were recognised as administrative centres for the regions) was accommodated by expelling bourgeois families who were excluded from the new order. For many people lost their jobs in 1948 and although they were not persecuted by imprisonment and forced labour (the fate of the regime's political enemies) they were, nevertheless, obliged to withdraw to the countryside, usually to the Baragan in the case of expellees from Bucharest. Additionally an allocation of six square metres of living space per person generated a surplus in the larger private houses whose owners had tenants billeted on them for very low rents. Some new housing was built, continuing the practice of small blocks with between four and eight apartments, with an emphasis on the capital city, industrial towns like Hunedoara and mining centres like Anina and Vulcan. When the pace accelerated after 1957 larger complexes were preferred. Initially the sistem cvartelor was preferred. This involved a rectangular building of three to five storeys focussing on a central square with roads and green space. Access might be gained at a point midway along each of the four sides of the rectangle which would break up the building into four wings or quarters. The system was used in Hunedoara in 1952 and applied elsewhere during the remainder of the decade.

Building work was initially tentative. In the capital city the communist government launched their plan de reconstructie socialista a orasului Bucaresti in 1952 and this was spelt out two years later in the Schita a planului general de sistematizare, subsequently revised in 1956 and approved by the local authority for the city in 1957 (Radu 1956, Titu 1956). The scale of new house building was distinctly modest, although some projects made a considerable impact locally and the Grivita district has been seen as a Particularly striking example of revolution in an urban landscape between 1958 and 1965. In view of wartime bombing and the association of the area with Gh. Gheorghiu-Dej, a leader of the strike of railway workers in 1933 and subsequently a leader of the governing Communist Party, there was scope for a powerful architectural effort to create a

social   gestalt   (Church 1979).   The army's
Bucharest studio for the plastic arts turned out
blueprints for monuments comprising purposeful
human figures complete with tensed arms, clenched
fists and sober faces. However even this phase of
development should not be exaggerated for it plays
only a minor role in an architectural ensemble
whose monumental character is still substantially
pre-socialist (including the railway terminus
building of the 1930s).

Apartment building in Bucharest averaged only
0.6 thousand per annum between 1948 and 1956 but
the rate increased progressively to 9.0 between
1957 and 1963, 14.3 between 1964 and 1969 and 23.8
between 1970 and 1976. Substantial modifications
to the city plan were made in 1962 following the
Eighth Congress of the RCP which was significant
in validating the Russian concept of microraion
involving large integrated urban complexes
(unitati urbanistice complexe) covering housing,
industry, services and recreational spaces. Hence
the housing complex of Titan around the lake and
park of Balta Alba, with employment on the eastern
industrial estate; also the development of new
suburbs close to the lakes of Pantelimon and
Plumbuita which extended the sequence of water
storages along the Colentina valley.   Drumul
Taberei has been built on the west side of the
city while Nitu Vasile has arisen to the south,
around the main streets of Sos. Giurgiului and
Sos. Oltenitei (Ciobotaru 1971).   Other schemes
are proposed.

HOUSING COMPLEXES: THE CASE OF CLUJ-NAPOCA

Since 1962 there has been a substantial
literary output focusing on the complex de locuit
(Radu 1962, 1965; Rau & Mihuta 1969) to achieve
greater efficiency in building (Ciobotaru 1974,
Gheorghita 1981). The importance of planning with
transport and services in mind has been emphasised
(Margarit & Fulicea 1964) and the need for high
architectural standards has been underlined by two
prominent authorities: D. Gusti (1965, 1974) and
C. Lazarescu 1972, 1977). Unfortunately recent
experience suggests that the architects have been
constrained and empirical studies have shown that
whereas urban growth in the 1960s and 1970s
involved an increase in the built-up area (Chitu
1969) as well as increase in density (Poghirc

et al. 1964) there is now greater insistence on growth simply by increased densities achieved by redeveloping some of the older suburbs. High density development reduces the per capita cost of services and minimises the loss of agricultural land. Both considerations have become increasingly important in spite of the need to retain flexibility against a time when better housing conditions can be afforded and, also, in spite of the fact that Romanian agriculture has ample capacity for increased output through intensification which could easily compensate for greater areas of building land.

The concept has been applied very widely since it was placed at the centre of socialist urban development (Lazarescu 1986). Greenfield estates have been developed with houses, shops, factories and workshops. Blocks are spaced according to height and also with regard to the need for access and for green spaces, playgrounds and sport/recreation facilities. Services must also be allowed for: education, health and culture, shops and catering facilities. Some scope is allowed for higher living standards in future: better services, increased living space per person and smaller school classes. In the provinces this new scale of housing provision is very evident. Layouts are based on waterways, like the Bahlui in Iasi, but other topographical features and historic monuments may also be taken into account. In Cluj, Grigorescu is planned with explicit regard for the river Somes to the south and Hoia hill to the north while Republicii in Baia Mare is planned to achieve a feeling of intimacy on a site clearly delimited by the Sasar river and Dealul Florilor. New streets in Deva offer a view to a prominent hillock capped by a ruined fortress, while the new centre at Piatra Neamt projects monuments of the Stefan cel Mare era.

Some old residential areas have been completely redeveloped, including town centres. The growth of towns frequently places overwhelming pressure on centres built with much smaller populations in mind and considerable renewal is inevitable. However the model is often applied only partially in old residential areas. Apartment blocks may be built along the main thoroughfares with traditional buildings left standing behind, e.g. Soseaua Giulesti, Calea Mosilor and Piata Salii Palatului RSR in

Bucharest, Piata Teatrului in Brasov and Calea Bucuresti in Craiova. Where the existing street system is unsatisfactory clearances may be made to provide space for apartment blocks to be built along new boulevards, as in Constanta. In smaller towns even less radical treatment is needed: relocation of industry to industrial estates (especially when pollution is serious), creation of new employment (e.g. institutions) to moderate commuter movements across the town, installation of services (water, drainage, gas and electricity) and repair of houses. Any rebuilding should take place section by section to minimise disruption. Of course towns that are no larger than the nominal population for a complex de locuit will require shopping only in the centre unless the town is dispersed along a valley or transport artery.

Even in small towns the emphasis on apartment blocks is becoming increasingly evident and it is virtually impossible for the private developer to find a site for a conventional town house. When the Iron Gates hydropower project led to the reconstruction of Orsova in the late 1960s there was a balance between apartment blocks and detached houses (Chiriac 1968). The same sort of balance was reported in an urban study of Slatina-Olt (Damian 1970). Some 300 properties were investigated: 210 were individual houses with courtyards (communal in 72 cases) and only 95 were apartments, most of them in tall blocks. Of course the individual houses were relatively primitive (only 19 had a bathroom though 150 had toilets and 179 had a kitchen) whereas virtually all the apartments had all three facilities. But they were better endowed with cellars, gardens and orchards, contained more rooms (36 per cent comprised three rooms or more compared with 24 per cent for the apartments) and were less affected by the need for two more households to share the same housing unit (32 per cent compared with 50). However, over the past decade the state has insisted more uncompromisingly on the apartment as the norm. This can be seen everywhere in the systematic demolition of traditional housing to make way for apartment blocks but it was witnessed most dramatically at Zimnicea on the Danube where a large proportion of the 33,000 houses lost in the 1977 earthquake were concentrated (Ceausescu 1969-1980 12, pp. 135-209). Although the government decided on immediate reconstruction with the help of building workers from all parts of the country (Ilie 1977) the town plan involved the near total exclusion of traditional town houses and was widely regarded as a psychological trauma or similar dimensions to the natural disaster.

While the increase and modernisation of the housing stock is to be applauded it is a matter for regret that so much traditional building is being destroyed. The partial reconstruction of town centres and the emergence of residential complexes on the outskirts is being followed up at present by near-comprehensive redevelopment of whole districts built up before 1945. While some properties may require replacement the lack of discrimination is now becoming a matter of great concern affecting many historic buildings, which could be retained without significant costs. Bucharest is now undergoing radical change with the building of a new boulevard to symbolise the victory of socialism but hundreds of substantial properties have been lost in the process. Several provincial cities have been almost completely rebuilt, for example Botosani and Pitesti. Only brief reviews are available to record the Medieval legacy in building and street systems (Greceanu 1981, 1982). In the Transylvanian cities such as Sibiu there is a greater awareness of the importance of conservation (Perenyi 1973, pp. 95-96), with various schemes of restoration along with movement of industry and redirection of traffic, but even here the immediate prospects are not good. It is possible that the authorities are unaware of the damage which is being done in order to maintain the dubious principle that agricultural land should not be sacrificed to urban expansion, necessitating therefore high density redevelopment of the existing built-up area. But the extent of the damage which continues in the face of growing opposition leaves the government open to the serious criticism that the changes are ideologically-motivated as part of a drive to loosen the ties of the people with its history which has lived on through several decades of socialism.

It is evident from Figure 4.2 that in the city of Cluj-Napoca the proportion of households resident in accommodation built since 1945 (overwhelmingly in blocks) is now substantial in all districts.

Figure 4.2: Cluj-Napoca: Age of buildings

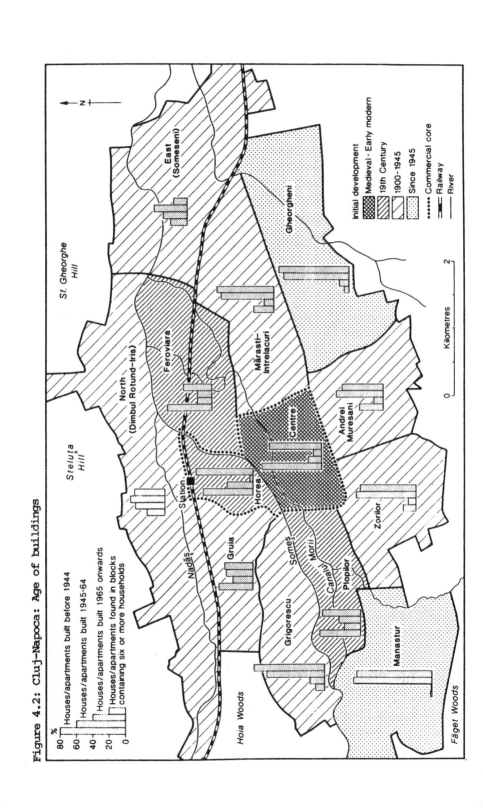

There are individual apartment blocks on restricted opportunity sites (common in the central area) and larger complexes which call for the demolition of existing housing (usually single-storey rural-type detached houses with substantial gardens). Of course apartment blocks are by no means restricted to the postwar period, although the very tall blocks with up to 10 storeys and the very long buildings (around 100 metres) are characteristic of this period alone. If blocks are regarded as buildings containing six or more apartments then it would appear that they account for proportions in excess of the share of post-1945 buildings by the approximate margins indicated on the map. As for other variations between districts these can only be discussed impressionistically because of the lack of reliable evidence. However it would appear that areas consisting of substantial houses built in the nineteenth and early twentieth centuries and situated in a pleasant environment reasonably close to the centre are regarded as particularly desirable, so that higher income families are able to compete effectively for rented or owner-occupied accommodation.

Generally speaking all pre-1945 accommodation seems to be held in high regard because the high densities characteristic of new housing complexes are avoided. The only difficulties are large rooms which give rise to high heating costs and make it difficult to rent small areas of surplus space. It is not unusual to find vacant apartments in some of the large peripheral housing complexes because the more affluent families will try and find older properties in the lower part of the city. Such tendencies may be exaggerated where there are perceived variations in shopping, transport and amenities. In other words while there may be only modest differences in living space per person and in possession of basic household equipment, variations tend to be more marked in terms of proportions of households where the head of the family is intellectual, with higher education. Social segregation is very apparent. There may also be variations in the level of car ownership. This is not surprising because when most options for investment are closed (e.g. property and stock markets) differences in income will tend to register in areas like housing and car ownership. But it is a matter for speculation to what extent the proportions vary and to what extent there are differences between towns in different parts of the country.

SITING PROBLEMS: THE CASE OF IAŞI

In allocating land for the construction of apartment blocks attention must be given to microclimate (wind direction and aspect) to ensure reasonable shelter and direct sunlight. Terrain is an important factor where slopes exceed five percent. On slopes of between five and eight percent long buildings need to be placed perpendicular to the slope and on even steeper slopes there are substantial increases in building costs, although greater dispersal of building to avoid broken ground and ravines (viroage) may result in higher servicing costs. But planning must always take careful note of abrupt slopes running down from the edge of low plateaux (platforme marginite de cornise). It is quite common for the more prominent blocks to be situated close to the edge in order to enhance the impressiveness of the townscape, as at Cluj (Gheorgheni), Galati (Tiglina) and Pitesti (Calea Craiovei) where ten-storey blocks on the edge of a broken plateau effectively mark the urban limit (though the blocks on the edge of the Danube bluffs at Tiglina have been considered somewhat monotonous). Blocks of four or five storeys (without lifts) are usual under normal geotechnical conditions. Where there are difficulties there can be a reduction in height or there can be more elaborate foundations and a greater volume of construction, perhaps compensated for an increase in the number of apartments (and reduction in average size). Since the very serious earthquake of 1977 it has been necessary to give greater attention to the seismic factor with appropriate modification to building systems and to the height and disposition of blocks particularly in Vrancea and other areas close to the Carpathian 'curve' including Bucharest (Boian et al. 1977, Constantinescu 1978).

As regards the physical basis for house building particular interest attaches to the studies in urban geomorphology carried out at Iasi University during the 1950s. Many of the leading towns of Moldavia were covered and Figure 9 indicates the synthesis that was achieved for Suceava (Martiniuc and Bacauanu 1964; Martiniuc et al. 1967). However, the most detailed attention was given to the city of Iaşi where severe physical constraints (see Figure 4.3) required the planning of a more compact settlement. The studies in physical geography were made in conjunction with a general plan de sistematizare of 1945-1946 and a more detailed program de sistematizare drawn up in 1954. Iaşi is, of course, an historic city, being the capital of Moldavia from the 1550s until Bucharest became capital of the unified principalities of Moldavia and Wallachia in 1862. Its national importance arose from its central position in Moldavia (prior to the loss of Bessarabia to the Russian Empire in 1812) and its commercial role related to the trade routes running south from Poland to the Lower Danube. But there is also a local marketing function enhanced by the contact between the Moldavian Plateau and the lower ground of the Prut valley and southern Moldavia (Bogdan 1904, Ungureanu 1976). The chosen site was the northern side of the asymmetrical Bahlui valley (15 kilometres from the Prut) where there was a well-developed terrace system above the flood plain (Martiniuc et al. 1956).

The terraces provided not only a dry site sheltered from north winds but also one endowed with springs (for sources at Ciric and Sapte Oameni sustained the first organised water supply at the Golia monastery's casa de apa in the early eighteenth century). But above all the terraces are relatively untroubled by the landsliding which affects much of the hill and plateau country of northern Moldavia. Figure 4.3 shows first the alluvial plain of the Bahlui and the dejection cone of the Nicolina. The whole of this area, up to two kilometres wide, was inundated during the catastrophic floods of 1932, although moderate flooding (arising through rapid run-off from impermeable ground) is frequent and occurred 15 times during the period 1921-1953. This is not an area where housing can be encouraged although with the regulation of the Bahlui and the construction of water storages on tributary streams in the vicinity of Iaşi (e.g. Ciric and Chirita) and further afield (the Belcesti was completed in 1974) it is much more feasible to contemplate industrial development, adjacent to the main road and railway (Gugiuman 1956).

Figure 4.3: Building constraints in Iaşi and in Suceava

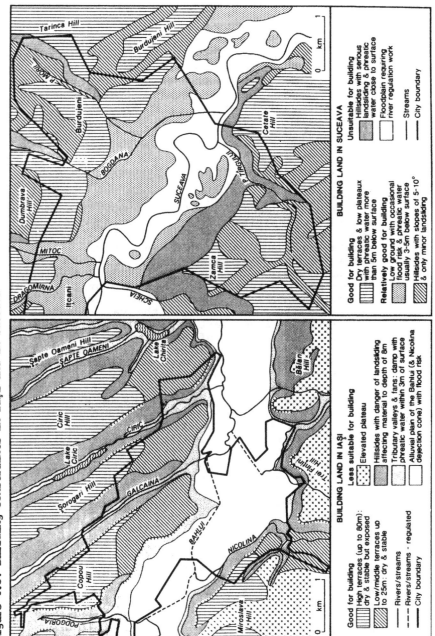

BUILDING LAND IN SUCEAVA

Good for building
Dry terraces & low plateaux with phreatic water more than 5m below surface

Relatively good for building
Low ground with occasional flood risk & phreatic water usually 3-5m below surface
Hillsides with slopes of 5-10° & only minor landsliding

Unsuitable for building
Hillsides with serious landsliding & phreatic water close to surface
Floodplain requiring river regulation work

Streams
City boundary

BUILDING LAND IN IAŞI

Good for building
High terraces (up to 80m): dry & stable but exposed
Low/middle terraces up to 25m: dry & stable

Less suitable for building
Elevated plateau
Hillsides with danger of landsliding affecting material to depth of 8m
Tributary valleys & fans: damp with phreatic water within 3m of surface
Alluvial plain of the Bahlui (& Nicolina dejection cone) with flood risk

Rivers/streams
Rivers/streams - regulated
City boundary

Immediately above the floodplain proper is a narrow strip of damp ground comprising the bottom lands of the tributary valleys and the fans developed by such streams in the main valley of the Bahliu. Such land is rich in underground water with a hydrostatic layer at a depth of 1-3 metres (compared with 0-2.5 on the floodplain) and even brick houses built in clusters are likely to suffer from problems of damp. However the terraces are much more suitable for building and the system extends from the lowest terrace at 10 metres (Frumoasa-Nicolina) to 25m (Palat-Tatarasi), 80m (Ciric/Copou/Galata) and finally 140-170m (Aroneanu/Breazu/Miroslava). The terrace system is restricted south of the Bahlui (with only a 10m terrace at the Nicolina-Bahlui confluence) but well developed on the northern sides especially between Galcaina and Copou streams where the Medieval town developed, protected from attack from the south by the Bahlui floodplain and the outpost forts Cetatuia and Galata on the south side of the valley (Bacauanu and Martiniuc 1966, Martiniuc and Bacauanu 1959). The first stage of development occupied the lower and middle terraces up to 25m. These terraces are of Upper Pleistocene age and have a covering of loess to a depth of 5-8m. Later growth, occurring between the eighteenth and twentieth centuries, occupied the upper terraces on the same Copou ridge (Barbu and Brindus 1967). These are of Middle Pleistocene origin and have a thicker loess layer (10-15m). The development plan must, therefore, concentrate attention on the terraces, maximising opportunities at Tatarasi and extending growth further east to the Ciric ridge between the Ciric and Sapte Oameni streams. The regulation of the Bahlui is also a fundamental element. It has been related to the improvement of navigation on the Prut and the construction of a navigable canal from Ungheni (at the Bahlui-Prut confluence to Iaşi). The scheme has also been associated with a diversion of water from the river Siret across the watershed at Strunga with hydroelectricity generation as the water descends eastwards to the Prut. However the plan of 1954, aiming at the expansion of heavy industry in Moldavia (Miftode 1978, Sandru et al. 1954) has not yet been fulfilled. But by concentrating on high-density residential development on the terraces Iaşi has experienced a rapid growth of population with relatively little increase in the built-up area.

SOCIALIST ROMANIA: THE RURAL DIMENSION

There is a continual need for new housing (despite a declining population) because many of the older houses are not particularly durable by virtue of the materials used and some of the traditional styles and arrangements are less acceptable in the context of modern living. However the effort is distinctly modest by comparison with the development in the towns partly because the figures relate only to entirely new houses. There is an unknown number of improvements to existing property which may amount to almost total rebuilding. The distribution of new private house building in the rural areas is very uneven. This calls for further investigation but it is possible that in some areas there is strong interest in building by families intending to retire (eventually) to a scenically attractive part of the country. Private building remains a significant activity despite the increasing constraints of the planning system and recent fears about state restrictions on property sale and inheritance. House building is an attractive way of utilising surplus funds and great pride may be taken in the construction which is usually done by family and friends, perhaps with some help from skilled neighbours and with much scope for reciprocity within the community. The official standard for living space of 12 square metres per person cannot be grossly exceeded but much effort can go into the fittings and furnishings which take many years to install. It is common to a house which is structurally complete but where only some of the rooms are occupied.

There have been changes in building materials with reductions in the use of wood and <u>chirpici</u>. In the mountains high quality construction timber is less readily available due to nationalisation of woodlands, while bricks and building blocks are made available by the state for authorised building. It is sometimes the case that timber and brick are combined where the former is obtainable. Brick is used for the bottom of the house supporting the weight of timber above. It is also common for the exterior walls to be coated with a mixture of cement and lime (secured by laths in the case of wooden walls). In the plains brick largely superseded <u>chirpici</u> although changes are not so obviously apparent because of the application of the lime/cement wash. There have also been quite radical changes in roofing materials with tiles and asbestos sheeting (sometimes metal sheeting) prominent where previously thatch and wood shingles were used. Design has also changed. Although artistic skills have sometimes been transferred from wood to metal, concrete and plaster, the post-war period has, generally, brought about a considerable simplication of the exterior and the disappearance of balconies. The tradition persisted longest in the southwest for many modern houses in Gorj and Mehedinti have a corridor running along the front of the house marked by a series of Brancovan arches. This arrangement is convenient in the summertime but is less appropriate in winter in the context of modern heating systems.

RURAL PLANNING

    The state is also exerting an influence through its rural planning programme. The authorities brought about a transformation in the countryside through collectivisation of agriculture and greatly improved access to non-agricultural employment. Most of the jobs were provided in the towns and hence the practice of daily commuting by train and bus emphasising the satellite relationship between village and town (Caloianu <u>et al</u>. 1964). At the same time there was a growth of rural employment in mining, forestry and food processing, so that many villages assumed an urban appearance (Onisor and Susan 1966). For many years the state did not interfere directly with rural housing and settlement patterns but during the 1970s <u>sistematizare</u> has become more explicit and <u>modernisation</u> of the countryside is to be accelerated by restricting new housing to designated building perimeters and by promoting many key villages (ultimately some 300) to urban status (Ioanid 1969, Matei 1960, Urucu and Candea 1984). 129 villages have been picked out for the first phase of new town development (Turnock 1986) and a detailed study has been made of one of these villages, Feldioara near Brasov (Sampson 1976, 1984) although no formal designation has yet been made. But the impact on housing is already considerable because there will be higher densities within the perimeters and a sharp reduction in the area of courtyards and orchards (Gheorghe and Gheorghe 1978) and new houses must have two storeys (although away from the main streets the rule can usually be satisfied by installing an attic and dormer window). Consolidation is also being accelerated directly by the state. In the past there has been very little state housing in the villages: only small programmes to provide accommodation for professional people working in research stations and industrial enterprises (though most professionals had their own country houses or else commuted daily from the nearest town). Now small apartment blocks are being built in selected villages on land taken from private owners with minimal compensation. The villages designated as new towns are witnessing much of this development but suburban villages like Apahida near Cluj-Napoca and Otopeni near Bucharest, are also affected.

As rural planning becomes more restrictive, through high-density development within designated building perimeters, many attractive buildings are lost either through demolition to make way for apartment blocks or through neglect arising out of a remote situation where building is no longer permitted. Fortunately there are a number of open-air museums where representative buildings are assembled. Supported by funds from Fundatia principele Carol Romulus Vuia organised the Ethnographical Museum of Transylvania in Cluj in 1927 with an open-air section on Hoia Hill in 1932. Then in 1936 Dimitrie Gusti founded the Village Museum in Bucharest. A third important development arose in Dumbrava Park, Sibiu where the idea for a Museum of Popular Technology (Bucur et al. 1986) emerged during Vuia's sojourn in the city during the Second World War (when the Vienna diktat of 1941 transferred North Transylvania, including Cluj, to Hungary). Several other museums play a valuable regional function in Arges (Golesti, specialising in fruit farming and viticulture), Banat (Timisoara), Gorj (Curtisoara), Harghita/Mures (Cernat and Reghin), Maramures (Baia Mare and Sighetul Marmatiei), Vilcea (Bujoreni) and Vrancea (Focsani) (Mihalache 1972). However these do not have open-air sections where whole buildings can be preserved. In any case what is more important than mere preservation is the perpetuation of local traditions in new buildings and although the authorities have paid lip service to such ideas, for the president himself has expressed confidence in architects for 'their sense of initiative and creative imaginative power' (Ceausescu 1969, p. 251), it has not carried sufficient priority.

CONCLUSION: PROBLEMS AND PROSPECTS

After a slow start the Romanian housing programme has maintained a level of effort roughly commensurate with the scale of population growth and redistribution. The plan has also played an important part in the improvement of living standards during the 1960s and 1970s because apartments have become larger and better equipped (Damian 1975). Yet the present situation is far from satisfactory as building rates have started to flag during the 1980s, a period of severe economic crisis. At the same time densities have increased and the understandable desire to avoid waste of land and excessive expenditure in servicing costs seems to be taken to illogical extremes. Seen in the context of redevelopment, involving the destruction of many well-built town houses and valuable historical monuments, the policy may be quite counter-productive. The greatly reduced scope for private house building and the abandonment of apartment sales by the state amounts to a significant reduction in consumer choice at a time when the general trend in Eastern Europe is running in the opposite direction. Increasingly Romanian families have little option but contemplate their future as tenants in a state-owned apartment block. In 1987 there were widespread fears over a prohibition on property sales which could be the first step to nationalisation of housing and a consequent simplification of urban development procedures. To be sure all these problems are most acutely felt in the towns but even in the countryside action over housing is constrained by the building perimeters and the increasing role of the state in the rural building programme. However it must be emphasised that the limited amount of statistical information and nationwide survey material make overall assessments difficult.

The future is very difficult to predict. The attack on individualism may be an inevitable side effect of the economic crisis or, more probably, it may be conditioned by ideological considerations: especially at a time of stress nobody must be appearing to profit from property speculation. But while the form of new housing may be uncertain it is almost inevitable that the growth of the larger towns will continue through both natural increase and in-migration. While commuting from the villages may well increase there is little sign yet of the improvements in transport that would persuade the majority of new urban workers to seek a permanent home in the country in preference to one in the town. Public transport is uncomfortable and the relatively small number of families owning a car have great difficulty getting petrol. Meanwhile the relatively rapid increase of the gypsy population may create more social problems within apartment blocks for it is usual for such families to be dispersed among other sections of the population through allocation of first floor apartments where structural problems are often encountered. Security is also becoming a matter for concern in the cities with crime against both people and property reported to be on the increase in virtually all neighbourhoods. These problems are not unique to Romania but the policies are quite distinct at present: rejecting the counter-urbanisation evident in Albania without allowing scope for the private sector which is evident in the other socialist countries.

REFERENCES

Apolzan, L. (1987) Carpatii tezaur de istorie: perenitatea asezarilor risipite pe inaltimi, Scrisul Romanesc, Craiova

Bacauanu, V. and Martiniuc, C. (1966) 'Cercetari geomorfologice asupra teraselor din bazinul Bahluiului', Anuarul stiintifice ale Univ. A.I.Cuza din Iaşi IIc, 12, 147-156.

Barbu, N. and Brindus, C. (1967) 'Solurile de pe teritoriul orasului Iaşi si a imprejurimilor sale', Anuarul stiintifice ale Univ. A.I.Cuza din Iaşi IIc, 13, 171-174

Bogdan, N.A. (1904) Orasul Iaşi: schite istorice si administrative, Tip. Nationala, Iasi

Boian, S. et al. eds. (1977) Contremural de pamint din Romania de 4 martie 1977, Ed. Academiei RSR, Bucharest

Bold, I. et al. (1974) Sistematizarea rurala, Ed. Tehnica, Bucharest

Housing Policy in Romania

Breharu, G. (1966) 'Contributii geografice la studiul dezvoltarii orasului Gherla', Communicari de geografie, 3, 253-265

Bucur, C. et al. (1986) Museums' complex of Sibiu: museum of folk technology, Brukenthal Museum, Sibiu

Caloianu, N. et al. (1964) 'Asezarile satelit ale oraselor din sudul regiunii Brasov', Analele Univ. Bucuresti: geografie, 13, 156-161

Ceausescu, N. (1969-1980) Romania on the way of building up the multi-laterally developed socialist society, Meridiane, Bucharest, 16 volumes

Chiriac, D. (1968)' Orasul Orsova in perspective', Terra, 1, 44-46

Chitu, M. (1969) 'Dezvoltarea teritoriala a orasului Ploiesti', Communicari de geografie, 8, 157-164

Church, G. (1979) 'Bucharest: revolution in the townscape art', R.A. French and F.E.I. Hamilton eds. op. cit., 493-506

Ciobotaru, I. (1971) 'Sistematizarea municipiului Bucuresti' Terra, 3(1), 51-58

_____ (1974) Sistematizare si urbanism: probleme de eficienta economica, Ed. Tehnica, Bucharest

Constantinescu, L. (1978) 'An extreme Romanian earthquake and its wider geonomical setting', Revue roumaine: geographie, 22, 179-206

Damian, N. (1970) 'Schimbari ale structorilor familiale in procesul de urbanizare', M. Constantinescu ed. Procesul de urbanizare in RSR: zona Slatina-Olt, Ed. Academiei RSR, Bucharest, pp. 209-254

_____ (1973) 'Efecte sociale ale locuirii in noile blocuri', Viitorul social, 2, 391-402

_____ (1975) 'Locuirea in mediul urban: indicator al calitatii vietii', Viitorul social, 4, 146-153

French, R.A. and Hamilton, F.E.I. eds. (1979) Socialist city: spatial structure and urban policy, Wiley, Chichester

Gheorghe, G. and Gheorghe, N. (1978) 'Gruparea satelor mici si a gospodarilor dispersate: analiza de caz', Viitorul social, 2, 276-282

Gheorghita, S. (1981) Economia constructiilor, Ed. Tehica, Bucharest

Greceanu, E. (1981) Ansamblul urban medieval: Botosani, Muzeul National de Istorie, Bucharest

_____ (1982) Ansamblul urban medieval: Pitesti, Muzeul National de Istorie, Bucharest

Gugiuman, I. (1956) 'Inundatiile Bahluiului si pericolul lor pentru dezvoltarea spatiala a orasului Iasi', Probleme de geografie, 3, 169-181

Gusti, G. (1965) Architectura in Romania, Ed. Meridiane, Bucharest

_____ (1974) Forme noi de asezare: studiu prospectiv de sistematizare macroteritoriala, Ed. Tehnica, Bucharest

Ianos, I. (1982) 'Locul si rolul oraselor mici in reteaua nationala de asezari', Terra, 14(3), 25-28

Ilie, G. (1930) 'Colonizarile in Banat in sec. XVIII-XIX', Analele Banatului, 3(2-3)

Ilie, V. (1977) 'Zimnicea de astazi si de miine', Terra, 9(2), 36-38

Institutul Central de Cercetari Economic (ICCE) (1983) Progresul economic al Romaniei socialist in conceptia presidente lui Nicolae Ceausescu, Ed. Politica, Bucharest

Ioanid, V. (1969) 'Probleme actuale privind sistematizarea rurala in Romania', Comunicare de geografie, 8, 209-216

Karateva, V. (1966) 'Dezvoltarea aglomeratiilor urbane', Studia Univ. Babes-Boylai: geologia-geographia, 11(1), 133-139

Lazarescu, C. et al. (1972) Arhitectura romaneasca contemporana, Ed. Meridiane, Bucharest

_____ (1976) 'Probleme actuale ale sistematizarii terit oriului national si asezarilor umane in Romania', Viitorul social, 5, 31-38

_____ ed. (1977) Urbanismul in Romania, Ed. Tehnica, Bucharest

_____ (1986) Arhitectura si viata oraselor, Ed. Tehnica, Bucharest

Margarit, N. and Fulicea, V. (1964) Orasul si circulatia urbana, Ed. Technica, Bucharest

Martiniuc, C. et al. (1956) 'Contributie la studiul hidrogeologic al regiunii orasului Iaşi', Probleme de geografie, 3, 61-95

Martiniuc, C. and Bacauanu, V. (1959) 'Harta geomorfologica a orasului Iaşi', Analele stiintifice ale Univ. A.I.Cuza din Iaşi II, 5, 183-190

_____ (1964) 'Problemes de geomorphologie appliquee dans la systematisation des villes en Moldavie', Revue roumaine: geographie, 8, 223-231

Martinuic, C. et al. (1967) 'Contribution a l'etude geomorphologique du territoire de la ville de Falticeni et de sese environs', Revue roumaine: geographie, 11, 127-139

Matei, M. (1960) Sistematizarea asezarilor satesti, Ed. Tehnica, Bucharest

Miftode, V. (1978a) Migratiile si dezvoltarea urbana, Junimea, Iaşi

_____ (1978b) 'Urbanizarea si migratiile intraurbane', Viitorul social, 2, 269-275

Mihalache, M. (1972) Guide des musees de Roumanie, Ed. Touristiques, Bucharest

Montias, J.M. (1967) Economic development in communist Romania, MIT Press, Cambridge, Mass

Morariu, T. et al. (1968) 'Contributie la studiul retelei de asezari din regiunile inalte sale Capatilor', Studia Univ. Babes-Bolvai: geologia-geographia, 13(2), 131-138

Onisor, T. and Susan, A. (1966) 'Asezari asimilate urbanului', Studia Univ. Babes-Bolyai: geologica-geographia, 11(1), 115-125

Panait, L. (1969) 'Dezvoltarea teritoriala a orasului Brasov', Comunicare de geografie, 7, 197-204

Perenyi, I. (1973) Town centre planning and renewal, Hungarian Academy of Sciences, Budapest, pp. 95-96

Poghirc, P. et al. (1964) 'Contributii geografice-economice asupra orasului Falticeni', Analele stiintifice ale Univ.A.I.Cuza din Iași IIb, 10, 147-158

Poncet, J. (1972) 'Les enseignments des inondations catastrophiques du printemps 1970 en Roumanie', Annales de Geographie, 81, 298-315

Popescu, B. et al. (1970) 'Inundatiile catastrofale din Mai-Iunie 1970', Hidrotehnica, 15, 611-644

Radu, L. (1956) Constructia oraselor, Ed. Technica, Bucharest

_____ (1962) Probleme de estetica oraselor, Ed. Tehnica, Bucharest

_____ (1965) Urbanismul, Ed. Tehnica, Bucharest

Rau, R. and Mihuta, D. (1969) Unitati urbanistice complexe, Ed. Tehnica, Bucharest

Ronnas, P. (1982) 'Centrally planned urbanization: the case of Romania', Geografiska Annaler, 64B, 143-151

Sampson, S. (1976) 'Feldioara: the city comes to the peasant', Dialectical Anthropology, 1, 321-347

_____ (1984) National integration through socialist planning: an anthropological study of a Romanian new town, East European Monographs, Boulder, Col.

Sandru, I. et al. (1954) 'O varianta a schitei-program de sistematizare a orasului Iași', Analele stiintifice al Univ. A.I.Cuza din iași II,1, 341-352

Titu, E. (1956) Introducere in probleme generale de constructia oraselor, Ed. Tehnica, Bucharest

Turnock, D. (1974) An economic geography of Romania, Bell, London

_____ (1980) The human geography of the Romanian Carpathians, Geographical Field Group, Nottingham

_____ (1986) The Romanian economy in the twentieth century, Croom Helm, London

_____ (1986) 'The rural development programme in Romania with particular reference to the designation of new towns', Leicester University Geography Department Occasional Paper 13

_____ (1977) 'Romania': A.H. Dawson ed. Planning in Eastern Europe, Croom Helm, London pp. 229-273

_____ (1987) 'Urban development and urban geography in
Romania: the contribution of Vintila Mihailescu',
GeoJournal, 14, 181-202
Ujvari, I. (1972) 'Consideratiuni geografice cu privire la
perspectivele gospodaririi apelor din tara noastra',
Terra, 4(4), 19-27
Ungureanu, A. (1976) 'Evolutia teritoriala a oraselor din
Moldova', Anuarul stiintifice ale Univ. A.I.Cuza din
Iaşi II, 22, 77-78
Urucu, V. and Candea, M. (1984) 'Aspecte ale modernizarii
satului romanesce', Analele Univ. Bucuresti: geografie,
33, 79-86

Chapter Five

HOUSING POLICY IN BULGARIA

Frank Carter

INTRODUCTION

Housing policies in the centrally planned
economies of the developed world appear to differ
as much amongst themselves as between them and the
mixed economies of the more advanced capitalist
countries. The Bulgarian housing situation tends
to support this view and is probably related to
two distinct reasons; first, any examination of
Marxist classical writings on the topic of housing
provides little general guidance for contemporary
planners and even less on the specific role of
owner-occupied dwellings within a socialist state.
Second, Bulgaria like the other centrally planned
countries of the Eastern bloc, inherited its own
historical housing problems, particularly
resulting from five hundred years of Ottoman rule,
as well as a standard of living and degree of
collective/state ownership which varies from its
socialist neighbours. It should be remembered
that socialism only came to Bulgaria as to the
rest of Eastern Europe, after the Second World
War, and the move towards socialist ideals in the
country has been perhaps at a slower pace than in
some of the other states in the Comecon bloc. It
is against this background that a more detailed
examination will be made of Bulgaria's housing,
its organisation and the contemporary problems
involved.

IDEOLOGY AND HOUSING: FROM MARX TO ZHIVKOV

For Marx, the problem of housing was closely
associated with the capitalist system; once over-
thrown, this problem would be easily solved.

170

Similarly, Engels believed that the housing situation reflected the political and economic system of the time, and that any housing shortages could easily be remedied by re-allocation of the present housing stock. Excessive overcrowding of the working class would disappear with the confiscation of property belonging to the wealthier classes, once political power had been placed in the hands of the proletariat. Furthermore, he saw the security of house tenure, and ownership of property, with its gardens and fields, as hindrance to the workers, especially with the development of large-scale industry; Engels was also critical about the role played by housing rents. Furthermore, he was reluctant to speculate on what housing policies should be adopted by a future socialist society, being satisfied that sufficient housing stock existed to give the working class of the day adequate accommodation for its spatial requirements and a healthy environment. In other words, Engels only used the housing situation as a tool for his criticism of the contemporary capitalist system, and declined to speculate on the way future leaders of a socialist society would organise the distribution of residential accommodation. (Engels, 1963, pp. 13, 21, 32, 98).

For future Marxist interpreters this was the only text on the housing question and having read and digested these words in the months prior to his assuming power in the Soviet Union, Lenin was to differ little in his views from Marx and Engels. Similarly, after the liberation of Bulgaria from Nazi domination in September 1944, the future socialist leaders had little to guide them on the ideological approach to the housing problem. Certainly Georgi Dimitrov, like Lenin, was faced with a situation demanding practical answers; housing was woefully inadequate, with much of the country containing mostly old, lath and plaster and unsanitary building, in badly planned towns and villages with their muddy or dusty streets that had little changed since Ottoman times. The expropriation of property from the former wealthier classes only left the new socialist state with a burdensome inheritance in the sphere of housing construction. Meanwhile, most of the poorer people in both towns and villages were living in gloomy, sunless and small houses, covered with thatch or roofed with stones, and often lacking sanitation.

Housing Policy in Bulgaria

Given this situation there was little time
for searching ideological debates on the type and
form of dwellings to be built, and the state set
aside financial support to erect whatever new
accommodation it could. Thus between 1944 and
1963, 905,000 dwellings were built of which over
two thirds (625,000) were in rural areas
(Roussinov, 1965, p. 202). By the early 1960s,
every third house in Bulgaria had been built in
the post World War II period. In the 1970s, Todor
Zhivkov, the country's leader made some
improvements in the housing situation. Up to then
priorities established in the Stalinist years had
meant that "accumulation" had invariably had a
greater share in the gross national product than
"consumption"; investment goods always had a
larger share of gross national product than
consumer goods which included housing. Even in
the 1970s there was little time to spare pondering
ideological questions about housing; urban housing
shortages prompted Zhivkov to declare in 1973 that
the residential situation was to be "radically
solved". Nevertheless, this prediction fitted
into the overall contemporary belief that housing
should be given a high priority in the country's
social and economic planning, because "The general
level of a country's material and cultural
wellbeing depends on its housing and living
conditions" (Todorov, 1971, p. 10). In a way this
supports Engel's earlier view, that such
priorities "lead directly to 'Utopia'" (Engels,
1963, p. 68).

In the 1970s and early 1980s the Bulgarian
planning authorities had an obsession for the
construction of high-rise buildings, even in
places where smaller edifices would have been more
appropriate. However, this was probably not on
ideological grounds, such as symbolic rejection of
pre-war capitalist one-family houses, or privately
rented accommodation, but the more immediate
practical expedient of finding homes for the
growing population and increased migration to the
larger urban centres. Even so, in 1986 some
notice was taken of criticisms frequently made in
the Bulgarian press against such high-rise
residential forms; this was mainly in opposition
to local party leaders and planners who wished to
see their urban centres reflecting the high
skyscraper outlines of modern world cities. They
were accused of paying insufficient attention to
the country's architectural heritage and the

172

adaption of such edifices to the overall environmental milieu. Zhivkov has proclaimed through his government, that as from July 1987 new village houses together with those in some selected centres, should be no higher than three storeys. Moreover, a strong case must be put forward now for the construction of five-storey or higher buildings anywhere in the country (Drzhaven Vestnik, 1986). The reasoning behind this move however, appears to be more aesthetic than ideological.

THE EARLY YEARS: 1945-1965

This period was one of rapid transition from a predominantly agricultural pre-war society to a modern industrialised state. Large-scale urbanization that emerged after the Second World War was part, as well as a consequence, of the profound social and economic transformations that took place after the war. The new socialist economy had its impact on patterns of employment with people moving to the cities and larger towns to work in manufacturing, education, the service and tertiary sectors. In 1946, a quarter of the population (24.7%) were already living in urban centres. Further, the intensified post-war urbanization process gave rise to serious housing shortages in cities, further aggravated by the need to replace war damage; for example, in the capital Sofia about a third of the housing stock had been destroyed or damaged. Moreover, there was a demand to provide modern residential facilities for those inhabitants affected by slum clearance (Ognyanov, 1981, p. 109).

During the 1950s, as in other socialist bloc countries, architecture and urbanism were considerably influenced by contemporary Soviet ideas. As a result in Bulgaria classicism became the main source of architectural inspiration, with styles, details and planning principles interpreted according to classical doctrines. Thus, planning construction was performed on a grand scale characterised by geometric regularity and rigid symmetry; such architectural gestures were seen as purveying strength which in turn implied optimism for the further glorification of the ordinary working individual; the latter's well-being was seen as the ultimate purpose of socialism. Every urban centre had its neo-classical buildings from the 1950-1958 period which imposed their massive allure and rigorous architectural style on the surrounding area and included such examples as the housing in the district around the river Perlouska, Communist Party Headquarters and shopping precinct (ZUM) in Sofia, and the Town Hall in Burgas. By the end of the 1950s, however, this type of architecture no longer corresponded to the changing attitudes of the time. It began to be severely criticized by the official authorities as being undemocratic, pompous, too rigid, outmoded, unsanitary and out of place. The result was a dramatic reversal of architectural and planning ideals in Bulgaria towards a more rational and functional style of building construction.

Whilst the architects and planners were embroiled in the pros and cons of classicism, housing during the immediate post-war period remained an acute social problem and a sensitive topic in the country's official publications. Certain firm data, however, relating to this early period may be gleaned from the two major censuses of 1956 and 1965 which included some information on the housing situation. For example, residential building construction according to certain criteria are seen in Table 5.1. The major trends apparent from this table are threefold. First, the number of occupants per dwelling increased from 3.8 from the 1945-1948 period to 4.24 for the 1961-1965 era, reflecting the failure of government policy to keep housing construction at a similar pace to population growth; second, and as a consequence of the above, floor space per occupant declined from 11.17 square metres in the former period to 9.96 square metres in the latter. Third, as a result of both these factors the average number of occupants per room grew from 1.55 to 1.80 for the two comparative periods. Finally, the overall totals for the period 1945-1965 reveal that nearly 85% of useful floor space was taken over by the living area, and 73.33% of the living area consisted of rooms as opposed to kitchens, toilets etc.

The 1956 census (T.S.U., 1959) gave the first detailed information on housing in the post-war period (Table 5.2).

Table 5.1: Bulgaria: Residential building construction by period 1945-1965

Housing Construction	Period Of Construction					Total
	1945-1948	1949-1952	1953-1956	1957-1960	1961-1965	
Number of buildings	141808	152672	163819	221820	221060	901179
Number of dwellings	157098	175997	198782	277387	322855	1132119
Rooms, kitchens etc. of which:	529431	596099	658424	952544	1034667	3771165
Rooms	387095	437253	480636	702654	758114	2765752
Useful floor space in '000 m2 of which	7956	9077	10044	14556	16235	57868
living floor space in '000 m2	6737	7644	8484	12360	13669	48894
Number of occupants	602907	694131	831691	1212478	1371502	4712709

Source: Statisticheski Ezhegodnik 1968, 1968, Table 9, p. 254

Table 5.2: Bulgaria: Housing facilities according to the 1956 Census

	With piped water only	With sewage system only	With electricity only	With piped water and electricity	With piped water and sewage only	With electricity and sewage only	With piped water, sewage and electricity	With piped water, sewage, electricity and central heating	Without any of these facilities	Total
Houses built	522	213	219398	59499	306	7546	63322	654	49658	401118
Of which: in towns	515	210	199370	58185	303	7356	62286	612	44998	373835
Of which: in rural areas	7	3	19938	1314	3	190	1036	42	4660	27193
Living space in m²	14172	5136	6529092	2345492	13787	255048	5157956	387373	987310	15695366
Of which: in towns	13978	5045	5838336	2282569	13414	241987	5025255	362801	873496	14656881
Of which: in rural areas	194	91	690756	62923	373	13061	132701	24572	113814	1038485

Source: Calculated from Tsentralno Statisticheske Upravlenie, Prebroivane na zhilishtniia fond v Narodna Republike B'lgaria na 1.XII.1956 g (Nauka i izkustve) Sofia, 1959, p. 220

Table 5.2 stresses the difference in housing facilities between rural and urban areas; over 90% of houses built, and the living space in them, were located in towns and cities. However, even in the urban areas individual dwellings with piped water, sewerage, electricity and central heating in 1956 formed less than 1% of the total, but over two-thirds had electricity. Piped water in homes was restricted to a mere 15%, whilst only 2% were connected to a sewerage system. Perhaps most alarming, one in eight dwellings had no basic facilities at all.

Further information from the 1956 census gives some indication of construction methods and the distribution of storeys.

By the mid-1950s, this table suggests that Bulgaria was a land of largely single-storey housing (82%), where the techniques of reinforced concrete construction were still in their infancy (30% of total dwellings). Nearly half the country's housing comprised wooden framebuilt single-storey dwellings (45% of total dwellings), whilst the five-storey or more buildings were almost exclusively constructed from reinforced concrete and mainly in urban areas (96%). The overall findings of the 1956 housing survey led to questions being officially posed on future housing construction methods (Toshev, 1959; Anon, 1956 (a); Anon, 1956 (b)).

Bulgaria, like most other eastern bloc countries had followed the Soviet example by introducing low household expenditure on housing. An enquiry into average annual expenditure by manual workers' families on housing and related items in 1957 in Bulgaria supported this view. Total expenditure devoted to such matters was 15.1%, of which 10.7% was allocated for buying furniture, general upkeep of the dwelling and equipment; a further 3.2% was spent on fuel and lighting. Engels would have been proud of the figure for rent - a mere 1.2% (Donnison, 1964-1965, p. 96; Donnison and Ungerson 1982).

In order to meet the increased demands for large-scale construction during the 1960s, and particularly for housing, architecture had to have a more simple style than previously, with little or no excessive decoration. This would allow for the utilization of pre-fabricated elements which could be rapidly and easily constructed. In sum, it has to comply with the demands of the contemporary construction industry, as did urban planning.

Table 5.3: Bulgaria: Housing construction and number of
storeys according to the 1956 Census

	Single storey	Two storey	Three storey	Four storey	Five or more storeys	Total
Overall	327,168	65,440	5,438	1,594	1,388	401,028
Of which: in towns	304,871	60,746	15,247	1,587	1,384	373,835
Of which: in rural areas	22,297	4,694	91	7	4	27,193
Reinforced concrete buildings	2,984	4,728	2,081	1,248	1,326	12,367
Of which: in towns	2,866	4,597	2,021	1,242	1,322	12,048
Of which: in rural areas	118	131	60	6	4	319
Solid-built dwellings	162,144	42,731	3,149	338	62	208,424
Of which: in towns	149,658	39,836	3,025	337	62	192,918
Of which: in rural areas	12,486	2,895	124	1	–	15,506
Frame built (lath & plaster) dwellings	162,040	17,981	208	8	–	180,237
Of which: in towns	152,347	16,313	201	8	–	168,869
Of which: in rural areas	9,693	1,668	7	–	–	11,368

Source: As in Table 5.2

Moreover, a new notion was infused into planning, namely that of open space, which allowed for freely standing buildings, where asymmetry replaced that of symmetrical design. Spatial and structural changes were meant to provide a more efficient and reasonable use of territory, particularly in urban areas; here building density and height was increased to meet pre-conceived norms and building regulations, together with providing green space and more public facilities. Unfortunately, the early 1960s were to see the introduction of the concepts of rhythm and repetition, in which houses and other buildings were subject to only two construction types, namely the slab and the tower. These ideas led to drastic changes in the suburbs and outskirts of cities and large urban areas, with new residential areas subject to building by industrial methods. This enabled construction to take place in a relatively short time, but unfortunately such new estates had the appearance of a landscape completely lacking in a sense of time, history or other clearly visible aesthetic factors.

By the close of this early period of Bulgaria's post-war housing development the pace of construction was increasing, as were the heights and lengths of buildings to the detriment of all else. This was particularly noticeable for the people who had to live in them, and the surrounding environment of new residential areas. The overall housing situation around 1960 is seen in Table 5.4.

The most noticeable change between 1957 and 1963 was urban/rural housing ratios. In 1957 urban areas had only a third of the total number of houses built that year, but by 1963 this had increased dramatically to a half, whilst the annual dwelling total constructed at these two dates was little changed. The surface area allotted to housing construction was augmented by a fifth between 1957 and 1963, as was the residential floor space. Useful floor space, however, grew by only 7%, emphasising the importance of living space in residential accommodation. Residential living space as a percentage of useful floor space grew from 68% in 1957 to 77% in 1963, i.e., nearly 10%. Useful floor space is defined as including the whole area within the walls of a self-contained flat, i.e., it includes kitchen, bathroom, passages etc., but excludes common hallways and staircases.

Housing Policy in Bulgaria

Furthermore, "useful floor space" provided in Bulgaria's new apartments during the early 1960s ranged from 56 to 60 square metres, compared with 47 to 50 square metres in the Soviet Union, 54 to 67 square metres in Czechoslovakia, and 33 to 48 square metres in Poland (Donnison, 1964-1965, p. 110). "Living space" in Bulgaria is defined as the total area of habitable rooms which have at least 4 square metres in area and a ceiling height of at least two metres over the major area of the room. This increased from 35 square metres in 1960 to over 46 square metres three years later.

Table 5.4: Bulgaria: Housing construction 1957-1963

	1957	1960	1962	1963
Total annual number of dwellings built	43,462	49,786	43,708	43,904
Of which: in urban areas	14,666	20,193	21,249	21,677
Of which: in rural areas	28,796	29,593	22,459	22,227
Total housing construction surface (in '000 m²)	2,901	3,277	3,352	3,479
Useful floor space (in '000 m²)	2,466	2,797	2,575	2,647
Of which residential living floor space (in '000 m²)	1,687	1,759	1,982	2,032

Source: <u>Statisticheski Godishnik</u>, Sofia, for relevant years

A different complexion is placed on these achievements when housing construction is compared according to sector. The emergence of co-operative house building schemes towards the end of the 1950s, together with the continued popularity of private housing construction (not differentiated in the statistics) shows how small state involvement was in the housing programme (Table 5.5)

Table 5.5: Bulgaria: Annual housing construction according
to sector 1957-1963

	1957	1960	1962	1963
Total construction by co-operative and private sector	38,741	44,385	36,384	36,925
Of which: in urban areas	11,152	15,436	15,967	16,230
Of which: in rural areas	27,587	28,949	20,417	20,695
Total construction by state sector	4,721	5,401	7,325	6,979
Of which: in urban areas	3,512	4,757	5,283	5,447
Of which: in rural areas	1,209	644	2,042	1,532

Source: As for Table 5.2

Given total housing construction for these
years (Table 5.5) it appears that the co-operative
and private housing sector was responsible for
over 80% of the houses built, varying from its
highest contribution in 1960 (89.16%) to its
lowest in 1962 (83.24%). In rural areas, state
housing construction remained below 10%.

The early post-war years had proved a
difficult time for Bulgaria's housing programme.
War damage and the adjustment to a new socialist
political system, with its emphasis on heavy
industry to ensure military strength, meant that
housing was not given the investment priority it
deserved. Much was left to the private sector to
provide what housing was needed, especially in
rural areas; state policy concentrated on
producing larger quantities of housing, but often
at the expense of reducing dwelling size and
quality. Thus, by the mid-1960s there was renewed
pressure on the state to build better housing in

an attempt to raise building standards rather than
increase output

THE LATER YEARS: 1966-1987

By the mid-1960s the fruits of socialist
endeavour and planning were beginning to become
apparent on the Bulgarian landscape. Not only had
there been changes in the rural landscape, but
also towns and villages were gradually being
transformed. Housing had become an important
social and state problem and large financial sums
were set aside for construction. State attitudes
had not precluded individual owner-occupation and
private ownership of houses was not viewed with
the same scepticism as that of land. This
probably relates to classical Marxist writings
which were rather ambivalent on the general role
of housing within a socialist state and even more
so with reference to owner-occupied dwellings.

By the late 1960s what was seen as
refreshingly simple architectural styles during
the earlier years now began to appear rather dull
and boring. The quickening pace of housing
construction, together with ever larger building
complexes meant that attitudes towards the human
element tended to be ignored. This was manifested
in the overall decline in the quality of the
living environment in new residential areas.
There was a need for the planners to produce a
more physically and functionally varied climate
for habitation based on detailed sociological
research. In the 1970s this was reflected in an
attempt to eradicate the monotony and anonymity of
residential areas by the use of colour, and a more
imaginative approach to housing design and
building configurations. Even this proved
insufficient with public and professional opinion
rejecting the spatial approach; human scale was
seen as uppermost in significance which demanded a
reduction in the number of storeys of dwellings
and more use of open space. Also there was a need
to return to more traditional housing styles with
their greater flexibility and use of sophisticated
building techniques. It was only in the early
1980s that the fruits of this movement began to
have effect with a reduction in building heights
in residential areas; city edifices were to be
limited to ten storeys, whilst in urban centres
with less than 30,000 inhabitants four storeys was

the maximum. This in turn encouraged small-scale building development and greater experimentation in types and forms of housing constructed. As a result the 1970s and 1980s were to experience an increasing interest in traditional house forms derived largely from an interpretation of shapes and planning principles coming from the country's own earlier national house styles.

The years 1965-1966 provide a natural watershed in housing development; more generally these years signalled a temporal demarcation for the Bulgarian economy between the Fourth and Fifth Five Year Plans; further, from a housing viewpoint they heralded the second major census in December 1965, which included specific questions on the dwelling situation. Detailed information was given on residential buildings both by type and structure (Table 5.6).

This table reveals the high proportion of solid and frame-built structures which still dominated Bulgaria's housing in the mid-1960s. Use of reinforced concrete structures and large-sized panels was still in its infancy both for residential buildings and individual dwellings. Nearly two-thirds of the rooms were located in solid-built edifices but a further third were in the lath and plaster frame-built category. A similar proportion of households and residential accommodation were in these two categories.

The number of storeys utilized for residential accommodation was also included in the census details (Table 5.7).

Table 5.7 emphasises the dominance of one- and two-storey housing in Bulgaria during the 1960s; they totalled 90% of all residences; further analysis reveals that single-storey residences totalled 60% in all the above categories, whilst two-storey dwellings covered on average, 31%.

Finally, the 1965 census gave some insight into Bulgarian housing facilities (Table 5.8).

This table, as with Table 5.2 for 1956 tends to emphasize the continued discrepancy between housing in urban and rural areas with rural housing having a higher percentage of the less favourable factors. For example, nearly three-fifths of rural housing had only piped water, three-quarters had only electricity, and 86% were without any facilities. As expected, the one positive advantage of rural residence was a higher percentage (65%) of living space.

Table 5.6: Bulgaria: Residential buildings by type and structure (1/XII/1965 Census)

Type and Structure	Buildings	Dwellings	Rooms, Kitchens etc.	Of which: Rooms	Useful floor space in m^2	Of which: living floor space in m^2	Households	Residents
Large-sized panels	762	13445	42268	29561	711123	571900	17385	55031
Reinforced concrete frames	9192	122140	397063	285708	6903639	5734139	183408	515985
Solid-built	1003738	1203014	4137280	3049827	63771453	53695237	1484846	4984693
Frame-built (lath and plaster)	682745	724641	2062097	1466164	30063897	25635070	783894	2522383
Total	1696437	2063240	6638708	4831260	101450112	85636346	2469533	8078094

Source: Statisticheski Ezhegodnik 1968, Sofia, 1968 Section XVIII, Table 10, p. 255

Table 5.7: Bulgaria: Residential buildings by number of storeys (1/XII/1965 Census)

	Residential buildings as percentages							Total in figures
	One storey	Two storey	Three storey	Four storey	Five storey	Six storey	Seven or more storeys	
Buildings	70.70	27.99	0.87	0.27	0.09	0.05	0.03	1696437
Dwellings	61.93	28.31	3.43	3.35	1.25	0.95	0.78	2063240
Rooms and kitchens etc.	57.86	32.40	3.33	3.39	1.27	0.97	0.78	6638708
Of which: rooms	56.83	33.52	3.34	3.35	1.27	0.94	0.75	4831260
Useful floor space in m^2	55.63	33.41	3.69	3.78	1.48	1.13	0.88	101450112
Of which: living floor space in m^2	56.81	32.44	3.62	3.70	1.47	1.10	0.86	85636346
Residents	58.60	30.58	3.98	3.70	1.39	0.99	0.76	8078094

Source: As in Table 5.6 (Table 11, p. 257)

Table 5.8: Bulgaria: Occupied dwellings according to facilities (1/XII/1965 Census)

	With piped water only	With electricity only	With piped water and electricity	With piped water and sewerage	With sewerage and electricity	With piped water, sewerage and electricity	With piped water, sewerage, electricity and central heating	Without any of these facilities	Total
Dwellings	4787	1219704	81108	576	125930	464568	24578	100476	2021727
Of which: in towns	1997	292591	46037	343	85289	411983	22761	14051	875052
Of which: in rural areas	2790	927113	35071	233	40641	52585	1817	86425	1146675
Living space in m²	166378	49772497	3850558	24772	4753147	21563359	1065215	3005203	84201119
Of which: in towns	57843	9686065	2024824	13573	2790506	18932352	1012360	307098	34824621
Of which: in rural areas	108525	40086432	1825734	11199	1962641	2631007	52855	2698105	49376498

Source: As in Table 5.6 (Table 12, pp. 257-258)

In all other aspects life in towns was more favourable. Of those dwellings possessing piped water, sewerage, electricity and central heating, 92% were located in urban environments, together with two-thirds of all dwellings possessing a sewerage system and electricity, and nearly three-fifths with piped water and electricity. Such factors help provide one reason for the rural-to-urban migration which was so strongly developed in Bulgaria during the late 1960s. Nevertheless, if one examines the surplus/deficit household and housing unit data from the census, the rural/urban dichotomy does not appear so noticeable (Table 5.9).

This table illustrates the lack of self-contained housing units, apartments, and homes for each Bulgarian in 1965. Housing construction was not keeping up with increased demand caused by a continued migration of people from the villages to towns and cities. Further, the increasing number of young people beginning families, the creation of separate households by members of families which formerly had lived together, and the destruction of older housing units at a higher rate than expected, all added to the problem. The 1965 Bulgarian census had been the first to inquire about migration experience of enumerated population, and referred to place of residence in 1956 (Carter and Žagar, 1977, p. 209). The real problem it reflected, however, was insufficient funding for housing from central government. Even by 1968 only a mere 6.45% of the Gross National Product was being allocated to housing (Stanev, 1969).

It was becoming increasingly obvious that more money had to be spent on housing in Bulgaria. In 1970 a programme was adopted by the government in an attempt to alleviate the severe housing shortage caused by rapid urbanization. Even so, in that year the state only built 13,390 dwellings, less than a third of total housing construction in 1970. The Bulgarian government, however, had one other possible means at its disposal, namely private initiative. Since the mid-1960s, there had been increasing official readiness to accept and encourage private housing developments; they were seen as a very successful way of removing surplus spending power from Bulgarians, who were experiencing rapidly growing incomes and quickly rising personal savings.

Table 5.9: Bulgaria: Household and housing unit data, 1965

	Households (in 000s)	Size of household	Housing units total (000s)	Housing units occupied (000s)	+ Surplus - Deficit (000s)	Households per 100 housing units	Population (000s)
Total	2527	3.2	2055	2019	- 472	123	8228
Of which: urban	1278	2.9	874	874	- 404	146	3823
Of which: rural	1248	3.5	1181	1145	- 67	106	4405

Source: As in Table 5.6

Furthermore, private house construction saved the government scarce investment capital and gave personal home builders a feeling of supplying an important social need, which helped relieve pressure on disguised inflation (Morris, 1984, p. 138). The most popular form of private housing construction was the co-operative which, by 1970, was found in most East European countries. Members of such co-operatives in Bulgaria were obliged to pay a substantial sum towards construction, but the state granted individuals loans. These could be repaid over a 25-year period, usually at very low or zero interest rates. The following table illustrates state involvement in housing construction at this time.

Table 5.10: Bulgaria: New state housing 1961-1979 (selected years as percentage)

1961-1965	1969	1970	1977	1978	1979
20	28	30	50	50	50

Source: Statisticheski Godishnik, Section IX for relevant years

This clearly indicates that by the early 1970s private housing construction was providing over two-thirds of the country's new accommodation. Much of the total was located in rural areas and was further encouraged in 1971, when the Bulgarian government declared that every citizen had the right to own a house, flat or villa. Nevertheless, two years later a property law on private ownership, limited the size of privately owned apartments to 120 square metres, and imposed penalties on ownership violations. This act was part of a much wider government package which emerged with the Country Planning Act of 1973. It suggested the need for a national territorial plan; and was based on the idea of reducing regional economic and geographical disparities. This was to be controlled through selected investment in order to stem the flow of rural-to-urban migration (Volle, 1979). In future housing policy was to be more closely connected with physical urban planning.

Certainly a report by Todor Zhivkov, the Communist Party leader, given at a Central

Housing Policy in Bulgaria

Committee meeting in December 1972, intimated
changes, particularly to remedy some of the acute
shortages experienced earlier. The urban housing
shortage was one of these and would be "radically
solved" within the next 10 to 15 years. He
proclaimed that there would be a separate dwelling
provided for each family, and a separate room for
each family member - a tall order indeed (Guardian
1973). Such optimism was to be applauded, but had
to be measured against the financial spending of
the average Bulgarian citizen. An analysis of the
consumption pattern for inhabitants at that time
revealed the following wages breakdown: food and
drink 50.1%; housing 8.8%; furnishing 3.6%;
clothing and footwear 13.9%; transport and postal
services 2.7%; social and cultural activities
3.8%; taxes and dues 4.7% and other costs
(including cigarettes/tobacco, hygiene etc), 12.4%
(Evgeniev, 1974, p. 35). Whilst over a half of
the monthly salary was spent on food (46.8%) and
alcoholic beverages (3.3%), only 8% went on
dwellings through housing and furnishing.

Housing availability was becoming
increasingly difficult in spite of Zhivkov's
optimistic pronouncements. Morton has calculated
one possible indicator of this situation by
comparing the number of marriages each year with
the number of new housing units built annually in
various socialist bloc countries. The accumulated
deficit in housing units between 1971 and 1975 was
-112,124 in Bulgaria. This was the third highest
in Eastern Europe after Poland (-433,683) and East
Germany (-282,557), and considerably higher than
Czechoslovakia (-71,904) Romania (-64,015) and
Hungary (-53,651) (Morton, 1979, p. 309). Whilst
this is by no means a perfect measure of
Bulgaria's housing deficit it does provide a yardstick for some kind of international comparison at
that time. This is particularly so when true
demand is hidden by the fact that local
authorities allocate available housing for which
no data are published. Further support for the
presence of a housing shortage was the Bulgarian
government's failure to publish all the data from
the 1975 national census: only a selection of the
statistical evidence has been made generally
available to the public. For example, data on
household amenities reveal that 67.8% had piped
water, and 99.9% electricity (Shoup, 1981) whilst
there were 11 telephones per 100 inhabitants
(Anon., 1981, p. 408).

There was obvious disquiet in government circles about the housing situation by the mid-1970s. The main task of the Seventh Five Year Plan (1976-1980) was to accelerate housing construction in order to meet the nation's needs. The guidelines to this plan stated that the aim was to build 400,000-420,000 new homes and improve the exterior and interior design of blocks of flats. Further, the housing fund was to pay special attention to the needs of families with several children, student families, and families on low incomes and in most need. Flats were to be built together with a system of service establishments, with necessary communication links, especially transport, telephones, water, electricity and central heating (Anon., 1976a, pp. 44). More detailed information on how this was to be achieved was given in the Theses of the Bulgarian Communist Party, not only for the Seventh Five Year Plan but up to 1990.

In sum, it reiterated Zivkov's claims of 1973, to build some 1.6 million dwellings by 1990. Special attention was to be paid to temporary accommodation for young families in large towns and cities; use was to be made of group-individual (co-operative) housing construction to provide 20,000 dwellings annually. Further, all new housing construction had to have suitable surroundings with quotas established for planning designs and building methods. New dwellings linked with national industrial enterprises were to enjoy proximity to workplace, unpolluted air, etc., whilst more funds were to be allocated to modernize and up-date existing housing stock. There was to be an increase in the relative share of one- and two-room flats, which under the Eighth and Ninth Five Year Plans was to be augmented for three- and four-room flats. Demands for improved architectural design and the building of blocks of flats with more floors, in which there would be a better combination of space for collective life and some public catering projects were called for in these plans. Exterior design of residential buildings would be diversified to give a better appearance to housing blocks. Finally, there would be an improvement in the housing rent system through the introduction of greater differentiation based on dwelling size and family status, (Anon., 1976b, pp. 39-42).

Housing Policy in Bulgaria

In spite of such laudable demands, reality proved otherwise. Figures for housing construction after the first year of the new plan, i.e., 1976, gave total completion figures of 63,000 dwelling units, i.e., 6,000 more than 1975, but 20,000 less than the targeted average if the Seventh Five Year Plan total was to be achieved. Furthermore, it should be noted that of the 1976 total, 9,700 dwellings were privately built (Rabotnicheske Delo, 1977, p. 1). Falling targets led to higher demands; for example housing construction in 1978 was normally planned to provide 82,600 new housing units; but in reality far fewer were completed in that year (Anon., 1980, p. 1). This meant that even larger increases would be necessary in 1979 and 1980 to make up for the poorer earlier figures. The question of quality as well as quantity was also critical, for finishing work often proved the Achilles' heel of housing construction. This part of the job was hurried and carried out by unskilled workers; dwellings could still be unfinished when the new tenants obtained occupancy (Feiwel, 1980, p. 91). In desperation some tenants completed this cosmetic work themselves, and as noted in one Bulgarian newspaper, "It is not easy to buy a new apartment and start repairing it upon moving in" (Technicheski Delo, 1977, p. 3).

During the latter half of 1978 it was becoming increasingly clear that housing targets set in the Seventh Five Year Plan would not be fulfilled. During the first eight months only 17,400 apartments had been completed, i.e., just over a quarter of the planned total, which had now been lowered from 82,600 to 60,245 (Rabotnicheske Delo, 1978(b), p. 1).

Part of the answer to these disappointing results may have lain elsewhere. The government's emphasis on increased housing targets had led to greater use of prefabricated sections, produced in newly established enterprises. These factories totalled 22 by mid-1976, and were planned to double in number by 1980 (Ikonomicheski Zhivot, 1977, p. 3). Their success in turn was dependent on adequate raw material supplies, particularly cement. Output of the latter had slightly increased during the early 1970s but had remained static during 1976; although in 1977 it showed a 7% increase to reach 4.6 million tons. Cement production was far below the planned target (8.5

million tons) for 1980. Further, a continual
manpower shortage in the construction materials
industry meant that planned targets could not be
met; this was especially so when industrial
projects always took precedence for cement and
other building materials.

Two other aspects of the housing situation in
the late 1970s should be noted. First, there was
an amendment to the rents law passed in 1969
(Drzhaven Vestnik, 1977a, p. 1). The 1969 law
had replaced an outmoded version from 1945; in
future house owners had more rights and provision
was made for ousting forcibly assigned tenants
from private housing, who paid very low rents. A
further amendment in 1973 on the property law now
made it illegal to rent state or low rent
municipally-owned housing to anyone (his/her
family) who had a second home in the countryside
near to their work place. A further amendment in
December 1976 encouraged use of second homes, and
other accommodation near to the larger urban
centres. Encouragement was given to people to
make them their permanent residence; this in turn
would relieve dwellings in urban centres for new
occupancy. It also encouraged people to build new
private houses in peripheral urban areas (Drzhaven
Vestnik, 1977 (b), p. 2). The aim of the 1977
rent amendment decree was linked to easing the
rather restrictive 1973 act, so that suburban
housing could now be more favourably rented by its
owners.

The second aspect concerned a new law on
housing construction co-operatives (i.e.,
condominiums). Prior to 1978 this important
sector of Bulgarian housing was hardly covered by
any legal backing. Although no official
statistics on the number of houses built by co-
operatives are available, it has been estimated
they totalled a quarter of all dwellings by the
late 1970s (Radio Free Europe, 1978, p. 3). The
idea of the new law was to place the establishment
and dissolution of co-operatives on a legal
footing; this covered cases of voluntary
withdrawal by, or death of a member, any changes
in construction plans, reduction in number of
flats etc. It also set a minimum of six members
per co-operative with a separate flat for each; on
completion the co-operative is disbanded after all
legal documents have been signed by the new flat
owners (Drzhaven Vestnik, 1978, p. 4). In future,
it was hoped that this legal basis would help to

solve in a speedier manner any disputes or conflicts which may arise in the co-operative housing sector.

In spite of all these legal refinements and set housing targets the housing situation in Bulgaria still left much to be desired by the end of the 1970s. This was partly due to other factors such as the accelerated pace of industrialisation, improved transport facilities and a high rate of population growth in urban centres which all placed greater pressures on the housing construction sector. "In all large towns housing construction is of the utmost necessity" (Kiradjiev, 1979, p. 71). This along with other housing aspects were of prime importance for Bulgaria's planners by the beginning of the early 1980s.

During the 1980s housing has remained a recognised social, economic and political problem in Bulgaria. The decade began with optimism reiterated in the theses for the Eighth Five Year Plan (1981-1985), which stated that improved living conditions for the population would be achieved by the construction of at least 400,000 dwellings, to be followed in the ensuing Ninth Five Year Plan (1986-1990) by a further 450,000 housing units (Dinev, 1981, p. 30). These figures appear to have been too optimistic. The actual law on the plan approved a more realistic target of 360,000 dwelling units (Radio Free Europe, 1983). Later official sources stated that the 1981-1982 targets were only about 4,000 housing units below schedule (Rabotnicheske Delo, 1983a). Such results were beginning to make Zhivkov's 1972 claim of ensuring a house or flat for every family in the country after 10-15 years as now a hope rather than a reality. Furthermore, in 1982, Gallagher noted that "Currently there is a housing stock of about 2½ million houses, but 100,000 of these are uninhabitable because they are sub-standard, and there is extensive over-crowding in the towns" (Gallagher, 1982, p. 124).

In Sofia, as in most other big cities, the perennial housing problem continued to be extremely difficult to solve. Year after year, housing construction plans have remained unfulfilled. Official statistics for the first two years of the Eighth Five Year Plan demonstrated this continuing lag in target performance. In 1981, it was announced that about 70,300 dwelling units were completed although the

actual plan target was never published, and in
1982 only 67,980 were built, i.e., 4,620 below the
annual target of 72,600 (Rabotnicheske Delo,
1981). Early in 1983 the government admitted that
a shortfall of 3,938 dwellings had been recorded
against the 1981 and 1982 planned completion
targets (Rabotnicheske Delo, 1983a). Such returns
may have prompted the government to pass a new
decree in 1983 with specific reference to housing
construction (Rabotnicheske Delo, 1983b). This
new initiative meant the government would play a
diminishing role in the housing sector; in future,
workers' collectives and state enterprises were to
be responsible for solving accommodation problems
for their employees utilizing "Their own forces
and means in the construction distribution and
management of housing" (Rabotnicheske Delo,
1983c). Unfortunately, this added yet another
social responsibility to the already overburdened
state enterprises, which over the last few years
had accumulated commitments for developing
auxiliary farms, supplying their own premises with
trade and service facilities and supporting
schools and kindergartens.

The problem of overcrowded cities also
provoked government attention. An offer was made
for young people to live in villages, which were
becoming rapidly depopulated by in-migration to
the cities. In 1982 the government began offering
incentives and privileges to young people who
would be willing to take up farming as a career.
The state campaign offered various benefits to
young people who would be willing to live in
economically undeveloped rural areas for at least
ten years. They would receive a lump sum of 4,000
levas (equivalent to about two years average
salary), together with a state apartment, a free
plot of land and favourable credit terms to build
a house on it (The Times, 1982). A year after
this campaign had started, a sociological survey
amongst the 2,686 new settlers of the most undeveloped regions (Strandzha-Sakar) indicated that
solving personal housing problems, financial and
social worries had been uppermost in their
decision to migrate there (Narodna Mladezh, 1983).

The perennial housing problem seemed no
closer to solution in 1984. Instead of the
planned 73,000 housing units being completed only
68,798 were built; this was even lower than the
74,000 dwellings constructed in 1980
(Statisticheskij ezhegodnik SEV, 1985). A decree

on living standards passed early in 1984 made specific reference to the housing problem (Rabotnicheske Delo, 1984a). Although aimed at encouraging higher birth rates, the size of housing units was seen as an important obstacle; closer links between family and dwelling size needed to be considered, with housing priority being allocated to families with two or more off-spring. A further incentive to newly-wedded couples was a state loan of up to 15,000 levas (i.e., over six years' average individual income). This would enable them to buy a home, with no down payment and 30 years for repayment. This house price could be reduced by 3,000 leva if a family had a second child within four years of the first, and a further 4,000 leva deducted on the birth of a third child with no time penalty. When it is considered that the price of a two-bedroom apartment in Sofia at that time was around 15,000 leva, and a one-bedroom flat 11,000, this was a substantial incentive (Drzhaven Vestnik, 1984).

Despite such incentives the overall housing situation remained unsatisfactory. Late in 1984 a report to the government by the State Planning Office admitted that housing construction had for many years fallen short of targets and could do so again under the Eighth Five Year Plan (Rabotnicheske Delo, 1984b). To date only 280,000 housing units had been completed since 1981; the report suggested that 78,900 dwellings would have to be built in 1985 to fulfil plan targets, a figure far higher than ever previously achieved. Even so it was only 1,100 housing units below the 360,000 dwellings target for the 1981-1985 plan.

With such poor results in mind, it must have been with some trepidation that the government planned the 1985 population census. This was to include questions on the housing situation. According to its aims and objectives the results would ensure comparability with previous censuses and give more detailed information on qualitative housing aspects. The latter would be analysed from questions on housing comfort, sanitary conditions and other facilities associated with personal and family life. Information would be forthcoming on housing conditions from residents and their personal data files (e.g., marital, professional and social status, age and other socio-economic details). Finally, detailed questions would be asked on non-urban activities such as the ownership of private plots, buildings, their quality and quantity. These questions were specifically aimed at villa ownership, since their construction was seen as an inevitable result of rapid urbanisation. It also illustrated the need of city dwellers to escape at weekends from the unhealthy urban environment for rural leisure and recreation activities (Totev, 1985). The complete results should have been ready for the XIIIth Congress of the Bulgarian Communist Party in April 1986, but were not available. As with the 1975 census, it is likely that only a selection of the census data will ever be published for general public consumption.

The shortfall in housing provision probably led to the appearance of a new decree in November 1986 (Drzhaven Vestnik, 1986). This law hoped to encourage large-scale participation by the public in housing construction. It deliberately encouraged the private construction of dwellings in suburban areas, whilst prohibiting new houses of more than three storeys. In future individual one-family houses with gardens would be condoned as well as providing incentives for people to join co-operatives for building condominiums. The government's underlying thinking here was probably to gradually shift housing responsibility away from the state sector. The onus of providing living accommodation in Bulgaria is slowly moving towards the co-operative and private sectors, with the government concentrating more on quality dwellings. For example, the size of new state apartments was increased on January 1st 1988. As from this date, the state will no longer build one-room apartments with a kitchen in the corner. Moreover, in Sofia even one-room apartments with a separate kitchen will no longer be allowed, although in other areas this will depend on the views of local authorities (Radio Free Europe, 1986).

In November 1986 further encouragement for people to move to the suburbs, especially around Sofia and the other large urban agglomerations, was made by a new law. People who now agree to construct dwellings for permanent residence in suburban areas, will not lose their city residence permits. This is especially important in places like Sofia, where a strict allocation of residence permits has been normal policy for several years. The aim has been to control in-migration flows to large urban centres. Also more incentives have been offered to co-operative building. Private individuals have been forming co-operatives for building purposes over many years in Bulgaria, but they have always been very strictly controlled by the state. Now (1986) co-operative ventures have received a fillip; any private individuals willing to build condominiums can obtain state advice on building designs, receive construction materials, and utilize labour provided by local authorities. Furthermore, state loans will ease financial worries. Prospective candidates can apply for loans of 15,000-18,000 levas; the latter sum is confined to those willing to build larger houses on the urban periphery. Such government incentives are timely, as the 1986 annual plan report once again revealed that the annual number of dwellings was 3,202 below the planned target of 57,400 (Rabotnicheske Delo, 1987a).

The post-war period outlined here suggests that Bulgarian housing may be on the threshold of a new phase of development. This would be dominated by greater private participation, increasing awareness of social factors and the demands of rising living standards. If this is true, the next phase will be more complicated and indirect than that experienced in earlier years, but results may justify this change of government attitude towards the housing problem. It may also be the time for a more imaginative approach on the part of the state towards greater architectural freedom, away from the stereotyped styles of previous decades; now the hopes and aspirations of private individuals may play a more prominent role in the future of Bulgaria's housing policy.

HOUSING INSTITUTIONS AND ORGANISATIONS

In the preceding pages an attempt has been made to give a broad outline of housing in Bulgaria, but naturally it has not been possible to detail all the relevant provisions included in the housing policy structure. Housing is part of the overall planning for raising the country's standard of living, and a whole network of bodies has been created to implement this programme. The Standard of Living Committee of the Council of Ministers was established both for the practical organisation and methodology attached to raising conditions in Bulgaria, not only for the immediate but also long-term planning needs.

The Standard of Living Committee is formed from senior government officials and acts as a co-ordinating body with supervisory functions; in other words it acts as the major motivating force behind improving living standards (Evgeniev, 1974). The task of actually implementing the Committee's plan directives falls to the various sectoral and regional bodies of the country's administration. These, through their managerial function, act as government agencies responsible for planning and cover all aspects of social and economic development. Whilst the government retains the monopoly right of fixing capital investment priorities including housing, the actual nuts and bolts of the organisation are carried out by local government through the People's Councils. Thus five-year and annual plans stipulate the amount of investment for housing construction, the building materials to be used and the number of state multi-dwellings to be constructed. Further, they determine how much credit will be available for co-operatives and private individuals through state loans, for building their own homes.

State housing construction and provision is largely in the hands of the People's Councils, which carry out the role of local government. In 1987 there were 28 regional districts, each responsible for planning at that level, together with 29 municipalities which implement local housing policy decisions. The People's Councils are responsible for constructing houses and employ state building enterprises for this task; the councils pay all building costs, and then let or sell the finished dwellings. House sales by the People's Councils are based on different priority categories, e.g., family size. Special consideration is given to newly married couples and families with several children (i.e., two or more). Each category is based on calculations made by the Scientific Institute of Territorial Planning which is responsible for assessing national housing demands. The various categories include single mothers, elderly and single people, and married couples with or without children; each category contains a consumer unit, which is allocated a separate dwelling. According to Gallagher, the planners take into account over twenty consumer unit types, which include "minimum space standards and provision of amenities as well as the external environment and local services and employment" (Gallagher, 1982, p. 124).

It should be noted that the present Bulgarian system of local government is in a state of flux. Zhivkov in an attempt to eliminate so-called parochialism and isolationism has hinted during 1987 at changes in the territorial organisation pattern. The problem identified by him as "territorial feudalism", has led to the integration of some administrative districts (e.g., Ruse and Razgrad in February 1987), two of the 28 regional units utilized since 1959 by the government for local economic management and political administration in Bulgaria (Rabotnicheske Delo, 1987b). In August 1987, the 28 regional districts were replaced by 9 consolidated regions. Further, Zhivkov also has plans for changing the role of the 29 municipalities (Otechestven Front, 1987) which are to become "self-managing"; this will encourage them to exhibit more of their own initiative with increasing contact between themselves for solving local socio-economic problems like housing. Nevertheless, planning is to remain the major instrument for managing the economy, but subject to a new approach. This will involve more contacts at all administrative levels based on greater equality and shared responsibility. The present territorial units will be replaced by larger units which are seen as providing better conditions for the development of productive forces, the territorial infrastructure and environmental protection (Central Committee of the Bulgarian Communist Party, 1987). Further discussion has continued on the country's territorial structure. In March 1987, party guidelines on regional planning were published (Rabotnicheske Delo, 1987c), which referred specifically to housing construction. Plans should not commence for new dwellings while housing stock is available in a settlement area. This means that empty houses vacated through rural-to-urban migration located in villages surrounding an urban agglomeration, should be occupied before new houses are built in more popular centres of the settlement area.

Besides the People's Councils, the other main source of government housing construction is organised by the state enterprises and other socialist organizations. Part of their income is set aside for the provision of workers' dwellings.

This gives enterprise managers an added advantage to hold on to their work force, particularly the more qualified personnel, through accommodation provision. During the 1960s some enterprises (e.g., metallurgy) engaged in ambitious accommodation programmes; later much of this early enthusiasm waned. The petrochemical works at Burgas on the Black Sea coast, was an exception to this tendency, and had provided about 4,000 dwelling units for its employees by 1979 (Morris, 1984, p. 138).

By the early 1980s new housing provision by enterprises was inadequate, and caused some tension between workers and management. Another factor was the growing financial problems of some municipalities. In principle, they were supposed to be financially self-supporting but some, especially in the more mountainous and remoter parts of the country, suffered financial difficulties. In 1982, all municipalities were reclassified according to their infrastructure level and stage of economic development; 142 of the 291 municipalities (i.e., nearly half) were in the fourth (developing) and fifth (economically weak) functional groups (Otechestven Front, 1982). Obviously housing finance was being severely tested in them. This may help explain why the government transferred some housing costs onto enterprises. In 1983, enterprises and other socialist organizations were told to build housing for their employees, utilizing their own workers and financial sources; furthermore, socialist organizations were given the right to set up their own special housing construction fund and allowed to transfer some money to it from other sources (Radio Free Europe, 1983a). The long-term effects have yet to be seen on the already overburdened enterprises and other socialist organizations; it does, however, fit in with the government's greater self-management drive at all levels of state control.

Private initiative is the other source of housing construction. By the mid-1960s, the government was openly encouraging such activity. More new houses came from the private than state sector (Figure 5.1). Not until the early 1970s did state housing surpass private levels, and it has remained higher since then. The housing co-operative is a popular form of private construction for building condominiums in Bulgaria, and has been so for over fifty years.

Housing Policy in Bulgaria

Figure 5.1: Bulgaria: Number of houses built according to
sector 1965-1984

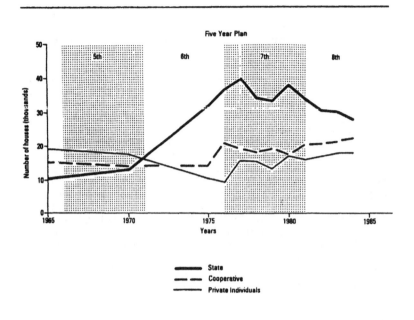

Source: Statisticheski Godishnik 1985 p. 247 XI Table 22

It is strictly controlled by state legislation,
but a group of six or more people are allowed to
form a co-operative to build houses or flats.
Since the early 1970s co-operatives have held
second place to the state sector in dwelling
construction and maintained it in spite of
government regulations. For example, the Law of
Housing Construction Co-operatives in 1978,
clearly stated in Article 9 that "persons may be
accepted as members of a housing construction co-
operative if they have been registered in the
files of those needing housing" (<u>Drzhaven Vestnik</u>,
1978). Prior to this law people saw joining a co-
operative as an easier way of obtaining accommodation by avoiding housing waiting lists, but since
1978 this incentive has been lost. Now only those
deemed as needing housing are eligible and this
may explain the temporary stagnation of co-
operative house building around 1980. Since then

condominium ownership has received higher state
planning priority, together with an offer of a
fifteen-year mortgage valued at ten times the
average annual income of a newly-wedded couple.
Certainly official statistics between 1980 and
1984 reveal how state housing construction
steadily decreased, while that of co-operatives
and private individuals increased (Figure 5.1).
Hopefully, this will help Zhivkov's 1972 target
prediction of one room for each family member by
1990 (Lampe, 1986, p. 163).

Private individual house building has always
been popular during the post-war years. This is
especially so in rural areas. The early post-war
years saw rural areas with greater access to
building material and labour, whilst later
individual families on collective farms were
allowed to keep their own houses and surrounding
outbuildings. By 1980 some 15,000 families have
received low-interest long-term loans for building
their own homes, much of which has been on
collective farms; it is popular for example among
some minority groups such as gypsies (Puxon, 1980,
p. 12). Even so, since the early 1970s there has
been a decline in privately built houses. This is
in spite of the 1971 Bulgarian Constitution which
confirmed a citizen's right to house ownership.
It may have been related to increased availability
of state accommodation, and the popularity of co-
operative housing activity. State housing
construction peaked around 1977; since then new
private dwelling construction has continued to
rise in popularity (Figure 5.1).

Ideally, the principle of supplying housing
as part of its social services has not been
adhered to in Bulgaria. Unfortunately, not enough
housing units have been supplied to satisfy
demand. Therefore, the Bulgarian government has
allowed co-operative and privately financed
housing construction to flourish. This has not
only helped offset demand, but also made the
consumer contribute a significant share of new
dwelling units and their later maintenance. As a
result Bulgaria is now reaching a point where the
state, co-operative and private individual housing
sectors are providing a similar percentage of the
new houses built in the country (Figure 5.1).

HOUSING MARKETS AND ALLOCATION MECHANISMS

Bulgaria's investment in housing construction has failed to keep pace with accommodation demand. The latter is largely a result of continued rural-to-urban migration, the demolition of older, often pre-war housing, more households with fewer people in them, and a rising number of young married couples beginning families. All these factors have grown faster than the planners had anticipated. The deficit in housing units vis-à-vis household demand is a significant factor in Bulgaria's housing market. It is most acute in urban areas where emphasis on rapid industrialization has attracted many migrants from rural areas, anxious to obtain higher wages in new, or modernized enterprises located in major cities and towns. The planners failed to allocate sufficient housing for this influx of people. By 1975, Bulgaria's average of 22.5 dwelling units per 1,000 inhabitants was the lowest in Eastern Europe; since then annual housing targets have had a 20% shortfall (Lampe, 1986, p. 193).

Figure 5.2: Bulgaria: Housing construction 1957-1986

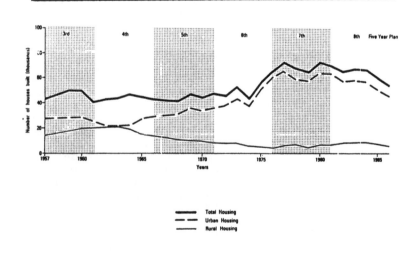

Source: Statisticheski Godishnik (relevant years)

During the early 1960s housing construction in urban and rural areas was roughly equal (Figure 5.2). Since then urban housing has steadily increased with a dramatic rise during the Seventh Five Year Plan. This was a response by the Party's Central Committee which saw solving the housing problem as "an important strategic task", (Anon., 1976a, p. 39). Decline in new housing provision during the Eighth Five Year Plan further aggravated the urban market; rural housing construction has remained stable, if much less, over the last two decades. However, it is in urban areas where the housing deficit is most serious. Figure 5.3 clearly illustrates the relationship between urban population and number of urban apartments. Whilst the former was approaching six million in 1985, the latter was still well below the two million mark. This reveals a considerable shortfall in urban housing provision, and means qualifying households in large towns wait several years for apartments. The overcrowding is worst in Sofia where people wait five to ten years for accommodation. This is in spite of the government's claim that between 1972 and 1982, 650,000 new housing units were built in Bulgaria for two million inhabitants (Lampe, 1986, pp. 193-194).

The most accurate way of estimating housing surplus or shortage is the relationship between the number of housing units and household demand. Such data is not always consistently available as in Bulgaria, but a possible surrogate may be calculated from the number of marriages each year compared to the number of housing units built. This gives some indication of a surplus/deficit in the housing market (Morton, 1979, p. 305). Admittedly, it is not as accurate because remarried divorcees who vacate dwellings are not included; it also fails to include new unmarried people entering the accommodation market, and dangerous dwellings demolished through necessity. In spite of these differences a glance at Figure 5.4 suggests that from 1957 to 1979 the number of new housing units never exceeded that of marriages. The greatest disparity existed during the Fifth Five Year Plan (1966-1971), but after 1975 the number of new houses built increased, narrowing the difference for the first time. During the Seventh Five Year Plan there was a small housing surplus. Thus 1980, 1981 and 1984 experienced housing unit surpluses of 4,582, 4,880

and 3,569 respectively, but in 1985-1986 another
shortfall situation occurred.

Figure 5.3: Bulgaria: Population and apartments in urban
areas 1946-1985

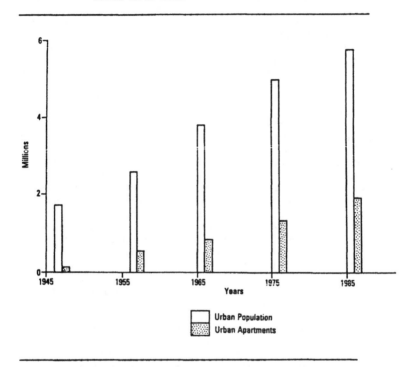

Source: Statisticheski Godishnik (relevant years)

With regard to the housing allocation mechanism,
the Council of Ministers issued a decree in 1978
concerning "Regulations related to the Sale and
Distribution of Housing" (Drzhaven Vestnik, 1978a)
which attempted to clarify former rules. The
decree stressed that local authorities were
responsible for the distribution of all new
dwelling units through the People's Councils. The
whole system was based on priority; elaborate
procedures were outlined relating to an
applicant's contemporary housing conditions and
his/her family situation. Allocation priority was
dependent on the number of children, those people

who had suffered over a long period from poor
accommodation, recipients of government service
awards, medals etc., veterans of the Second World
War, and those who had worked in a voluntary
capacity on building sites. Special housing was
allocated to those people recognized for their
"activities in quelling fascism and capitalism"
together with enterprise workers who had helped to
build accommodation for their own staff.

Figure 5.4: Bulgaria: Marriage and housing unit data
           1957-1986

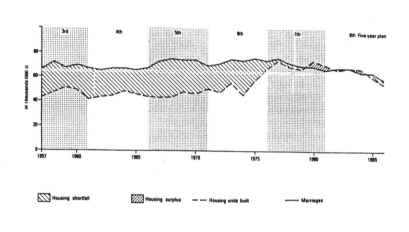

Source: Statisticheski Godishnik (relevant years)

     This latter point was further refined in 1983
in an effort to increase the pace of housing
construction (Rabotnicheske Delo, 1983a). This
encouraged workers' collectives and socialist
organizations (e.g., enterprises) to utilize their
own labour for the construction, distribution and
management of housing, for their own employees.
Elsewhere the overall responsibility for housing
construction and allocation to labour collectives
remained with the District People's Councils; they
designated three types of housing project, namely
that utilizing enterprise funds and its own labour
force, that using the latter only for finishing

purposes, and finally projects where the enterprise was allocated some dwelling units after its labour force had helped to build either its own, or some other housing scheme.

Another housing allocation incentive applied to white-collar employees. Ever since the early 1970s these workers had received favourable house purchase loans if they had been employed for one and a half months of paid vacation annually working in agriculture or any form of building construction (Radio Free Europe, 1973). Later in the 1970s this idea was broadened to include anyone wishing to buy an apartment (i.e., usually the younger generation) who would be willing to help on the finishing work of dwellings (Ikonomichesko Zhivot, 1976; Drzhaven Vestnik, 1977a).

Once allocated a dwelling unit, the new owner has to pay something towards its construction costs. This form of government-subsidised housing is popular but involves joining long waiting lists because of unsatisfied demand, and therefore means a considerable time lag. The logic behind this idea is that the value of waiting time equals factor cost, a large part of the latter being labour. In Bulgaria, on average, people pay only 22% of housing cost compared with 34% in Romania, 21% in Czechoslovakia and Poland, 17% in East Germany, and 14% in Hungary (Alton, 1983, pp. 13-18). It should be remembered that these subsidies may in time be eroded as privately bought co-operative housing receives new government support and emphasis. The Bulgarian government has deliberately expanded the private housing sector in place of its public counterpart.

Usually payment for a dwelling is made from a person's housing savings account held in the State Savings Bank. This bank is specifically designed to cater for personal deposits compared with others responsible for industrial investment or trading activity abroad. Housing savings accounts saw considerable growth in the late 1950s and 1960s. For example, between 1956 and 1968 such accounts grew sevenfold in number and deposited money, twenty-three times (Anon., 1969, p. 165). In the 1970s consumer loans for house purchase were on a long-term basis with up to 25 years repayment at a low interest rate (2.5%). This represented anything between 25-30% of total cost, providing the dwelling was 120 square metres or less (Volle, 1979, p. 122). Although the consumer on average pays only 22% of the cost, this can vary between 20-30%. The state fixes house prices nationally, but there are regional variations; for example, higher state housing subsidies are paid in areas needing more workers, or in under-developed regions like Strandzha-Sakar.

The government subsidy covers the differential between actual price and the cost of a new housing unit. In 1981, for example, Gallagher states that "the average selling price was 170 leva per square metre, but the construction company cost was 200 leva per square metre" (Gallagher, 1982, p. 124). State subsidies are only given on new housing. Any subsequent sale by the owner is subject to a state-fixed price formula. The State Savings Bank also gives loans for home improvements, e.g., central heating installation, and for building houses, but any do-it-yourself projects are under surveyance by building inspectors employed by the bank to see that standards are kept and construction work is being done. Bank loan repayments are debited at income source to avoid fraud or default.

The sale of owner-occupied accommodation is also subject to strict state rules. It is compulsory to sell your dwelling unit through the People's Council; this body acts as estate agent, and receives one percent of sale price for its trouble, although no transfer or other legal costs are charged. Furthermore, the People's Council has to agree to the transaction, and appoints its own new occupant on the basis of accommodation need. The dwelling's vendor has to provide the People's Council with evidence of sale, e.g., insufficient space through additional children, and provide proof of alternative house purchase. Moreover, the State Savings Bank will insist on any outstanding loans being paid off on the original property, prior to giving any further financial assistance. All these rules are designed to prevent illegal practices such as gazumping, or profit from capital gains (Gallagher, 1982, p. 125). A state-built co-operative apartment of 70 square metres sells for 10,000-20,000 leva: the average annual household income in 1980 was 6,000 leva (McIntyre, 1988, p. 54). Loans up to 12,000 leva at 2 percent interest over 30 years are available from the State Bank (Ministry of Construction and Architecture, 1982).

Housing Policy in Bulgaria

Rented accommodation remains comparatively cheap. Roussinov states that "in 1939 those working people who are fortunate enough to get better lodgings, spent an average of 23 per cent of their monthly salaries on rent" (Roussinov, 1965, p. 202). Donnison has shown that around 1960 the average annual expenditure of Bulgarian manual workers on rent was less than 2% (Donnison, 1967, p. 122), whilst in 1965 Roussinov maintained that "The new flats are mainly supplied to the working people, who pay minimum rents, averaging 4.6 percent of monthly salaries (or 1.5 per cent of total family expenses)" (Roussinov, 1965, p. 202). In 1979, Volle stated that rents were fixed at 0.19 leva per square metre of useful surface area for an average quality apartment, equivalent to 5% of a family's income. This compared favourably with owner-occupation as the sale price of family accommodation was 2-3 times that of the household's average annual income (Volle, 1979, p. 123). Tenants also enjoyed greater mobility according to Gallagher, but at the same time had equal status before the law to their home-owner counterparts. For example, tenants can appoint who succeeds them when they die. Furthermore, the state retains the burden of property upkeep, and dwellings may be located in the same residential block or area as privately owned housing. Rented accommodation is popular with lower salaried people. State employees on contract work in different parts of the country are provided with housing when away from home (Gallagher, 1982, p. 124).

From this brief resume of the housing market and its associated allocation mechanisms it can be seen how it is subject to tight government control. Restrictive legislation, the significant role played by district and local People's Councils, and financial controls stipulated by the State Savings Bank, prevent speculation, in theory at least, in the country's housing market.

PROBLEMS AND POLICY ORIENTATION

The perennial housing problem in Bulgaria, especially in larger urban centres, has proved in the past to be extremely difficult to solve. Continued shortfalls in planned housing targets is a constant cause for concern, whilst the reasons for the failures are numerous and complex.

Shortages seem a major factor; insufficient numbers of trained construction workers is ever present, probably because the more able and hard working are sent abroad to the Soviet Union, or to the developing Third World countries to obtain hard currency earnings. There is also a continual shortage of building materials and poor delivery dates, together with the more general problems of inadequate state organisation and planning.

Stemming from these basic shortcomings are other interrelated problems. The first of these and a major issue is connected with construction. More generally, the whole Bulgarian construction industry seems to be suffering from lethargy. This fact prompted Grisha Filipov, a member of the government's Politburo and chairman of the National Assembly's socio-economic development commission, to warn the party in June 1987 that the number of unfinished projects remained impermissibly high. As a result, more effective methods would have to be found to increase construction, and weed out people who failed to observe government decisions. Even the construction of priority industrial facilities was considerably behind schedule (Tanjug, 1987; British Broadcasting Corporation, 1987a).

If the demands of industry which always takes priority are not being fulfilled, then it is probable that housing is in a similar position. Even so, official housing statistics from the first half of the 1980s were encouraging. The type of houses constructed were over two-thirds from reinforced concrete, and a further quarter 'solidly built' compared with the inferior frame-built lath and plaster. Similarly, the percentage of number of rooms per dwelling changed over this period (Figure 5.5) with a noticeable decline in one-room apartments compared with a substantial increase in three-room dwellings. This complements guidelines for the Ninth Five Year Plan (1986-1990) which promises that the number of three-room apartments will increase by half. Furthermore, since 1980 average usable floor space (in square metres) has risen steadily (Figure 5.6) after annual falls during the Seventh Five Year Plan (1976-1980). Finally, residential living space per capita (in square metres) has also shown continual increase between 1965-1985, rising from 10 square metres to 16 square metres (Statisticheski Spravochnik, 1986).

Figure 5.5: Bulgaria: Dwellings according to number of
rooms 1978-1985 (percentages)

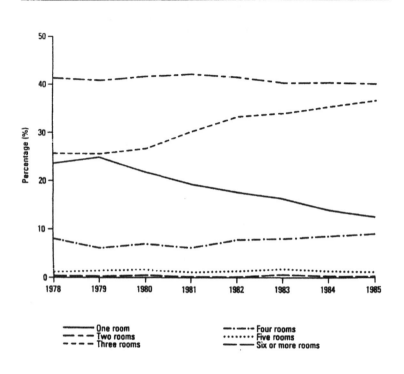

Source: Statisticheski Godishnik (Section IX)
(relevant years)

In spite of this official data, there are problems
associated with poor quality construction work,
leading to future maintenance difficulties and
discontented occupants. Lack of financial support
often means houses and flats are poorly finished
with a drab, and colourless appearance. Problems
also exist when some enterprises try to build

their own housing. Some have little or no experience in the building sector and the quality of construction is amateurish, because it is done by employees with no professional qualification in construction. Admittedly, one way of solving the housing crisis is through enterprises providing financial investment and construction facilities but this may result in unnecessarily high costs and low quality performance. Furthermore, such action may also deter enterprises from completing normal plan targets because both labour and finance have to be diverted to support supplementary housing projects.

Official housing statistics can also be misleading, as Natalija and Vladimir Kostov pointed out. During the Seventh Five Year Plan, 352,000 dwellings were built, i.e., 100,000 more than in the Sixth Five Year Plan. The statistical yearbooks give some indication how this was achieved; Figure 5.6 shows that the Sixth Five Year Plan had an average usable floor space of about 49 square metres. Guidelines for the Seventh Five Year Plan indicated this was insufficient and should be increased to 80 square metres. In reality, the opposite trend took place, declining by 1980 to 44.78 square metres, i.e., 5 square metres less than the previous plan period. The statistics do not give the number of dwellings constructed with 150 or 200 square metres, but they do help to explain how the extra 100,000 apartments appeared at the expense of size differential (Kostov, 1983, p. 91). Similarly, official statistics give no indications of problems relating to interior/external finishing and bad quality fittings; further, the shortage of building materials such as cement means external wall rendering often suffers long delays before completion. Part of the answer lies in government investment decisions. These are not directed towards obtaining high-quality production and/or finishing in the building industry. Little capital investment is earmarked for introducing modern building techniques and complaints have been made that construction organizations are slow and cumbersome. This in turn helps explain why targets are not achieved in the five year housing plans (Gallagher, 1982, p. 125).

Housing Policy in Bulgaria

Figure 5.6: Bulgaria: Average usable floor space 1957-1985
(square metres)

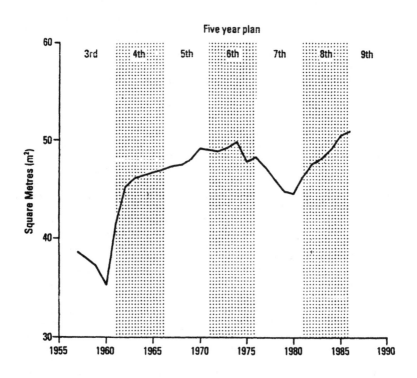

Source: Statisticheski Godishnik (relevant years)

The whole housing mechanism seems to be bedevilled by poor organization and planning. Shortcomings in Bulgaria's plan fulfilment were stressed at the July meeting of the Council of Ministers in 1987. Growth rates had fallen short of plan requirements in both qualitative and quantitative terms; there appeared to be a lack of urgency in the move towards self-management and remedial measures had proved unsatisfactory (British Broadcasting Corporation 1987b). Thus in the housing industry the proliferation and entrenchment of state organizations has led to the second problem. It is often a case of the left hand not knowing what the right is doing, due to lack of co-ordination between various ministries, economic bodies, local People's Councils and site managers. One may therefore find that construction agencies have to deal with central government committees, e.g., the Finance and Construction Ministry, the State Savings Bank, Architecture and Planning Committees, as well as regional authorities, e.g., District People's Councils, local municipalities and building research institutes. It is not surprising therefore that sometimes confusion reigns supreme; for example, in Sofia alone there are over 100 building organizations and enterprises grouped into 25 different economic bodies, which in turn are answerable to nine different government ministries.

As a result negative factors such as the bribery of officials can flourish in such situations, particularly as housing demand is never in tandem with supply. Corruption is therefore the third problem in Bulgaria's housing situation. Whilst most evidence of such activity is covered up by the state's press, reports do occasionally expose individual offences of bribery committed by state officials (Pogled, 1985; Starshel, 1985). Criticism of the housing industry in 1985 included information of a scandal in Sofia involving several state housing commission officials who had accepted bribes over a four-year period totalling some 45,000 leva, from illicit dwelling allocations (Rabotnicheske Delo, 1985). Such revelations at that time were part of a campaign to expose corruption and other misdemeanours, with the publication of people's names involved. Other urban centres recorded corruption of officials at this time. Both in Varna and Plovdiv housing officials were accused of failing to give preference to newly-wedded couples and those with large families, in housing allocation, in which bribery and nepotism were suspected motives (Radio Sofia, 1985; Radio Free Europe, 1986). Though a short-lived campaign, this attitude may be revitalised with the current emphasis on 'glasnost' in the country.

Another form of corruption concerns the popular growth of second homes. Many have been constructed legally with state permission for summer residence, particularly in favourable mountain and coastal areas; others have been built illegally without reference to urban planning, often on arable land, private plots, or auxilliary farms attached to non-agricultural enterprises (<u>Rabotnicheske Delo</u>, 1978; 1981a; 1981b; <u>Zemedelsko Zname</u>, 1978; <u>Otechestven Front</u>, 1979; <u>Pogled</u>, 1981; <u>Starshel</u>, 1981). Further a large number of these second homes have been officially designated as too large for occupancy for the average 50 days per year. This is interpreted by the state as an excessive waste of manpower and construction costs which is particularly deplorable from a government viewpoint, considering the number of deserted houses available in some areas. In contrast, the government's appeal to second home and villa owners on the periphery of large urban areas to use their property as permanent dwellings, and alleviate central city congestion by commuting to work, has received little support. Poor public transport and other services have encouraged urban dwellers to remain in the city and only travel to their property on the periphery at weekends. Such attitudes could be reversed if the transport network with the outlying city areas was improved, and sufficient shopping facilities, adequate water supply and other basic services were provided around the city's edge.

The government also has encouraged use of abandoned villages - the fourth housing problem. Some regions, such as Strandzha-Sakar and the more mountainous areas, have a plethora of such settlements, which on occasions constitute 70-90 percent of total dwellings in a district. It is probable that the old administrative district system is partly to blame through the promotion and growth of bureaucracy and local inertia. They have done little to resolve the future of abandoned villages which are particularly prevalent in the Stara Planina Mountains (187 in Gabrovo district, 100 in Veliko Turnovo district) (Radio Free Europe, 1983) with over a further 96 in the Sofia district. Unless government policy solves this problem, it has been estimated that about a thousand Bulgarian villages will have been abandoned by the end of this century (<u>Nova Makedonija</u>, 1983). A state propaganda campaign could be launched to try and convince people that deserted villages do provide ideal recreational sites for second home occupation, because their restoration can be achieved at a third the cost of building new weekend residences. Perhaps a start would be a state second home organization, responsible for the location, purchase, re-

condition, furnishing and sale of old abandoned
village dwellings. Indirectly this could also
help to preserve earlier architectural styles, as
well as repopulate (if only intermittently) some
of the more deserted parts of the country.

Resuscitation and preservation of earlier
architectural styles leads to a fifth housing
problem facing the Bulgarian government, namely
that of urban design. At present there is an
increasing state concern for architectural
innovation and conservation. During the immediate
post-war period and throughout the 1950s,
architectural styles were rather rigid, heavy and
based on monumental planning methods reminiscent
of the Soviet Stalinist period (Bichev, 1961;
Tsapenko, 1953; Zlatev, 1948). Fortunately, after
1960 a new architectural approach emerged
(Tangurov, 1972), aided by an extensive housing
and public building programme; indigenous
architects were invited to utilize earlier
Bulgarian styles in the construction of coastal
resort complexes, planned housing estates etc.,
together with re-designing older, dilapidated
residential areas. The state encouraged
architects to blend elements of national
architectural styles into newly planned projects;
unfortunately, the effort required to do this was
not sufficiently rewarded. As a result the drab
appearance of many housing blocks constructed in
the 1960s and 1970s continued, which has received
considerable criticism of late. There has been a
demand for improved design during the 1980s. The
state has responded with new laws on height of
high-rise blocks, and some attention has been paid
to the significance of architectural heritage and
contemporary town planning (Stantcheva, 1980).
However, there is still plenty of scope for inte-
grating former building styles into the present
architectural landscape (Danailov, 1978;
Obretanov, 1978; Stamov et al., 1972).

A sixth housing problem concerns social
attitude. Bulgaria's birth rate remains low and
lack of adequate housing is seen as one of the
main factors in a family's decision to restrict
its size. In future, housing units must be better
adapted to the number of children per family, and
larger accommodation should be given to young
married couples. Pressure on the government is
also coming from an increasing divorce rate and
growing demands from single-parent families; also
the number of people living longer has derived

217

from an improved state health service, which in turn has placed a growing demand on the need for sheltered accommodation for the elderly. Rural-to-urban migration has tended to break up traditional family bonds and customs, where parents/grandparents/in-laws all lived in one large house with their young married children. This is now very much a thing of the past in many areas.

Lack of sensitivity towards social attitudes by the government was also revealed in the most recent national population census (Tsentralno Statistichesko Upravlenie, 1985). It may have been designed with honourable intentions in mind regarding the documentation of building and housing stock throughout the country. Unfortunately, some of the questions posed could have been interpreted as infringing on civil liberty rights, with detailed questions on housing and living conditions. In such a census one may expect normal household questions on presence/absence of water, heating and other communal services, but some sections demanded data on personal belongings. These included possession of washing machines, televisions, video recorders and other consumer durables; perhaps the most sensitive question referred to compulsory registration of all villas and second homes. This had never been asked for in previous censuses. Owners who had used building materials obtained illegally or who had built without state building permission felt particularly vulnerable. Other owners may have interpreted this question as a forerunner for registration of private property which at some future date could be used for compulsory re-allocation to poorer families, or young married couples in the hope of stimulating the birth rate. Also, there were others who could have interpreted such questions as data collection on the illegal practice of property sub-letting. This provides untaxed earnings for a significant part of the Bulgarian black economy (Radio Free Europe, 1985).

The seventh housing problem facing the government concerns the balance between private and public sectors. The obvious advantages of encouraging privately built houses has relieved pressure on the state economy, by diverting attention away from the government's continued failure to reach planned housing targets. As Gallagher notes, "Over the last five years 320,000 houses were built by the State and 80,000 by private individuals" (Gallagher, 1982, p. 126), i.e., a fifth. Furthermore, such private initiative leads to quicker occupancy, particularly in rural areas; here once one or two rooms have been built, the family moves in and continues to complete the rest of the dwelling whilst in occupancy. Government policy therefore must not make state rented accommodation too attractive, if it wishes to further encourage the private sector. At present a rented state apartment from the local authority guarantees quality maintenance, low rent, and no liabilities if one wishes to move to another area. According to Gallagher, "there is growing evidence that many people would prefer to rent" (Gallagher, 1982, p. 125). If rented accommodation becomes more popular, private construction will decline and greater pressure will be placed on the state to fulfil its housing quotas and planned targets. This, in turn, will place heavier demands on state subsidies for housing, in order to keep consumer prices artificially low as part of the government's social policy. In turn, this will put extra pressure on state financial sources, diverting them from Bulgaria's prestigious programme of industrialization.

Finally, the government has to consider housing problems in the more underdeveloped parts of the country. In 1982, a government decree aimed at encouraging rapid socio-economic development of settlements, namely, fourth and fifth functional level settlements, i.e., the lowest and poorest, together with frontier areas (particularly Strandzhar-Sakar region near the Turkish border) (Rabotnicheske Delo, 1982). The idea was to elevate the living standards of these regions to those found in municipalities of the three higher or developed categories. This in turn would discourage out-migration and encourage re-population of these underdeveloped regions; according to the decree this would affect a fifth of Bulgaria's population and nearly half of its territory (i.e., 142 of the country's 291 municipalities). In these poorer areas one of the incentives was that of rent-free housing, besides that of permanent (not seasonal) employment, an infusion of agricultural technology, better communications, and basic services. Housing, particularly for the younger generation, was seen as a major attraction of this scheme, according to a sociological survey of the Strandzhar-Sakar region in 1983 (Trud, 1983a).

In theory the scheme was most attractive, but in reality housing posed on it a severe restraint. People would only migrate to these underdeveloped regions if sufficient modern housing accommodation was available. This meant importing people with building and construction skills into these regions; they in turn demanded suitable dwelling facilities, which in spite of abandoned villages, were in short supply. The government hoped to attract such building workers to these areas, but most were reticent to leave employment in the better developed parts of the country. Therefore housing construction remains only in the planning stage in many poorer parts of the country (Radio Free Europe, 1983b). Legal constraints sometimes limit in-migration to these areas; for example, Ahtopol is a town with surplus housing as a result of out-migration. Unfortunately, the 1983 rent law excluded Ahtopol municipality and others (e.g., Burgas) from allocating temporarily unoccupied or private dwelling accommodation to in-migrants (Trud, 1983b). This illustrates how on occasions, state laws frustrate initiative at the local administrative and planning level. Moreover, even if such housing constraints did not exist, many of these poorer areas have insufficient employment possibilities for the resident population, let alone new settlers, because capital investment into new enterprises has been made in many underdeveloped regions.

Even if new settlers arrive in areas free of legal constraints, and they obtain employment, problems with housing remain. For example, in parts of the Strandzhar-Sakar region rapidly constructed pre-fabricated dwellings possess leaky asbestos roofs and look out of place against local architectural styles, whilst insufficient capital is being invested in the lengthier process of restoring abandoned village family houses. Sometimes financial investment is misdirected; for example, in the Burgas district, two-thirds of the capital allocated to building construction in 1983 was invested in one new hotel, in spite of housing demand for in-migrants (i.e., 600,000 leva from 900,000 leva allocation) (Kooperativno Selo, 1984). As a result, reduced housing finance led to inadequate allocation in this district of individual living space with three-child families given one-room apartments of 6 square metres. The

Burgas and Yambol districts have particularly severe housing shortages with eight applicants for every dwelling place in 1984 (i.e., 775 out of 6,683 applicants in Burgas district; 591 out of 4,396 in Yambol) (Radio Free Europe, 1984). Finally, new housing projects located beyond the periphery of original villages inhibit contact and integration, particularly where there is an age differential between young settlers and older village inhabitants.

PROSPECTIVE DEVELOPMENTS

Bulgarian housing has progressed a long way since that described by Vakarelski for the earlier part of this century (Vakarelski, 1965). Future developments have to be seen against the wider trends of prospective policies. Todor Zhivkov, the country's leader, is now appraising the values of 'perestroika' which have led to some realignment of certain ministries and various councils. Further clarification has come from the special party conference in Sofia on 28 and 29 January 1988, in which hints were made that the state machine may be reformed through streamlining. Restructuring has already seen the number of administrative regions reduced to a third, the relocation of around 100,000 middle-ranking bureaucrats, and early retirement for many older officials within the next year (The Times, 1988). The Bulgarian plan for economic reform will indirectly affect housing; for example, enterprises will be granted more autonomy to set prices, salaries and targets, with less interference from government ministries. Project managers will now have more financial choice from several different banks, whilst price reform, together with more attractive joint ventures between domestic and foreign partners are planned. However, a recent survey in Bulgaria's provincial areas has indicated that the average person does not fully understand these reforms (The Economist, 1988). Thus private housing construction may decline until the full implications of such developments are known.

Amendments are to be made to the constitution to aid transformation to a self-managing society. Bulgarian housing policy should benefit with its emphasis on self-help in order to assist the collective aims of society. Presumably this means further technical and financial assistance will be given to private individuals by state housing organizations, and this will reduce overcrowding problems experienced in many residential blocks on urban housing estates and inner cities. The underlying government belief that many Bulgarians wish to acquire their own dwellings may fit comfortably into this new wave of political thinking; self-management and self-help can fortify state methods of housing finance, leading to a growth in Bulgaria's owner-occupation sector of society. However, people themselves will be left to decide between the attraction of state-rented accommodation and the advantages of home ownership.

Demand for either will partly depend on the current demographic situation. A long-term post-war housing deficit has affected the country's social life. One important issue, clearly connected with insufficient housing is a falling birth rate. Many families are no longer having two or three children, especially where housing accommodation is in short supply, such as the large cities of Sofia, Plovdiv and Varna. Overall demographic trends are disappointing; the national birth rate has fallen from 21 per thousand inhabitants in 1944 to 13.6 per thousand in 1984, although admittedly there are considerable variations at local level. Areas of high demographic growth often coincide with those inhabited by a Turkish minority. The Bulgarian constitution gives much wider representation to any ethnic group if it attains 10% of the country's total population, and the Turks are now rapidly reaching that mark. Their presence is particularly noticeable in some urban centres such as Kirdzhali, Shumen and Razgrad, where they predominate; unfortunately, many large Turkish families inhabit increasingly cramped and inadequate dwellings and growing tension between them and the local Bulgarian inhabitants is sure to follow. In 1985 a government attempt to 'Bulgarize' Turkish names, on the grounds that many of these people were originally Bulgarians forced to accept Ottomanization, may have been an attempt to artificially reduce the Turkish element of society as the 10% quota reaches ever nearer (Kassam, 1985; The Economist, 1985).

CONCLUSION

The previous pages have traced Bulgaria's housing development over the post-war period. Housing, from the early days of Bulgaria's inception as an East European socialist state to the present, has proved a perennial problem in spite of state planning targets and encouragement of private initiative in the dwelling construction sector. Lack of success in the earlier post-war years may have been connected with the birth pangs of a country struggling to transform itself from a predominantly agricultural past to a future industrial economy. Preferential investment to finance industry at the expense of other sectors of the economy like housing, meant the latter could not cope with the growing rural-to-urban migration, shortage of raw material allocation for dwelling construction, and insufficient labour. Supplementary factors associated with a centralized bureaucracy, the temptations of corruption and stifling of local administrative initiative, added further confusion within the housing sector. Self-help housing through co-operatives and private owner-occupation partly relieved this bottleneck in Bulgarian society but housing demand has continued to grow, in spite of a falling birth rate over the post-war years. Given these drawbacks, one has to be optimistic about the future, particularly with the blossoming of ideas on self-management. The transitional period towards self-management is fraught with difficulties, not least in the housing sector. Yet Zhivkov has proclaimed that "The idea of these changes is to create still better conditions for economic growth, for the development of the social sphere and of culture, fully taking into account the requirements of the scientific and technological revolution" (British Broadcasting Corporation, 1987b). Let us hope that such pronouncements help solve one of the country's most pressing problems - that of housing.

REFERENCES

Alton, T., (1983), Money Income of the Population and Standard of Living in Eastern Europe 1970-1982, International Financial Research Inc., New York, p. 143.
Anon. (1956) (a), Planirane na b'lgarskoto selo, B'Igarska akademiia na naukite, Sofia, p. 228.

_____ (1956) (b), 'Tipovoto proektirane na sgradi v B'lgariia; sbornik materiali', B'lgarska akademiia na naukite, Sofia, p. 248

_____ (1969), 25 Codini Sotsialisticheska B'lgarija, D'rzhavne Upravlenie za Informatsija pri Ministerskija S'vet, Sofia, p. 210

_____ (1976) (a), Guidelines of the Eleventh Congress of Bulgarian Communist Party on the Socio-economic Development of the People's Republic of Bulgaria Under the Seventh Five-Year Plan (1976-1980), Sofia Press, p. 57

_____ (1976) (b), Theses of the Central Committee of the Bulgarian Communist Party on the Further Implementation of the December Programme on Raising the People's Living Standards under the Seventh Five-Year Plan and up to 1990, Sofia Press, p. 60

_____ (1980), 'Social Policy: Housing gains', News from Bulgaria Bulletin No. 1, Sofia Press, p. 19

Bichev, M. (1961), Architecture in Bulgaria, Sofia, p. 304

British Broadcasting Corporation, Summary of World Broadcasts, EE/8609/B/3 (2/7/1987)

_____ Summary of World Broadcasts (a) (30/6/1987) Monitoring Report (EE/8607/i), London

_____ Summary of World Broadcasts (b) (16/7/1987), Monitoring Report (EE/8621/i), London

Carter, F.E., and Žagar, M., (1977), 'Post-war internal migration in South Eastern Europe' in Demographic Developments in Eastern Europe, Ed., L.A. Kosinski, Praeger Publishers, London, New York, pp. 208-244

Central Committee of the Bulgarian Communist Party, (1987), Information Bulletin, Vol. XLI, No. 10, Sofia, p. 120

Danailov, B., (1978), 'Novi tvorbi na moinumentalnot o iskustvo', Izkustvo, Vol. 28, No. 7, Sofia, pp. 16-19

Dinev, A., (1981), 'Rozvoj Bulharské lidové republiky v osmé pětiletce (1981-1985)', Planované Hospedarství, No. 9, Prague, pp. 21-30

Donnison, D.V., (1964-1965), 'Housing policies in Eastern Europe', Transactions of the Bartlett Society, Vol. 3, pp. 91-119

_____ (1967), The Government of Housing, Ch. 4., 'Housing in Eastern Europe', Penguin Books, pp. 112-150

Donnison, D.V., and Ungerson, C., (1982), Housing Policy, Ch. 8, "Housing in Eastern Europe", Penguin, Harmondsworth

Drzhaven Vestnik, (a), No. 33 (26/4/1977)

_____ (b), No. 4 (14/1/1977)

_____ (c), No. 38 (13/5/1977)

_____ (a), No. 55 (14/7/1978)

_____ (b), No. 24 (24/3/1978) (Decree No. 1, dated 27/1/1978)

_____ No. 24 (23/3/1984), Sofia

_____ No. 92, (28/11/1986) (Decree No. 65; 6/11/1986)
The Economist, (a) (2/3/1985), pp. 58-59
_____ (b) (14/12/1985), p. 55
_____ (12/1/1988), pp. 44-45
Engels, F., (1963), The Housing Question, Lawrence and
    Wishart, London
Evgeniev, G., (1974), 'The rise in living standards in
    Bulgaria: Some problems', International Labour
    Review, Vol. 110, No. 1., pp. 29-42
Feiwel, G.R., (1980), 'The standard of living in centrally
    planned economies of Eastern Europe', Osteuropa
    Wirtschaft, Vol. 25, No. 2, Stuggart, pp. 73-96
Gallagher, P., (1982), 'Housing in Bulgaria', Housing
    Review, July-August pp. 124-126
Guardian, (2/4/1973)
Ikonomicheski Zhivot, No. 18 (18/4/1976)
_____ No. 20 (11/5/1977)
_____ No. 3 (19/1/1979), Sofia, pp. 4-5
Kassam, M., (1985), 'Bulgaria's Muslims under the hammer',
    Arabia (Jumada Al-Thaniyhah, 1405), pp. 10-13
Kiradjiev, S., (1979), 'Tendetsii v cinamikata na
    naselenieto na golemite gradove v B'lgarija',
    Problemi na Geografijata, No. 2, Sofia (B'lgarska
    adademija na naukite), pp. 67-72
Kooperativno Selo, (10/4/1984)
Kostov, N., and V., (1983), Sotsializm't v B'lgarija. Opit
    za Raznosmetka (Replike i helezhki po aktualni
    v'prosi) (Peev & Popov Publ.), Paris, p. 174
Lampe, J.R., (1986), The Bulgarian Economy in the Twentieth
    Century, Croom Helm, London & Sydney, p. 256
McIntyre, R.J., (1988), Bulgaria: Politics, Economics and
    Society (Marxist Regimes Series), Pinter Publishers,
    London and New York, p. 201
Ministry of Construction and Architecture, (1982) 'On the
    settlement situation and related trends and policies',
    National Monograph submitted to the Economic
    Commission for Europe, Sofia, p. 56
Morris, E.P., (1984), Eastern Europe since 1945, Heinemann
    London/Exeter, p. 211
Morton, H.W., (1979), 'Housing problems and policies of
    Eastern Europe and the Soviet Union', Studies in
    Comparative Communism, Vol. XII, No. 4, 1979, pp. 300-
    321
Narodna Mladezh, (11/8/1983), Sofia, p. 3
Nova Makedonija, (29/1/1983), Skoplje, p. 3
Obretanov, A., (1978), 'Harmonichna i krasiva zhizena
    sreda', Izkustvo, Vol. 28, No. 8, Sofia, pp. 2-9
Ognyanov, L., (1981), 'Along the road to socialism', in
    Modern Bulgaria : History, Policy, Econ-Culture, Ed.
    G. Bokov, Sofia Press, Sofia, pp. 103-127

Housing Policy in Bulgaria

Otechestven Front, (13/1/1979)
_____ (19/5/1982)
_____ (28/8/1987)
Pogled, No. 4, (26/1/1981)
_____ No. 25, (24/6/1985)
Puxon, G., (1980), Roma : Europe's Gypsies (Minority Rights
    Group, Report No. 14), p. 20
Rabotnicheske Delo, (14/4/1977)
_____ (a), (14/10/1978)
_____ (b), (25/7/1978)
_____ (12/1/1979)
_____ (a), (13/2/1981)
_____ (b), (16/2 and 10/4/1981)
_____ (c), (17/8/1981)
_____ (18/5/1982)
_____ (a), (30/3/1983)
_____ (b), (24/8/1983)
_____ (c), (14/2/1983)
_____ (a), (25/4/1984)
_____ (b), (27/11/1984)
_____ (16/9/1985)
_____ (a), (30/1/1987)
_____ (b), (27/2/1987)
_____ (c), (18/3/1987)
Radio Free Europe, Bulgarian Situation Report, No. 30,
    (17/8/1973), p. 3
_____ Bulgarian Situation Report, No. 15, (18/8/1978), p. 3
_____ Bulgarian Situation Report, No. 10, (12/9/1983), p. 1
_____ Bulgarian Situation Report, No. 3, (4/3/1983), p. 10
_____ Bulgarian Situation Report, No. 14, (28/12/1983),
    p. 14
_____ Bulgarian Situation Report, No. 8, (8/6/1984), p. 8
_____ Bulgarian Situation Report, No. 13, (17/12/1985),
    p. 6
_____ Bulgarian Situation Report, No. 12, (22/12/1986),
    p. 3
_____ RAD Background Report /81, (E. Europe), (12/6/1986),
    p. 9
Radio Sofia, (26/11/1985), 19.30 hrs
Roussinov, S., (1965), Economic Development of Bulgaria
    after the Second World War, Sofia Press, Sofia, p. 251
Shoup, P., (1981), The Eastern European and Soviet Data
    Handbook 1945-1975, Columbia University Press,
    New York, p. 381
Stamov, S., Angelova, R., Caneva-Decevska, N., &
    Kolarova, V., (Eds.), (1972), The Architectural
    Heritage of Bulgaria, De Capo Press, London, p. 224
Stanev, S., (Ed.), (1969), Dvadeset i pet godin
    sotsialisticheska B'lgarija, Sofia, p. 186

226

Stantcheva, M., (1980), 'Museums in the street...
architectural heritage and contemporary town planning
in Sofia', Museum, Vol. 32, No. 3, pp. 107-118,
(UNESCO) Paris
Starshel, (20/11/1981)
_____ (18/10/1985)
Statisticheskij ezhegodnik stran chlenev SEV, (1985),
Moscow, p. 68
Tangurov, Y., (1972), The Architecture of Modern Bulgaria,
Sofia, p. 243
Tanjug Yugoslav Press Agency, (29/6/1987), Belgrade
Tecknicheski Delo, (20/8/1977)
The Times, (2/8/1982), p. 4
_____ (30/1/1988), p. 3
Todorov, Y., (1971), Housing Construction in Bulgaria,
Sofia Press, Sofia
Toshev, D., (1959), Vuprosi na zhilishtnoto stroitelstvo
v NR B'lgaria, B'lgarska akademiia na naukite,
Sofia, p. 198
Totev, J., (1985), Osnovna tsel zadachi na prebrejavaneto
na zhilishnija fend v NRB prez (1985) g., Naselenie,
Vol. 3, No. 1., Sofia, pp. 80-86
Trud, (a), (23/6/1983)
_____ (b), (9/3/1983)
Tsapenko, M., (1953), Arkitektura Bolgarii, Moscow, p. 190
Tsentralno Statistichesko Upravlenie, (1959), Prebroiavane
na zhilishtniia fond v Narodna Republika B'lgaria na
1.XII. (1956) g., Nauka i Izkustvo, Sofia, p. 220
_____ (1985), Prebroiavane na zhilishtniia fond
v Narodna Republika B'lgaria na 1.XII. (1985) g.,
Nauka i Izkustvo, Sofia, p. 300
Vakarelski, K., (1965), Etnografia B'lgarii, Ossolineum,
Wroclaw, 1965, pp. 90-113
Volle, J-P., (1979), 'Croissance urbaine et organisation
regionale en Bulgarie', Revue Géographique de l'Est,
No. 1-2. Nancy, pp. 101-133
Zemedelsko Zname, (30/3/1978)
Zlatev, T., (1948), B'lgarska bitova arkitektura, Vol. 1,
Sofia, p. 345

Chapter Six

HOUSING POLICY IN THE SOVIET UNION

Gregory Andrusz

INTRODUCTION

The XXVII Party Congress, held in 1986, passed a resolution to provide each family with its own separate flat or house by the year 2000. Speaking at the end of 1988, the deputy chairman of the State Construction Committee (Gosstroi) declared that since the housing stock now stood at 4,300 million square metres of floor space, a further 2,800 million square metres would have to be erected by the end of the century in order to meet this goal (Rozanov 1988). Put in other terms, the government has set the target at increasing housing construction by one and a half times and to erect 40 million new flats and individual houses by that date - in other words doubling the existing housing stock (Aganbegyan 1988a: 17). Abel Aganbegyan (one of Mr. Gorbachev's principal economic advisors), for whom the "housing problem (is) the worst social problem" faced by the country, has elevated housing - or rather the successful resolution of the perennial housing problem - to the status of keystone of perestroika. Although this might be an exaggeration, good grounds do exist for ascribing such prominence to housing in the restructuring process. Yet, the 'old' Party Programme, published in 1961 during the Khrushchev administration, had also grandiosely declared that by 1971 the housing shortage would have come to an end and by 1980 every family would have its own fully-equipped flat. Moreover, in the course of the second decade of building communism (defined, millenially, as the period 1970-1980) people would gradually cease to pay rent. Today, however, communism is no longer seen as lying just beyond the horizon and its advent has been more than just deferred. Because housing targets have not been met in the past, many people today express

through their letters to the Central Committee of the Communist Party and the Council of Ministers, at public meetings and to the press their serious anxiety that the goal of providing each family with its own separate accommodation by the year 2000 will not be achieved. (Postanovlenie 1987: no. 28, art. 96).

At the end of 1988 senior government officials were still remarking on the acuteness of the housing problem (Murphy 1988: 19,11; Rozanov 1988: 9,12).

At the XXVI Party Congress in 1981 Brezhnev repeated what had been said many times before: "As is well known housing occupies the most important place in our social programme" (Lenin 1981: 60; Postanovlenie: 26 February 1983). It was during Brezhnev's term of office in January 1981 that the Supreme Soviet ratified the Principles of Housing Legislation for the USSR and the Union Republics which came into force on 1 January 1982 (Vashets 1981; Tolstoi 1984; Gribanov 1983; Osnovy zhilishchnogo zakonodatel'stva 1986).

In his report to the XXVII Party Congress five years later, Mr. Gorbachev also spoke of "the social importance and acuteness of the housing problem". He placed a little more emphasis on tenures which have underperformed, namely co-operatives and housing for owner occupation. He also drew attention to an ethical issue of "radically improving the way in which housing is distributed" and, rather courageously, broached the question of rent: "Proposals for fair charges in the system of house rents by gearing them to the size and quality of all occupied living space merit attention" (Gorbachev 1987: 60). Most of his references to housing closely mirrored those to be found in the 'new' Party Programme, adopted at the Party Congress in 1986, including a vital statement: "The practice of encouraging people to contribute funds for the improvement of housing and living conditions ... will be broadened" (The Programme 1986 : 43).

A decree published in April 1986 "On the principal directions for accelerating the solving of the country's housing problem" covered most of the issues which subsequently were to be the subject of more detailed legislation and public debate: unwarranted demolition of habitable dwellings; injustices in the distribution of accommodation; the prioritisation of Siberia and the Far East in the allocation of state capital investment funds for housing construction; utilising the population's own income and savings for house building and for covering part of the cost of interior finishing by allowing the tenants of newly built blocks of flats to pay for decoration to their choice; systematisation of flat exchanges and increasing the volume of building materials for direct sale to the population (Postanovlenie 1986: no. 19, art. 100).

The two objectives, to provide each family with its own separate dwelling unit at a low rent, have been reiterated by the Party at successive Conferences and are embedded in various government resolutions passed over the past thirty years. It might transpire that the successful achievement of the first will be at the expense of the second goal of providing accommodation at a low cost to tenants and owners. At this stage, suffice it to say, that Aganbegyan, a 'market-oriented' reformer, still considers that "we shall be maintaining our country's preference for very low rents" (Aganbegyan 1988a: 17). A number of competing diagnoses of the housing problem and recommendations for its rectification have been freely circulating since the mid-1970s, although some were confined to academic and professional readerships. Today these ideas have been augmented by newer and more radical ones, which are being expressed more frequently and have a greater opportunity for adoption as policy. In this regard, 'socialist pluralism' is no misnomer.

The present chapter, written in the midst of what one day may be interpreted as the most important revolution in the history of Russia, will attempt among other things to discern the extent to which housing is perhaps the keystone of perestroika.

PERESTROIKA

There are a number of preconditions that have to be met for perestroika to fulfil its promises. One of them includes a radical reform of prices. This will almost certainly mean that over time the highly-subsidised prices for basic food products and housing (rent) will have to be increased.

The economic stagnation of 1981-1985, when there was practically zero economic growth, and the general economic crisis which attended it, saw a growth in speculation, bribery and exploitation of personal position. The Brezhnev era, now known as the period of stagnation (zastoi), is regarded as one when "social justice was systematically violated". At the same time, plans to increase welfare spending were not fulfilled; hidden price rises occurred; trends towards the erosion of wage differentials and arbitrariness in the fixing of remuneration scales together with privileges for certain categories of workers did enormous harm (Aganbegyan 1988a: 2-4, 16-17). A putative prescription for this pathological state would be for price rises and wage differentials to increase. Such a strategy, operating in tandem with reforms in the political realm, could enhance social justice.

In most advanced industrial societies the principle of social justice possesses what is tantamount to a hallowed status in political rhetoric. Although the concept is integral to socialism it has now assumed a quite new prominence. A glossary of the terms constituting the philosophy of perestroika explained that:

> According to Marx, Engels and Lenin, under socialism, the first stage of communism, social justice still has an historically limited character with inequality remaining in so far as the distribution of consumer goods and services is determined in accordance with the quantity and quality of labour expended, not in accordance with 'need'. That is, distribution adheres to the principle: 'From each according to his ability, to each according to his work'. In practice this means that social justice presupposes a struggle against wage-levelling, unearned income and the granting of unjustified privileges. It also implies increasing the incentive role of payment for labour (Politicheskoe obrazovanie 1987: 10: 123).

The most difficult task is to identify a method for defining the 'value of work' so that wage (and other reward) differentials can be devised which will command nationwide acceptance as being 'socially just'. It would seem that the Government considers this to be achievable through what may be called the 'monetisation' of the economy. This has two immediate corollaries: first, income differentials assume a greater importance as a method of 'activating the human factor'; and second, a greater emphasis is to be placed on income tax. (The period 1936-1968 witnessed a gradual shift from reliance on indirect taxation towards profit taxes and to a lesser extent personal income taxes. Between 1968 and 1983 the Soviet budgetary revenue coming from direct taxes declined from 45.6 percent to 38.1 percent. It seems likely that this decline will be arrested and reversed in the near future "and higher rates of income tax instituted".)

Taken together, all these changes rely on an expansion in the role of banking, credit and finance in the economy. Legislation introduced in 1987 dealt with the financing and crediting of capital construction (<u>Postanovlenie</u> 1987: no. 7, art. 31) and on improving the banking system as a means of raising economic efficiency (<u>Postanovlenie</u> 1987, no. 37, art. 121). The latter pointed out that credit had virtually ceased to fulfil its function as "an instrument for expanded reproduction and for the rational use of productive potential". A "fallacious practice has developed whereby credit is being granted without proper foundation and being used to cover losses and lengthy delays in meeting targets". As a result negligence and bad management flourishes while "deteriorating discipline in monetary payments and money circulation has made it difficult to balance the supply of material goods with finance". The organisation of credit fails to meet the contemporary needs of the economy including full cost accounting, self-financing and changes in investment policy. Changes have therefore to be instituted in order that credit may become one of the most important measures for securing the "inclusion of commodity-money relations within the framework of the planned administration of the economy". The negative effects of the existing (but in process of change) command economy - which "represses democracy, initiative and the creativity of workers (and) does not make workers interested in the final product of their labour" (Volkonskii and Koryagin 1985; Proshkin and

Povarich 1987; Ryvkina 1988; Gordon 1988) - can, it is alleged, be countered by making transfers to self-financing and self-management, thereby enhancing the role of prices, finance and credit. The enormous paradigm shift occurring in Soviet political thinking creates a dilemma for the Government which has to reconcile the moral issue of distribution with the practical desideratum of increasing the rate of economic growth. These changes are reflected in, and give rise to, vexing questions for housing policy.

As far as housing itself is concerned the concept of social justice may be seen to apply in three different ways. First, it applies to the organisation of the production of accommodation: work collectives have been given the right to determine wages which now depend on a working collective's final output. These policies, it is believed, will create greater opportunities for higher productivity in the construction industry and better quality dwellings (Vashets 1987: 93). Second, wider differentials mean that those with higher incomes can pay more for housing. Third, it relates to the crucial matter of fairness in the distribution of accommodation.

The primary changes in institutional structures and practices required to ensure that social justice is achieved are themselves of considerable significance. First of all, as a preliminary remark employers have historically (even prior to 1917) provided rented accommodation for their workforce. With the introduction of the first five-year plans in 1929 the share of industry in housing construction and management rapidly increased. Now that the rights of workers' collectives and trade unions are to be expanded, the whole question of the allocation of accommodation and other consumer goods at the point of production could be radically affected. The law already allows enterprises to make grants out of their social fund to employees so that the latter may become owner-occupiers or members of housing co-operatives.

Second, a precondition for debate on social justice to take place is openness (glasnost'). These two principles of social justice and openness develop in an interrelated fashion with the democratisation of the political system. The institutional mechanism for putting these precepts into practice are the local soviets, whose revitalisation is central to the policy of

perestroika. Their growing importance has a
direct impact on housing policy in that, first,
they can be the vehicle for expanding the
production of housing - state, co-operative and
individual (private) - and, second, they are
ultimately responsible for ensuring social justice
in the distribution of living space within their
administrative jurisdiction. The authority and
powers of deputies to local soviets are also being
strengthened which will enable them to be more
responsive to complaints raised by their
constituents on abuses in the allocation of accommodation.

The constitutional guarantee of shelter is
intimately connected with the simple reproduction
of the labour force, for some research suggests
that better housing conditions contribute to
higher birth rates (Broner 1980: 43-50). In view
of the falling birth-rate in republics other than
those in Central Asia and Transcaucasia, accurate
information on the causality of this relationship
is crucial to the formulation of the Government's
pro-natalist policy. The emergence of a serious
nationalities problem especially since the riots
in Kazakhstan in 1986, in Nagorny Karbaakh in 1988
and demonstrations in the Baltic republics have
further intensified regional antagonisms over
variations in the standard of living in different
parts of the country. In so far as housing serves
as an important indicator of living standards, the
disparities in per capita living space between
republics might serve to provoke ethnic enmity.
Furthermore, the falling birth rate in the largest
cities is causing concern among demographers. In
Moscow, the natural population growth fell from
2.4 to 1.8 per 1000 between 1970 and 1984 (TsSU
RSFSR 1985:7). This has meant that during the
1970s the Moscow authorities issued temporary work
and residential permits to 40,000-50,000 people
annually, so that for the whole period 1970-1985
the city's population grew by 700,000 through in-
migration. The implications of these demographic
processes for housing and social policy are well-
known in western capital cities. No doubt these
newcomers to Moscow are to be found among the
344,800 families and single persons on the city's
waiting list for accommodation (Goskomstat 1988:
32:5).

However, the functions of housing are not
restricted to population reproduction. It assists
in developing the worker's capacity for highly

productive work. Soviet research has identified positive correlations between improvements in housing conditions on the one hand, and increases in labour productivity and reductions in labour turnover, on the other (Postanovlenie 1980: no. 3, art. 17). Thus "housing affords the individual the opportunity to renew his strength and to rest so that in the final analysis it raises labour productivity and becomes a crucial factor in mobilising the labour force in the production process" (Zhelezko 1986: 81). This instrumentalist approach to housing, which suggests that a distinct association exists between housing conditions and labour productivity, has been an explicit feature of Soviet housing policy since the 1920s. Indeed, the reciprocity between improvements in production and general welfare (in which accommodation rates very highly) represents "an axiom of the socialist way of life" (Lenin 1982: 12-13; Zagorul'kin and Kolesnikov 1983: 32). It is in this context that the Fund for Socio-Cultural Measures and Housing Construction (FSKM i ZhS) - described below - has a role to play in providing, among other things, "the favourable conditions for the reproduction and better use of the labour force and its stabilisation". Legislation in 1978 and 1981 encouraged the funds' administrators to use it to provide support for individual and co-operative house building (Postanovlenie 1978: no. 17, art. 102; Postanovlenie 1981: no. 29, art. 170).

Lastly, it is within the tranquillity of the home that the worker can renew the strength expended at work and also "shed the weight of 'stress' inherent in modern society" (Poltorygin 1983: 34; Zhilaya yacheika 1982). Reservations remain on the actual 'economic return' from housing expenditure, for "the issue of the real effect which house building has on the national economy has generally speaking been little studied" (Dmitriev 1987: 72). On the other hand, better quality housing does appear to raise the population's capacity for work. In so far as it reduces morbidity, spending on improving living conditions can be translated into increases in output and in savings made on social insurance and health care (Dmitriev 1987: 73).

Housing Policy in The Soviet Union

CHANGES IN SOVIET HOUSING STANDARDS

The house-building record of the USSR has been impressive, yet the situation is still regarded as acute. The deputy chairman of the State Planning Committee (<u>Gosplan</u>) for the RSFSR confessed in an interview that large numbers of individuals continue to live in barrack accommodation, which in 1982 was officially defined as "one and two storey residences designed for temporary tenancy with an expected life of 10-20 years and consisting of shared kitchen and bathroom facilities". A further substantial number were also admitted to be living in log huts (<u>balki</u>, trailers and caravans (<u>vagonchiki</u>) and basements (<u>podval</u>)). In 1979 about one million people (inhabiting 9 million square metres of living space) were housed in barrack and basement accommodation. The period from 1976-1980 witnessed a concerted effort to rehouse them. As a result, in the space of the five years, 8.7 million square metres of barrack and basement housing 'disappeared'. By 1986 the remainder fell to 211,000 square metres which declined by a further 79,000 square metres in 1987. The outcome of this change was that at the beginning of 1988, in the RSFSR at least, only 29,500 square metres remained, providing shelter for 2,500 persons. Furthermore, over 45 million square metres - equivalent to about 2 per cent of the total housing stock providing shelter for about 3 million people - is officially registered as dilapidated and in need of emergency repair work. Sixty-five per cent of this atrocious housing belongs to ministries, constituting part of the 'departmental' housing stock (see below). The plan is to demolish only 18 per cent of this stock during the current (1986-1990) planning period. Grants will be made to the owners to undertake extensive modernisation of most of the remainder (Zapal'skii 1988). Interviews, documentaries and discussions conducted in the media have over the course of 1987 and 1988 brought state and society together to review openly and bitterly the state of housing.

In view of this situation it would therefore seem at first sight to be paradoxical that recent years should have witnessed a decline in the number of new flats erected from 2,591,000 units in 1960 to a low of 1,991,000 units in 1985, since when it has begun to rise again, reaching

2.3 million in 1987 (see Table 6.1).
The fall in output is above all a consequence
of the rise in the cost of providing accommodation
which occurs with improvements in space standards
and the quality of the fittings being installed in
flats.

Table 6.1: USSR: Number of flats built by sector

	1960	1965	1970	1975	1980	1985	1986
Total no. of flats built (in thousands)	2591	2227	2266	2228	2004	1991	2100
State, co-op organisations and house-building co-ops (thousands)	1331	1511	1723	1778	1667	1638	1743
%	51.4	67.8	76.0	79.8	83.2	82.3	83.0
Individuals (thousands)	1260	691	489	377	247	226	240
%	48.6	31.0	21.6	16.9	12.3	11.4	11.4
Collective farms (thousands)	-	25	54	73	90	127	117
%		1.1	2.4	3.3	4.5	6.4	5.6

Source: Goskomstat, Narodnoe khozyaistvo SSSR za 70 let.
Tubileinyi ezhegodnik. Moscow: 'Finansy i
Statistika', 1987, p. 510

This is reflected in the fact that the post-war
peak in 1960 of 2,591,000 units contained 109.6
million square metres of living space compared
with 2,100,000 units and 119.8 million square
metres in 1986. A shortage of good building land
compelling developers to in-fill on land on which
it is more difficult to build also contributes to
higher construction costs. In addition it is for
obvious reasons more expensive to build in Siberia
and the Far North, areas which are earmarked for
substantial increases in house-building activity.

Table 6.2: USSR: Capital investment in housing construction by sector: millions of rubles in comparable prices

Year	Total cap. investment in housing construction	% of overall capital investment in the economy	State and co-operative organisations		Of which: house-building co-operatives		Individuals		Collective farms	
			mln.R	% of total	mln.R	% of total	mln.R	% of total	mln.R	% of total
1918-1940	11356	18.4	6740	59.4	-	-	4616	40.6	-	
of which 1940	1228	16.4	853	69			375	30.5	-	
1956-1960	45203	23.5	33183	73.4	-		12020	26.6	-	
1960	10810	22.7	8066	74.6	-		2744	25.4	-	
1961-1965	52714	18.9	40808	77.4	1827	3.5	9541	18.1	538	1.0
1965	11245	17.5	8511	75.7	815	7.3	1690	15.0	229	2.0
1966-1970	70379	17.7	55284	78.5	4622	6.6	8230	11.7	2243	3.2
1970	15793	17.1	12555	79.5	1074	6.8	1636	10.3	528	3.3
1971-1975	89069	15.8	71907	80.7	4808	5.4	8560	9.7	3794	4.2
1975	19215	15.0	15530	80.8	872	4.5	1821	9.5	992	5.2
1976-1980	101943	14.2	83071	81.5	4416	4.3	8418	8.3	6038	5.9
1980	21123	14.0	17348	82.1	847	4.0	1561	7.4	1367	6.5
1981-1985	127627	15.1	100360	78.7	5898	4.6	10138	7.9	11231	8.8
1985	28081	15.6	21480	76.5	1439	5.1	2468	8.8	2694	9.6
1986	30875	15.9	23850	77.2	1505	4.9	2787	9.0	2733	8.9

Source: Goskomstat, Narodnoe khozyaistvo SSSR za 70 let. Tubileinyi statisticheskii ezhegodnik. Moscow: 'Financy i Statistika', pp. 328-329, 514

There has, moreover, been a decline of investment in housing construction as a proportion of
overall capital expenditure. It fell from 23.5
per cent in the 6th five year plan (1956-1960)
to 14.2 per cent in the 10th five year plan (1976-
1980), since which date it has gradually begun to
rise again: 15.1 per cent in the 11th five year
plan (1981-1985) and 15.6 per cent and 15.9 per
cent for 1985 and 1986 respectively (see Table
6.2).

As Tables 6.3 and 6.4 reveal, the current
trend in overall housing construction is now
upward.

The immediate effect of this decline in
construction is visible in housing waiting lists.
Seventy per cent of 175 enterprises questioned in
one recent sample survey, reported that a worker
would have to be employed at a factory for over 10
years before receiving a flat and an engineer over
eight years. A further 25 per cent put the period
at 6-10 years and only 5 per cent of housing
department managers surveyed declared the period
to be as short as 4-5 years (Alekseev 1987: 181).
The daily and periodical press is replete with
such information. One hero of socialist labour, a
delegate to a conference of Stakhanovite veterans,
referred to conditions at one cotton mill where 26
per cent of its 4,500 employees were on the
waiting list for accommodation. This is a
familiar picture throughout industry, especially
in the light manufacturing sector (Pravda 1985).

At the beginning of 1987, nationwide,
12,660,000 households comprising 22.3 per cent of
all Soviet households were on housing waiting
lists. Regional differences are again
considerable, ranging from 10 per cent in Estonia
to 31.6 per cent in Belorussia. The North-South
(Baltic-Central Asian republican) divide is not
evident in this particular index of housing need.
In fact, the three Slav republics (the RSFSR,
Ukraine and Belorussia) were all above the
national average, while Uzbekistan (11.8 per
cent), Kirghizia (16.7 per cent) and Turkmenia
(16.7 per cent) were below it (Goskomstat,
Narodnoe khozyaistvo za 70 let...1987: 519). This
is a slightly anomalous feature since on most
housing indices the Central Asian republics fare
less well than most others. One explanation might
be that waiting lists are generated by new
household formation, partly as a consequence of
divorce and partly because of the desire within

the European part of the country for individuals and couples to live separately from their parents in their own homes. In Central Asia on the other

Table 6.3: USSR: Total housing construction
1960-1986 (end of year)
(million m^2 overall (useful) living space)

Year	1 Total built (i) mln.m^2	2 State and co-operative mln.m^2	% (2/1)	3 House-building co-operative (ii) mln.m^2	% (3/1)
1960	109.6	55.8	50.9		
7th 5YP					
1961-1965	490.6	300.4	61.2	13.4	2.7
1965	97.6	63.2	64.8	6.5	6.7
8th 5YP					
1966-1970	518.5	352.5	68.0	33.6	6.5
1970	106.0	76.6	72.4	7.7	7.3
9th 5YP					
1971-1975	544.8	407.3	74.8	32.6	6.0
1975	109.9	83.3	75.8	5.8	5.3
10th 5YP					
1976-1980	527.3	413.8	78.5	27.4	5.2
1980	105.0	84.0	80.0	5.1	4.9
11th 5YP					
1981-1985	552.2	436.5	79.0	32.8	5.9
1981	106.4	84.5	79.5	5.3	5.0
1982	107.9	85.6	79.3	5.5	5.1
1983	112.4	88.8	79.0	6.7	6.0
1984	112.5	89.1	79.2	7.5	6.7
1985	113.0	88.6	78.4	7.8	6.9
12th 5YP					
1986-1990 (est.)	630.650				
1986	119.8	94.4	78.8	8.2	6.8

Notes: (i) total built (1) is the aggregate of cols. 2, 4, 5

(ii) house-building co-operatives are calculated here as a percentage of total housing

hand divorce rates and family breakdown are less frequent while attitudes towards living in the extended family remain traditionalist.

Table 6.3 (continued)

Year	4 Individual		5 Collective farm	
	mln.m^2	% (4/1)	mln.m^2	% (5/1)
1960	53.8	49.1		
7th 5YP				
1961-1965	184.9	37.7	5.3	1.1
1965	33.0	33.8	1.4	1.4
8th 5YP				
1966-1970	153.8	29.7	12.2	1.3
1970	26.7	25.1	2.7	2.5
9th 5YP				
1971-1975	120.8	22.1	16.7	3.1
1975	20.6	20.6	4.0	3.6
10th 5YP				
1976-1980	91.4	17.3	22.1	4.2
1980	16.1	15.3	4.9	4.7
11th 5YP				
1981-1985	80.3	14.5	35.4	6.4
1981	16.2	15.2	5.7	5.4
1982	16.0	14.8	6.3	5.8
1983	16.2	14.4	7.4	6.6
1984	15.5	13.8	7.9	7.0
1985	16.3	14.4	8.1	7.2
12th 5YP				
1986-1990 (est.)	630-650			
1986	17.4	14.5	8.0	6.7

Sources: Goskomstat 1985, 435; Goskomstat 1987, 508

Table 6.4: USSR: Construction of urban housing by sector, 1960-1986 (end of year)

Year	Overall total (million m^2 overall living space)	of which Local soviets state enterprises etc. million.m^2	%
1950	20.7	14.3	69.1
1951-1955	129.8	91.0	70.1
1956-1960	241.7	181.0	74.9
1960	59.0	44.6	75.6
1961-1965	291.6	227.4	78.0
1961	56.1	43.7	77.9
1962	58.9	47.5	80.6
1963	58.4	46.8	80.1
1964	57.5	43.5	75.7
1965	60.7	46.2	76.1
1966-1970	335.5	265.5	79.1
1966	63.4	49.0	77.3
1967	66.1	51.8	78.4
1968	66.1	52.5	79.4
1969	68.6	55.3	80.6
1970	71.3	57.0	79.9
1971-1975	377.4	312.4	82.8
1971	72.9	60.1	82.5
1972	73.5	61.0	83.0
1973	77.6	63.6	82.0
1974	77.1	64.0	83.0
1975	76.3	63.8	83.6
1976-1980	378.7	318.5	84.4
1976	75.9	(69.5)	(91.6)
1977	77.2	(70.9)	(91.8)
1978	76.6	(70.4)	(91.9)
1979	72.7	(65.8)	(90.5)
1980	76.3	64.2	85.2
1981-1985	384.8	317.9	82.6
1981	75.0	63.6	84.8
1982	75.7	63.5	84.0
1983	76.9	64.5	83.9
1984	77.2	63.9	82.8
1985	77.1	62.6	81.2
1986	82.9	67.3	81.2

Sources: Andrusz 1984, Table 1.4, p. 21; Goskomstat 1985; Goskomstat 1987b, pp. 509, 515

Table 6.4 (continued)

Year	of which House-building co-operatives		Private individuals	
	million.m^2	%	million.m^2	%
1950	-	-	6.4	30.9
1951-1955	-	-	38.8	29.9
1956-1960	-	-	60.7	25.1
1960	-	-	14.4	24.4
1961-1965	13.4	4.6	50.8	17.4
1961	-	-	12.4	22.1
1962	-	-	11.4	19.4
1963	1.8	3.1	9.8	16.8
1964	4.8	8.3	9.2	16.0
1965	6.5	10.7	8.0	13.2
1966-1970	33.6	10.0	36.4	10.9
1966	6.7	10.6	7.7	12.1
1967	6.5	9.8	7.8	11.8
1968	6.4	9.7	7.2	10.0
1969	6.2	9.0	7.1	10.4
1970	7.7	10.8	6.6	9.3
1971-1975	32.6	8.6	32.4	5.6
1971	6.8	9.3	6.0	8.2
1972	6.5	8.8	6.0	8.2
1973	7.1	9.1	6.9	8.9
1974	6.3	8.2	6.8	8.8
1975	5.8	7.6	6.7	8.8
1976-1980	27.4	7.3	31.2	8.3
1976	N.A.	N.A.	6.4	8.4
1977	N.A.	N.A.	6.3	8.2
1978	N.A.	N.A.	6.2	8.1
1979	N.A.	N.A.	6.3	8.7
1980	5.1	6.7	6.0	8.0
1981-1985	32.8	8.5	29.3	7.6
1981	5.3	7.1	6.1	8.1
1982	5.5	7.3	5.8	7.7
1983	6.7	8.7	5.7	7.4
1984	7.5	9.7	5.8	7.5
1985	7.8	10.1	5.8	7.5
1986	8.2	9.9	6.8	8.2

Furthermore, as Table 6.11 shows, building for private ownership continues to run at a high level in these republics, suggesting that members of these cultures might not register on local authority waiting lists, preferring to continue living with their parents or other relatives until they built their own homes.

The data cited above were reported in the annual statistical handbook for the first time in 1987. Subsequently, as an indicator both of the progress of glasnost' and of differences between the length of waiting lists in different cities, a popular weekly, Argumenty i fakty, published a table showing the number of people on the waiting list for housing in the capitals of republics and other large cities (see Table 6.4A).

Although Moscow has the longest waiting list, in proportionate terms Ufa, the capital of the Bashkir Autonomous republic in the Urals, with 36 percent of families is the worst off. As has been stated above, Soviet citizens are aware of the seriousness of the housing situation. Since the number of dwelling units constructed in these cities is published at the end of each year, local residents will now be in a position to compare the length of the waiting list in their city with the number of units erected. They will also be able to compare the relative levels of housing deprivation in different cities and the extent of the success of the authorities in reducing the waiting lists. With the publication of further information of this nature, it will be possible to conduct much more detailed analyses of the housing situation in the Soviet Union.

In order to deal with this shortage, the state plan envisages the erection of 630.4 million square metres of housing space in the period 1986-1990. A more recent optimistic variant puts this figure at 650 million. Even the lower target means constructing 16 per cent more than in 1971-1975, 20 per cent more than in 1976-1980 and 14 per cent more than in 1981-1985.

The size of the urban housing stock was 2.8 times larger in 1986 than in 1960 while the population witnessed a 1.7-fold increase, with the result that overall living space per capita rose from 8.8 square metres to 14.3 square metres. In 1983 this urban population was housed in 45.2 million flats and 6 million hostel beds, which means that the country's 37,000 hostels (obshchezhitie) accommodate about 5 per cent of the working and student population (Meerson and Tonskii 1983: 13).

Table 6.4A: USSR: The waiting list for improved housing in capitals of Union Republics and other large cities, 1988

City	No. of households on waiting list	% of households on waiting list
Alma-Ata	49,700	15
Ashkhabad	23,600	26
Baku	68,700	26
Vilnius	36,300	21
Gor'ky	123,100	27
Dnepropetrovsk	74,200	20
Donetsk	75,300	22
Dushanbe	31,300	22
Erevan	42,000	16
Kazan	112,900	34
Kiev	208,400	26
Kishinev	69,500	32
Kuibyshev	114,300	29
Leningrad	282,900	20
Minsk	134,600	28
Moscow	334,800	12
Novosibirsk	111,600	25
Odessa	80,400	23
Omsk	108,400	31
Perm	106,800	32
Riga	75,700	26
Rostov-on-Don	74,100	23
Sverdlovsk	130,600	31
Tallin	25,400	16
Tashkent	60,100	12
Tbilisi	59,000	19
Ufa	118,800	36
Frunze	31,800	17
Khar'kov	113,400	23
Chelyabinsk	109,500	31

Source: Goskomstat soobshchaet 1988

The justification for such a large hostel population - that population growth and the opening up of new areas and the development of the economy in general "makes it difficult to provide everyone with their own flat" - may be

sustainable. It is probably less true that "young
people do not always want at first to have their
own separate accommodation because at that stage
in their lives when they have not yet formed their
own family units they do not want a settled life"
(Rusakovskii 1983: 4). Although this form of
accommodation is unlikely to disappear in the near
future, unless there is a substantial improvement
in the standard of comfort which they offer, the
number of people living in such accommodation
could possibly fall. Standards in fact vary
considerably. National norms laid down by
legislation in 1976 are exceeded in some cases by
enterprises drawing upon their Fund for Socio-
Cultural Measures and Housing Construction to
provide better furniture, fittings and facilities.
In other instances, especially at the large
building sites, hostels are overcrowded, with
those responsible for their management turning a
blind eye to rule-breaking over minimum per capita
living space standards. Furthermore, while
existing regulations in most republics strictly
forbid families from living in hostels, this
proscription is perforce frequently evaded. This
has led some organisations and enterprises to
build family hostels where each family has its
own room or "whole dwelling unit". At present the
newest designed hostels offer only 6 square metres
of dwelling space per person, which is
considerably below the housing space norm.

In the absence of an expansion of the
privately rented sector - about which very little
information is available - hostels will continue
to meet a mobile labour force's need for shelter.
The average age of their residents is 24, with
one-third under the age of 20 and 12 percent over
30. Again, on average, the educational level is
quite high, but 'specialists' are outnumbered by
two-and-a-half times by those with only
uncompleted secondary education behind them. (The
subject of hostels could make an interesting post-
graduate dissertation.)

A new policy initiative presents a proportion
of the young adult population with an alternative
to hostel and privately rented accommodation. The
Government has intensified its efforts to
encourage younger people - partly through
financial incentives - to embark on self-build
schemes with the assistance of their employers.
Mr Gorbachev in his Report to the Party
Conference referred to "those who are backing the

construction of youth complexes" as "doing the right thing". In fact he was referring to a policy initiated in 1985 intended to promote the construction of housing complexes for young people (Postanovlenie 1985: no. 22, art. 111). According to the decree's preamble, young people in a number of major cities and in the countryside have taken up the challenge - out of necessity, it might be added - by erecting not only residential complexes and renovating older housing but also school, arts' and sports' buildings. Apparently they have reduced the construction time, improved the organisation and quality of the work and economised in outlays on materials. By 1988, 180 such complexes were in the process of being built in 110 cities in the Russian republic. By 1990 they will account for up to 10 per cent of total housing construction using centralised state investment (Postanovlenie 1988: no. 6, art. 20).

In 1926 - when the socialised housing sector accounted for 41.8 per cent of the total stock - the average overall living space stood at 8.2 square metres per person (Sedugin 1983: 11). Table 6.5 below shows the average per capita provision of housing space (in square metres) in 1985.

Table 6.5: USSR: Per capita living Space by tenure and location, 1985

Tenure	Urban		Rural	
	a	b	a	b
local soviets	15.7	10.0	13.7	9.4
state enterprises	14.6	9.3	13.5	9.5
house building co-operatives	17.3	10.9	18.6	12.7

Note: column (a) represents overall living space
      column (b) represents actual dwelling area

Source: Blekh 1987: p. 13

It should be noted that these figures differ from those cited in the principal national statistical handbook. The difference might be explained by the inclusion in Table 6.6 of figures for the owner-occupied sector. The other significant discrepancy between the two tables is that according to Table 6.6 in contrast to Table 6.5, rural inhabitants are better housed than their urban counterparts, at least in space terms. While Table 6.5 also clearly indicates the higher space standards in the co-operative sector - a subject discussed below - Table 6.6 and 6.7 equally vividly demonstrate republican variations.

Table 6.6: USSR: Average per capita housing provision
by republic, December 1986
(square metres of overall living space)

	Urban	Rural
USSR	14.3	16.1
Estonia	17.8	26.6
Latvia	16.9	23.4
Georgia	15.9	19.6
Lithuania	15.7	21.9
Ukraine	15.4	19.4
RSFSR	14.5	17.1
Belorussia	14.0	20.4
Armenia	13.1	15.0
Moldavia	12.8	20.3
Kazakhstan	12.6	13.6
Azerbaidhzan	11.9	8.9
Kirghizia	11.3	11.3
Uzbekistan	11.2	10.8
Tadzhikstan	11.1	7.6
Turkmenia	10.1	10.3

Source: Goskomstat 1987b: p. 522

In aggregate terms the "socially necessary minimum" housing space has been achieved. Nonetheless, substantial regional variations continue to exist. As Table 6.6 shows, whereas the urban average in the Baltic States ranges from 15.7 to 17.8 square metres per person, in the five Central Asian republics the range is 10.1 to 12.6. This disparity continues to grow because of differential investment rates: in 1986 per capita capital investment in the former contained the highest figure for the whole country, 139.5 roubles (against the national average of 109.6 roubles) while in Tadzhikstan, Kirgizia and Uzbekistan the respective figures were 55.8, 64.6 and 70.3 (Goskomstat 1987b: 514). Similarly large variations occur within the RSFSR, especially between the European part of the republic and Siberia. Recent studies of Siberia have statistically demonstrated the poor living conditions found there. Nine out of the eleven regions into which Siberia is divided were rated as experiencing 'relatively poor' and 'below average' housing conditions. Not a single region was considered to be 'relatively good' (the highest of the four categories) compared with 40 per cent of the regions in the rest of the RSFSR. Fifty-five per cent of the Siberian regions had an average per capita norm of less than 10 square metres of overall living space. Apart from low space standards, the housing is extremely poorly provided with amenities: only 70 per cent of the state and co-operative stock - almost universally better provided than owner-occupied housing - had mains water and sewage disposal, and 60 per cent a bath or shower (Khakhulina and Trofimov 1985). (The respective figures for these three amenities nationally in 1986 were 92, 90 and 84.) A typology of urban areas in West Siberia revealed that housing conditions in the nine areas were almost all worse than elsewhere in the republic. The study also demonstrated the area-specific nature of housing problems (Artemov and Khakhulina 1984). This particular research highlights the need for substantial investment in Siberia (and the Far East) for housing construction, especially if the government is to fulfil its intention of attracting and securing labour to develop the vast natural resources located in these regions (Rumer 1984). More generally, the methodology demonstrates the growing sophistication of empirical research, which is itself an acknowledgement by politicians that social scientists have a critical role to play in revealing the tremendous diversity of living standards, norms and traditions in the Soviet Union's multi-national society.

Table 6.7: USSR: Average size of flats built in the private sector by republic (square metres of overall living space)

	1976-1980	1981-1985	1980	1985	1986
USSR	63.1	68.4	64.8	72.1	72.5
RSFSR	53.1	60.3	55.0	65.2	66.1
Ukraine	67.6	68.5	67.0	69.9	70.2
Belorussia	62.2	66.3	63.4	68.8	70.5
Uzbekistan	61.9	70.4	63.6	77.4	98.8
Kazakhstan	73.9	80.8	79.5	85.2	75.6
Georgia	100.0	95.0	100.4	97.2	97.8
Azerbaidhzan	48.9	62.6	56.4	67.8	70.2
Lithuania	81.9	94.1	86.9	97.4	99.7
Moldavia	67.5	74.0	71.9	79.9	79.5
Latvia	88.9	103.5	93.3	105.4	107.6
Kirghizia	69.9	75.3	71.4	82.5	80.1
Tadzhikstan	54.1	52.6	52.9	52.7	50.8
Armenia	85.1	88.2	85.6	84.1	94.3
Turkmenia	88.0	87.5	88.0	84.5	88.9
Estonia	87.0	95.6	83.8	100.0	99.4

Source: Goskomstat 1987b: p. 513

At the same time, these per capita living-space statistics are extremely crude and it comes as no surprise that "the utility and meaningfulness of this index as a measure of living conditions far from adequately expresses the extent to which the housing problem has been solved" (Dmitriev 1987: 70). For instance, even in the Baltic republics, which according to this means of computation enjoy the best nationwide housing conditions (see Tables 6.6 and 6.7), "a very high proportion of flats, dating from the pre-war (1939) period, are poorly equipped and fail to correspond to family size and gender composition". The implication is that the deficiency of this index is at least as great elsewhere, hence its need for revision.

During the 1940s and 1950s the basic unit of measurement used in the design of housing was 1 square metre of dwelling area (zhilaya ploshchad) with the coefficient 'k' expressing the relation-ship between 'overall (useful) living space' [obshchaya (poleznaya) ploshchad' zhilishch] and dwelling area. (This important point needs explanation. 'Overall (useful) living space' includes living rooms, bedrooms, kitchens, corridors, bathroom and storage space. In order to calculate actual 'dwelling area', i.e. just living and bedrooms, the larger figure in the state sector is today multiplied by a coefficient of 0.60 (0.52 for a one-roomed flat to 0.65 for a four- or more roomed apartment). In the owner-occupier sector the republican range is from 0.56 (Latvia) to 0.79 (Armenia).) Table 6.8 shows the growth in average flat size over the past decade.

Table 6.8: USSR: Average size of flats built in the state and house-building co-operative sectors, 1976-1986

	1976-1980	1981-1985	1980	1985	1986
State built flats con-sisting of:					
1 room	32.4	34.1	32.9	35.0	34.4
2 rooms	48.2	50.0	49.2	50.6	50.7
3 rooms	62.5	65.7	63.5	66.8	66.7
4+ rooms	77.9	82.2	79.5	83.1	84.0
average	51.6	54.3	52.5	55.3	55.4
House-building co-operative flats consisting of:					
1 room	33.1	34.6	33.5	35.1	35.6
2 rooms	49.0	50.5	49.5	50.9	51.2
3 rooms	64.7	66.5	65.4	66.6	66.5
4+ rooms	83.4	88.4	85.6	87.5	87.0
average	53.0	55.5	54.1	55.8	56.2

Source: Goskomstat 1987b: p. 512

Table 6.8A: USSR: Number of square metres by which
co-operative flats exceed state flats in size

Size of flat	Date	
	1976-1980	1986
1 room	0.7	1.2
2 rooms	0.8	0.5
3 rooms	2.2	0.2
4+ rooms	5.5	3.0
average	1.4	0.8

Source: Derived from Table 6.8

This led designers and builders to seek to
maximise the size of the dwelling area by
increasing room sizes to the detriment of the
quality and convenience of layout. With the
growth in the size of the housing stock and the
policy of allocating self-contained flats to
individual families, this index became an obstacle
to improving housing design. As a result a decree
of 1969 altered the unit of measurement for
capital investment to one square metre of overall
living space (Postanovlenie 1969: no. 15, art.
84). However, procedures and practices in force
for over half a century have proved difficult to
change, so that the distribution of housing and
the determination of rents continues to be based
on an index of 'dwelling area' (Tonskii 1982: 31).
Recent years have seen a call for 'room density'
to replace metres as the unit of calculation.
Then in November 1988 the Chairman of the State
Committee for Construction announced that
beginning in 1989 the state plan for housing
construction will be calculated in terms of the
number of individual apartments to be built, not
square metres of living space. Indeed, a gauge of
the extent to which housing standards have
improved is the changing formula applied when
allocating tenants to flats. Where 'r' equals the
number of rooms and 'n' equals the number of
people, at the end of the 1950s and the beginning

of the 1960s new flats were tenanted on the basis of r = n - 2 (i.e. a 2-roomed flat would normally be occupied by a family of 4). By the end of the 1960s the formula r = n - 1 was applied, and by 1976-1980 over one-quarter of new urban flats were being allocated on the basis of 'one room, one person' (r = n). In the early years of the next decade each adult member of a family is expected to have his/her own room. This is probably an optimistic prediction.

In newly built dwelling units and in housing controlled by local soviets the average room density for the country at large stands at 1.4 persons per room, and for the urban housing stock as a whole the figure is 1.6, having improved from 2.8 persons in 1960 to 1.9 in 1970. The average flat density in the socialised sector is 3.3 persons and 3.2 in local soviet accommodation (Blekh 1987: 13). In comparison with 1967 when only 61 per cent of urban families lived in self-contained apartments, by 1975 the figure had risen to 76 per cent, to 80 per cent in 1982 and had reached 'about' 85 per cent in 1986 (Goskomstat 1987: 517). The remainder live in communal flats (where bathroom, kitchen and hallway are shared), hostels and in privately rented accommodation. Account has also to be taken of those who continue to live in 'complex' (extended) family homes composed of several conjugal couples of different ages, who are in fact in need of their own homes. Our knowledge about this and other matters concerning the housing stock and living conditions should be considerably enhanced when the results of the (January) 1989 decennial National Census become available. (In contrast with the 1979 Census form which had sixteen questions, the next Census will contain 25, with the last seven dealing exclusively with housing standards (Goskomstat 1987a).)

The proportion of three-roomed apartments has risen slowly from 31 per cent in 1971-1975 to 37 per cent in 1986 (Goskomstat 1987b: 511). Although family size is declining as the two-generational family (with 3.3 members) has become established, the fact that 57 per cent of all new flats consist of only one or two rooms does not suggest that the formula r = n will be achieved in the near future. High divorce rates and a growing number of single-parent households might justify the continued building of such a high proportion of smaller flats.

In 1958 the average area in a newly built apartment was 41 square metres. This increased to 44.5 square metres in 1964 and stood at 55.4 square metres in 1986 for dwellings erected by the state (and 56.2 square metres in co-operative apartments). More significant in some respects has been the corresponding increases in the difference between the 'dwelling area' and the 'overall living space', signifying an enlargement of the bathroom, kitchen, corridors and storage space. This changing ratio is a recognition of the need for more ancillary (storage) space attendant upon a universally higher level of ownership of consumer goods. Thus, in the period 1976-1980 when the average flat size was 51.6 square metres, 63 per cent (i.e. 32.5 square metres) was designed as the actual dwelling area, leaving 19.1 square metres for bathroom, kitchen and general ancillary space. By 1986, the coefficient had been reduced to 60 per cent while the overall space had risen to 55.4 square metres, which means that the ancillary space now amounts to 22.2 square metres (Goskomstat 1987: 512-513). One of the complaints of those favouring rent reform derives from the fact that rent is still based on the actual dwelling area. The evident anomaly in this practice is discussed below. Finally, discussion of housing standards would be incomplete without a brief commentary on architecture.

Architecture, it is frequently maintained, exerts a constant and pervasive influence on the public consciousness and is a "positive instrument in the moral and aesthetic education of the population, contributing to the formation of a communist Weltanschauung". In September 1987 the Government once again criticised Soviet architecture and severely censured the architectural profession partly on the grounds that "building production has begun to dictate the activities of architects" (Postanovlenie 1987: no. 45, art. 149; Andrusz 1984: 147-158). The highly centralised nature of standard design work, the predominance of large panel construction and a neglect of demographic and national-cultural factors had all created a situation where many cities and settlements are "monotonous and expressionless".

The contents of this twenty-page decree cannot be discussed in any detail here, so suffice it to say that the State Committee for Civil Construction and Architecture (Gosgrazhdanstroi) has been strengthened and the new emphasis on architecture reflected in its new name, the State Committee for Architecture and Town Planning (Goskomarkhitektury). An indicator of its political importance might be that its chairman is the First Deputy Chairman of the very important State Committee for Construction (Gosstroi). The incumbent of this post in 1988 was Boris Yeltsin, the controversial former member of the Politburo.

The principal objectives of the legislation are to extend the range of types of dwelling unit, devoting more attention to their size and layout in order to cater for different family sizes, single person households, young people and the elderly. Notwithstanding the fact that since the late 1960s policy makers have referred to the need to preserve old buildings, demolition has continued, both because these structures have not been regarded as worthy of protection on architectural grounds - a high-rise aesthetic overwhelmingly predominated - and because economic arguments prevailed over social considerations.

The new decree suggests that the social, historical and cultural merits will come to take precedence over the often philistine and less sensitive wishes of officials - not least because of the tourist revenue-earning potential from the restoration of settlements with historical centres and monuments. Another important influence, according to the decree, will come from the public who are being invited to play a more active role in discussions and decisions on architectural and town planning issues as "the principles of democratisation and glasnost' come to be felt in this sphere". With the 'pluralisation' of the polity, as more interest groups come to exert their influence, party officials and economic managers will possibly face greater resistance from architects. The decree wants to improve the prestige and enlarge the role of the profession by expanding its members' rights and promoting their responsibilities for the architectural quality of their designs and the final product. At the same time the number of students studying architecture, especially the art of restoration, is to be vastly increased. The production of equipment and materials specific to restorative work is to be funded in part by deductions from the profits made by organisations engaged in tourism - an increasingly important sector of the economy. Institutional changes will enhance the role of both the architectural and town planning departments within the framework of the local soviets and of the chief architect. An All-Union Scientific Research Institute in the theory of architecture and town planning is to be created as soon as possible with its headquarters in Moscow and branches in Leningrad, Kiev, Tashkent, Tbilisi, Novosibirsk and other as yet unspecified cities. Evidence of the Government's commitment to attract people of talent into this field is to be found in the fact that employees in these and other planning and research institutes operating under Goskomarkhitektury are to be placed on the premier salary category for research workers. Legislation on the architectural profession was quickly followed by a decree which shared a similar concern. On this occasion the object of censure were those ministries, enterprises and research and design offices which totally neglected the "aesthetic component in residential and industrial environments" and the "importance of design in raising the quality and the competitiveness of articles and equipment" (Postanovlenie 1988: no. 1, art. 1).

Attention is drawn to this legislation in order to emphasise the necessity of understanding the scale of the changes which are taking place in housing policy and the centrality to perestroika of well-designed and equipped accommodation situated within an attractive neighbourhood. In the 1920s peasants refused to deliver their produce to the market because there were too few goods to buy with their money. In the 1980s Soviet citizens are not going to be enthusiastic about new incentive systems which offer them rouble riches if they have no space in which to use or keep the goods that the Government fervently hopes will shortly become available on the market. The privatised, consumer and car-ownership oriented, status-conscious nuclear family exists and is being fostered by the Government. For it to flourish, to be satisfied and motivated to work requires solving the qualitative as well as the quantitative dimension of the housing problem.

THE COST OF HOUSING

There can be few government officials or housing specialists who in 1988 could display any real confidence in the attainability of the goal set by the Party to "provide each family with its own separate flat or individual house by the year 2000". Should it succeed in its aim, much will remain to be done, as has been suggested above, to the quality of the housing stock. A very high proportion of housing in the countryside both lacks the most basic of amenities and is in a poor state of upkeep. As a result demolition rates exceed those found in towns. Demolition amounted to about 9 per cent of the total amount of living space constructed throughout the country between 1981 and 1985. During the period 1981-1985 the average annual outlay on structural and other major (capital) repairs amounted to 3.3 - 3.6 milliard roubles - equivalent to over 50 per cent of the annual state subsidy to housing (Dmitriev 1987: 74). The tendency is for this figure to grow (Tonskii 1982: 53). Because of especially high levels of obsolescence in the countryside, extra resources have to be directed to replenish the stock. Yet, in 1975 the cost to the state of building one square metre of living space in villages was 14.9 per cent higher than in the towns and in 1980, 8.6 per cent higher. Then, with the movement during the 11th five year plan (1981-1985) to erect 'cottage type' dwelling units with gardens and outbuildings for rural inhabitants, the cost estimates rose by 18.6 per cent and by 1985 surpassed urban construction costs by 17.4 per cent. For the USSR as a whole, both town and country, the cost estimates for erecting housing space by the state have risen over the past fifteen years (in constant prices) by a huge 48.5 per cent (Dmitriev 1987: 71).

Soviet specialists have long been explicit in their recognition that despite the considerable improvements in housing standards, to which the statistics testify, compared with other industrialised nations, these standards are still low. To a much greater extent than in other societies, however, the Soviet government faces serious ideological and political obstacles to introducing consumer charges to pay for this 'service'. Since the rate at which rental payments are charged has not risen since 1928, the proportion of today's family budget spent on rent typically amounts to about 1 per cent (and, on average, 3 per cent if charges for communal services are included). As a consequence rent covers less than one-third of the costs of running and maintaining the housing stock, which the state has to subsidise at an annual cost of 5-6 milliard roubles, with one author, in an idiosyncratic unpublished document, making the figure 9 milliard roubles. The figure of 5 milliard was 'sighted' in 1975 - although it may have been in circulation earlier - and continued to be mentioned in 1988. While there are <u>prima facie</u> grounds for believing that this statistic has the dubious status of being notional (rather than fictional), it does nonetheless indicate - perhaps seriously understimating - the magnitude of the subsidy.

The decree of October 1937, which totally reorganised the pre-war administration of housing, considered that the best way to improve the running and management of the housing stock was to place each house-management office on a <u>khozraschet</u> (cost accounting) basis. Subsequently, a number of organisational changes were introduced to improve the efficiency of the socialised housing sector. However, revenue has failed to cover expenditure and by an ever increasing margin. A survey undertaken by the Central Statistical Administration of a number of towns and central departments (<u>vedomstva</u>) revealed that on average for the country as a whole, between 1966 and 1980, the expenditure of organisations running housing and municipal services (for instance, cold and hot water supplies, sewage disposal, central heating etc.) rose by 34.8 per cent for each square metre of dwelling space. Over the same period their incomes (taking into account fixed tariffs and rents on non-residential premises) increased by 24.6 per cent (Andrusz 1986).

While at a low level of economic development shelter is necessary in order to reproduce labour, as the economy expands and shelter becomes more differentiated, people can make choices as to the quantity and quality (including the locational aspect) of housing they wish to consume. The burden of subsidy for a socially necessary but heterogeneous service, which is becoming ever more onerous, requires the state to adopt policies which will reduce costs or transfer them onto the consumer. The principal strategies available to the government to make savings are: (i) rationalise the ownership and management of the state housing stock, (ii) reduce construction costs, (iii) modernisation, (iv) encourage co-operative house building and individual house construction, (v) change the system of charging rent.

RATIONALISATION OF THE STATE HOUSING SECTOR

As a general proposition, housing differs from other consumer goods and services in several ways. In particular, because its capital cost is high relative to family income; it can rarely be purchased directly out of income nor, in most cases, can it be wholly financed from an individual's savings. Thus, housing falls into two main categories: (i) housing which individuals wish to purchase and own, and therefore borrow money in order to do so; or (ii) housing in which public or private agents invest capital, with the intention of renting to others.

Soviet housing policy was established under the exigencies of revolution and civil war between 1917 and 1920. Land was nationalised and housing space was redistributed according to certain criteria of need and to a definition of a minimum requirement and maximum entitlement of space per person. Two new sectors of housing tenure were created - the municipalised and nationalised, the former being owned and operated by the local soviets and the latter by economic agencies and public institutions. The introduction of the New Economic Policy (NEP) in 1921 witnessed changes in housing policy. First, the state divested itself of the function of administering small houses. Second, the experiment in providing rent-free accommodation - which was introduced in January 1921 - was short-lived, for in April 1922 the payment of rent was restored. When the worst crisis was over (1923) and new building began again, it was the private sector which took the initiative and was counted upon to provide additional housing. Between 1918 and 1928 the state and co-operative sectors accounted for only 11.7 per cent of total housing construction. Legislation passed during the 1920s anticipated attracting both the savings of individual workers into building houses for themselves and the larger resources of private entrepreneurs for building to rent. De-municipalisation comprised an integral part of this policy (Andrusz 1984: 29-45). In its search for ways of financing house building and management, in 1924 the government introduced another quasi-private tenure form, the housing co-operative. Its conception during the NEP was far more important theoretically and ideologically than in terms of its practical impact on the housing shortage. Second, its abolition in 1937, its restitution during the Khrushchevian inspired 'thaw' in 1962, and the strong sponsor-ship which it has received from Mr Gorbachev suggests that, in the history of Soviet housing and domestic policy generally, the co-operative housing sector - examined in detail below - has served as a reliable barometer of change in Soviet society.

In 1988 the four tenure categories which had emerged by 1924 still form the four pillars of house ownership in the Soviet Union. Major changes in their relative importance serve as quite accurate seismographic registers of tectonic shifts in Soviet society. These tenures are: (1) local soviet (the former 'municipalised' sector); (2) state institutions, enterprises, trade unions (formerly, the 'nationalised' sector, and now the so-called 'departmental stock'); (3) house-building co-operatives; and (4) individuals. Unfortunately, disaggregated statistical data on housing in all four categories are not published systematically. Therefore, while the Central Statistical Administration now provides a breakdown for the 15 republics in terms of (i) state (categories 1 and 2); (ii) co-operatives; (iii) individuals - only fragmentary information is available on the relative balance between local soviet housing and that comprising the 'depart-mental stock'. Statistical handbooks for regions (oblast') and cities occasionally provide more detailed information on the breakdown by tenure.

The principal reason for bracketing various sorts of landlord (tenure types) together in category 2 as parts of the 'departmental stock' is that housing in all these cases is provided by the owners for those with whom the 'owners have productive relationships'. Functionally, departmental housing is clearly distinguished from housing belonging to local soviets whose task it is to allocate housing to individuals living witnin their administrative jurisdiction, irrespective of where they work. The historical

factors which gave rise to the distinction between
the local soviet and departmental housing stocks
are regarded as no longer operative by many Soviet
authors who therefore argue that all state housing
should be transferred into the hands of local
soviets. This demand, often emanating from the
soviets themselves, continues to face resistance
from the 'departmental' owners. The tug-of-war
taking place between these institutional bodies
for possession of a set of property rights remains
one of the central issues of present-day Soviet
housing policy and reflects a much broader debate
in Soviet politics.

Since Mr. Gorbachev became General Secretary
he has vociferously championed the cause of the
local soviets. In doing so he has built upon
foundations laid by his predecessors, Lenin,
Khrushchev and Brezhnev. The demand and requirement that the local soviets, qua the 'local organs
of power', whose position is enshrined in
successive Constitutions, should be the sole
owners of public housing within their
administrative jurisdiction is both functional and
moral. Functionally, it would lead to a more
rational and efficient housing construction and
management policy. Morally, it could result in a
more just distribution of accommodation both for
those in need and those who merit receiving better
housing - whether in terms of quality, size or
location. A distinction nonetheless has to be
drawn between democratic theory - which
frequently displays a lesser concern for the
prudential than the priority of the concept - and
the practice of local soviets whose malfeasance
has hitherto made them no obvious paragons of
democratic or political virtue. The action of
publicly mentioning and discussing in the media
scandals and misdemeanours involving the
distribution of accommodation by local soviets is
a necessary component of the catharsis required in
the political culture in order to revive this
emasculated institution which is the fulcrum of
any regenerated democratic polity in the country.
In the short and medium term, the rejuvenation of
the local soviets might be conducive to greater
social justice and economic efficiency. In the
longer term, it might be speculated, these local
authorities will have to divest themselves of some
responsibility for at least the management of the
vast housing stock accumulating in their hands.

The ramifications of the most recent housing
policy innovations strongly suggest this
possibility. Since these newer strategies are
discussed below, suffice it to say here that,
speaking at a meeting in London in November 1988,
Aganbegyan was quoted as saying: "We ... allow
people to purchase state flats. This is not
extensive but the first steps have been taken. I
would like to see more extensive steps in this
direction" (Aganbegyan 1988b).

Although at the end of 1986 the state
sector as a whole controlled 72.7 per cent of the
total urban housing stock (<u>Goskomstat</u> 1987b: 517)
and was responsible for 81.2 percent of all new
building in towns, the private and co-operative
sectors continue to fulfil important functions in
meeting the demand for accommodation and in
determining the spatial distribution of social
groups.

The housing decree of July 1957 initiated the
transfer of housing from state enterprises and
institutions to local soviets. Further decrees
were passed on the subject of transfer in 1967,
1971, 1978 and 1986. Numerous governmental and
official statements on the need for the transfer
to take place have been made over the past thirty
years. By the mid-1970s in the RSFSR, the local
soviet share amounted to 24 per cent, while 49 per
cent was classified as 'departmental', with the
private sector contributing 23 per cent and house-
building co-operatives 4 per cent (Lebedev 1980:
80). The figure had risen to 40.7 per cent by the
beginning of the 1980s. In the country at large
the local soviet sector increased its proportion
of the state housing stock from 39.2 per cent in
1975 to 39.6 per cent in 1979 and was still
somewhat less than 50 per cent in 1986 (Fetisov
1980: 56; Blekh 1987: 19). These global figures
of course conceal large variations between cities,
regions and republics in the extent to which local
soviets control housing. Concentration of public
(state) housing in the hands of local soviets is
highest in Armenia (over 80 per cent) and lowest
in Kazakhstan (38 per cent). In the largest
cities, such as Kiev, Alma-Ata, Kalinin and
Smolensk, the figure falls within the range 50-70
per cent and is over 70 per cent in Moscow (Blekh
1987: 19).

The issue of accelerating the transfer of
housing from departmental to the control of local
soviets was central to another decree issued in

1987 (Postanovlenie 1987: no. 27, art. 92). On this occasion, a time limit was set: republican Councils of Ministers and relevant ministries were required during the course of 1987 to draw up plans and schedules to effect the complete transfer by 1993 at the latest. However, not all 'departmental' housing will be handed over to the local soviets and it was left to the central (All-Union) ministries and departments to agree on a list of cities and urban settlements whose housing and utilities are not subject to transfer. It is uncertain whether information on this list will become publicly available.

When housing and communal amenities are transferred to local soviets this is supposed to happen simultaneously with the transfer to them of: the organisations and departments running, maintaining and repairing the stock; all relevant technical documentation for the normal functioning of the housing; and the 'material base' - workshops, stores, garages - which hitherto had been used by the enterprises owning the housing specifically for the purpose of managing the housing stock and infrastructure. In some cases, where these services and materials do not exist, then the enterprise transferring its accommodation has to find the capital investment necessary for their provision. Furthermore, should dwellings newly erected using the direct labour method and now the property of the local soviet require repairing within two years of their being commissioned, this has to be charged to the enterprises as the negligent party in allowing the faults to occur. Alternatively, they can pay the local soviets to rectify the defects (Buzhkevich 1988: 3, 32). The essence of the matter is that local soviets have for decades been poor relative to the industrial sector, receiving far fewer resources than the latter. They have had constitutional authority but no power. This probably reflected negatively on the calibre of the personnel working in local area management. Those with recognised ability would prefer to work in more prestigious sectors where they could exert influence. Those in local government have tended to prefer a quiet life which has been demonstrated in their reluctance to take on more responsibility (Fetisov 1980: 56-62). This disinclination has not been wholly unjustified, given the difficulties with which they would be confronted when trying to fulfil plan targets without adequate resources. Thus, <u>perestroika</u> simultaneously means a psychological restructuring as well as a restructuring of institutions, their interrelationships and cost effectiveness.

In the view of the Soviet authorities the cost of managing this huge public housing sector can be reduced through administrative re-organisation, chiefly by creating larger administrative units. Following legislation in 1959 housing management offices were 'placed on a firmer financial basis'. In contrast to the smaller housing managements which were constantly in debt to other organisations that had carried out repair work and provided materials and transport, the larger offices (partly by cutting administrative staff expenditure) began to operate profitably at least on a current account basis. This policy applies mainly to local soviet-controlled housing for, whereas the management offices of the latter take care of blocks of flats in one particular district, housing owned and controlled by enterprises is quite frequently distributed throughout the city and sometimes more widely still. In the early 1980s nationwide, the departmental sector with twice as much living space to run and maintain had seven times as many housing-management departments. One consequence is that the operating costs of running the departmental stock are relatively high and annually incur huge losses.

Constantly increasing maintenance costs, the persistence of too many small, unprofitable housing management offices, the inefficient use of resources and the poor quality of repair work carried out, occasioned the issuing of decrees in 1978 and 1986 containing proposals for the establishing of new methods of administering the housing sector. These were to include the setting up in the large cities of a single building repair service unit which would contract to carry out current and capital repairs for both the state and co-operative sectors. Such administrative changes, while likely to produce some of the desired effects, will overall make a marginal impact on the level of efficiency in managing and maintaining the housing stock and on the enormous disparity between costs and revenue.

But while pressures are accumulating within the logic of the development of Soviet society for this transfer of resources - for future investment and the existing stock - to local soviets, another 'socialist' logic is operating to partially offset this process. The enterprise Fund for Socio-Cultural Measures and Housing Construction, which has already been mentioned, can now be examined a little more closely.

Finance for housing in the departmental sector comes from two sources: the central state budget and internal funds allocated from enterprise- and organisation-generated revenue. However, light industry and agriculture, for instance, consistently receive far fewer resources than do enterprises in metallurgy, coalmining and heavy engineering. This is evident from Table 6.9 which shows the amounts of overall living space (in square metres) built by the state per employee in seven different economic sectors.

Table 6.9: USSR: State investment in housing by selected economic sectors

Sector	1965	1982
	square metres built per employee	
Energy & electrification	1.95	2.49
Chemical and Petro-chemical	0.90	1.36
Ferrous metallurgy	0.50	0.81
Building materials industry	0.15	0.54
Light industry	0.09	0.22
Food industry	0.09	0.19
Agriculture	0.01	0.17

Source: Alekseev 1987: 187

The author citing these figures, whilst acknowledging that the highest state planning authorities have to take into account the specific developmental needs of different sectors of the economy at different historical periods, concluded that nothing justifies such huge disparities. That they exist can only be explained by the fact that planners still lack sufficiently well-defined objective criteria for financing housing construction. A wide range of factors - such as the volume of capital investment in particular industrial sectors, the need to demolish old housing, in-migration and labour turnover - determines the allocation of resources for house building between ministries and eventually to individual enterprises. Yet, properly assessed, these factors tend to be quite subjective, being determined by the intuition, bias and personal predilections of senior officials in planning departments (Alekseev 1987: 187).

The outcome is that certain privileged sectors, which each year receive substantial allocations for house building, co-exist with 'second-rate' sectors which are wholly underfunded. So, in each town and district enterprises may be classified as <u>rich</u> and <u>poor</u> depending on the amount of housing that they construct. In order to combat this inequality, one recommendation is that the state should allocate resources for new house building according to a standard norm for all public sector workers, say 180 roubles a year, to achieve an annual growth of housing space of one square metre per worker. In this way all workers - who after all have equal rights as owners of the public means of production - will have a right to an equal proportion of the wealth which is created and allocated from the central consumption fund without there being any hard and fast link with their personal labour. This would ensure the equal right of all citizens to accommodation as laid down in the Constitution.

This would appear to be a commendable and credible policy. Yet, a moment's reflection reveals difficulties in its implementation. For instance, sectors owning and managing obsolescent housing stock subject to demolition and from which tenants have to move would have to be prioritised in the allocation of resources. Second, as already mentioned, sectoral differences aside, considerable republican and regional variations exist whose origins lie in established national traditions and in the types of social policies pursued by local soviets in these regions (Alekseev 1987: 188). Should existing differences in provision be taken into consideration, with the norms being raised in some cases and lowered in others? Or, should differences be disregarded in favour of the principle of the equal right of all individuals throughout the country to an equal proportion of resources allocated from the centre? The critical issue, however, is that the use of centrally distributed funds alone will not solve the housing shortage: additional resources must come from the surplus product of enterprises themselves. The vast majority of enterprise directors questioned in a particular survey considered that 60-80 per cent of the total sum allocated for housing should come from the state and the remainder (20-40 per cent) from their own internally generated resources (Alekseev 1987: 194).

The implication of this is that the standardisation and equalisation of the distribution of resources for housing which ensures social equality and social justice has to be combined with the opportunities for each worker and work collective to improve the quality of their living conditions. In this way workers will have an incentive to increase their labour productivity. At the same time, planning authorities can, by regulating the value relations between different branches of the economy through the use of prices, improve the living conditions of workers employed in those sectors that have to develop rapidly.

The vision of the mediating and overseeing role of the central state as the principal agency for achieving equality has deep cultural origins as well as being expressive of socialist ideology. This collectivistic or paternalistic conception also extends to the enterprise itself. It is considered right by some that workers expect enterprises and organisations for whom they work to provide them with good housing. "This approach is preferable for it corresponds better to the basic principles of a collectivist society than the present practice of co-operatives" (Alekseev 1987: 197). In other words, government policy should consist of two components. The first ensures that all workers have an equal opportunity to improve their accommodation through a national unitary expenditure norm on new house building. The second allows enterprises to adjust this norm in order to take account of a worker's individual contribution to output (Alekseev 1987: 198).

This particular prescription thus favours the distribution of rewards - both monetary and non-monetary - through the place of work. According to this view, the Constitutional right of each worker to receive state accommodation is guaranteed, while at the same time it promotes another basic principle of socialism which is to

267

enhance the interests of society as a whole, the
work collective and the individual worker. These
rewards are paid for out of the <u>socio-cultural
fund</u> which comes out of the overall sum allocated
for incentive payments. Although labour
productivity - and thereby reduced production
costs - is the most important index for assessing
a work collective's performance, the efficiency
achieved depends not just on the collective itself
but also on such factors as the reliability of the
equipment, the technological process and the
continuity of supply of materials and components
in a particular plant (Kalmantaev 1988). Needless
to say, these 'objective' factors (externalities)
promoting or constraining productivity and
therefore earnings constitute a serious source of
contention and complaint by workers who are being
adversely affected by the economic consequences of
the restructuring process. The drawbacks of
making increases in labour productivity the
determinant of the size of this fund are now
accepted. Refinements to the system of
calculating the fund's size have failed to
overcome the problem and so, not surprisingly,
only 7.8 per cent of managers considered the fund
adequate to finance the planned volume of house
building.

Since, according to the vast majority of
managers, their own generated resources cannot
solve the housing problem, they remain heavily
reliant on capital investment by the central
Government. On the other hand, managers have
frequently been rebuked for "not operationalising
all hidden reserves" (<u>Postanovlenie</u> 1987: no. 28,
art. 96). When the General Secretary and like-
minded reformers speak of the power of inertia and
conservatism as factors retarding the pace of restructuring, housing offers them a good example of
refractory behaviour by segments of the state
sector. Industrial ministries and republican
governments who continue in the old way of "not
using resources allocated by the state for housing
construction and the provision of amenities" have
been reprimanded for "a practice which is
completely intolerable" (<u>Postanovlenie</u> 1987: no.
28, art. 96).

REDUCTIONS IN HOUSING CONSTRUCTION COSTS

An assessment of the successes and failures, merits and demerits of the Soviet Union's housing policy has to take into account the scale of the housing programme, the keystone of which is the widespread application of standard designs and pre-fabricated methods of construction. The use of standardised components, the mechanisation of the industry, improvements in planning, organisation and the co-ordination of activities and a better qualified workforce have all contributed to the enhanced rate and efficiency at which new dwelling units are erected.

In terms of overall living space in towns, the proportion of all state and co-operative housing erected using large, four-metre wide panels rose from 1.5 per cent in 1959 to 38 per cent in 1970 and 61 per cent in 1980. This figure was expected to reach 67 per cent in 1985 and in the region of 75-78 per cent by the end of the century (Postanovlenie 1981: no. 29, art. 169; Meerson and Tonskii 1983: 9-12). The attractiveness of panel construction rests on the assumption that it leads to substantial reductions in building time and labour costs. This expansion has been at the expense of brick-built houses: the use of bricks in the socialised sector declined from 48 per cent in 1970 to 30 per cent in 1980 and was envisaged to contribute only about 10 per cent of all new urban house building by the year 2000. This policy has now been arrested and brick production is to be increased.

Progress made in the use of pre-fabricated building technology and standard designs has been closely associated with the advance towards the goal where the house-building combine (domostroitel'nyi kombinat - DSK) will form the basic production unit for housing. These combines not only manufacture all the main components making up a block of flats but also transport them to building sites using their own custom-built vehicles. They then assemble the flats, complete all the necessary interior finishing work and hand them over to the client.

Further dissemination of this technology, more spacious apartments, a greater interest in architectural design and a perceived need to use land more intensively have all contributed to the change in policy towards building heights. Thus, whereas in 1965, 5 per cent of all new state and

co-operative building was of 9 or more storeys, by
1980 the figure was 51 per cent.

In 1986 the average height of residential
buildings in towns was 6.2 storeys, while in
Moscow it was 8.9 storeys, which represented a 35
per cent increase on 1971 (Blekh 1987: 24, 25).
In the case of Moscow, Leningrad and Kiev
construction has shifted completely to blocks of
9, 16 and more storeys, while in a number of large
cities they comprise over 80 per cent of new house
building. In contrast, the overwhelming majority
of blocks of flats in small and medium-sized towns
(i.e. those with under 100,00 inhabitants) are
still of 5 storeys. In 1971 the 5-storey walk-up
block represented 62 per cent of all public sector
housing in towns compared with the 7 per cent
contribution of 1- and 2-storey units. The
corresponding figures for 1980 was 38 and 6 per
cent. As a rule, the large-scale construction
of 9-storey blocks is to be confined to cities in
the following situations: (i) those with over half
a million inhabitants; (ii) those in the
population range of 250,000 with difficult
terrain; (iii) where there is a combination of a
shortage of vacant space and limited opportunity
for territorial expansion; (iv) where the central
districts of cities in the 250,000-500,000
population category are being redeveloped. In all
other cases the 5-storey structure has to remain
the norm. Between 1971 and 1980 buildings of 6-9
storeys rose proportionately from 19 to 40 per
cent of new building in towns. Those of 10 or
more storeys rose from 3 to 11 per cent (Borovkov
1984: 46). Building heights are currently the
subject of renewed controversy (Fesenko 1986),
reflecting both increasing professional concern
with the aesthetic of the skyline and popular
preferences for low-rise accommodation.

The conventional wisdom holds that the DSK
and pre-fabrication offer a number of advantages
compared with traditional building methods. First
of all, the production of building materials and
construction can be unified within a single
administrative framework. Second, greater
continuity is achieved in housing construction; in
particular the industry is less susceptible to the
vagaries of the weather. Third, they permit an
improvement in the quality of the materials used
and of the completed buildings and in working
conditions, with building and assembly work coming
to assume characteristics of ordinary factory

production. (On the issue of quality, in 1977 only one percent of new houses built in Saratov were awarded a rating of 'excellent' and 58 per cent a rating of 'good'. In the following year matters deteriorated further: a mere 0.7 per cent were rated as 'excellent' and 33 per cent as 'good'. The remaining 66.3 per cent were accepted as 'satisfactory' - a category which the author of the article considered "does not merit comment" [Fetisov 1980: 61].) Fourth, waste products are better utilised and recycled - a subject currently receiving considerable attention from the Government. Lastly, they make it possible to reduce both the amount of paperwork and the size of the clerical staff and other white-collar employees. While all these improvements have increased the efficiency of building, the problem of costs remains acute.

The cost of erecting one square metre of living space in 1965 was 119 roubles; it rose to 167 roubles in 1971-1975, to 175 roubles in 1976-1980 and to 199 roubles in the 11th five year plan (1981-1985) (Sarkisyan 1983: 183). Put in another way, the average cost of building a flat increased from 5,000 roubles in 1960 to 8,700 roubles in 1980 (Tonskii 1982: 51). These increases are due to a number of factors. As building heights rise so do the initial capital investments: for instance, the need for 2 lifts in blocks with more than 10 storeys makes a 16-storey block 7-10 per cent more expensive to construct per square metre than one of 9 storeys. Current expenditures are also higher: in a 16-storey block running costs per square metre are 25-30 per cent greater and in tower blocks of over 25 storeys 60 per cent greater than in a 9-storey block (Tonskii 1982: 89; Sarkisyan 1983: 182). In 1988 the average cost of producing 1 square metre of housing was 200-250 roubles (Rozanov 1988).

Steps taken to restructure the building industry - entailing among other things improvements to the 'economic mechanism' and the transfer of building contractors and other concerns involved in the building trade to full cost accounting and self-financing - are intended to open new opportunities for qualitative improvements in capital construction (Postanovlenie 1988: no. 1, art. 2). Further rationalisation allowing enterprises or larger associations to combine the functions of designing and constructing accommodation is seen as another way of improving the efficiency of housing construction (Postanovlenie 1988: no. 18, art. 53). The principal criterion for evaluating an enterprise is whether it has fulfilled its contractual agreement to erect housing according to the full specification within the project's allotted time. Should it succeed 100 per cent in doing so, its 'material incentive fund' is increased by 15 per cent. It is certainly too early to predict the outcome of the new financial arrangements, especially their impact on the costs and quality of the housing built. Any future judgement on the success or failure of this particular policy will have to take into account the more far-reaching housing strategy of which it is part.

On the other hand, the cost reductions associated with these changes are countered by the operation of factors making housing construction more expensive, namely: the increased height of blocks of flats; improvements in the size and layout of buildings and individual apartments; the higher standard of furnishings, fittings and decoration; and the increased proportion of construction taking place in the northern and eastern regions where the physical environment makes building difficult.

The idea of dwelling units being produced like many other commodities for mass consumption by assembly-line technology appealed on ideological as well as "rational, scientific, technical" grounds when research and development on the large-scale production of housing began. It would be difficult to deny that the strategy adopted has not had a measure of success: between 1965 and 1985, 164.5 million people moved into new homes, built by themselves or by the state (Goskomstat 1987: 516). The huge social and economic costs of this success in providing shelter for the population will become an increasing burden for future generations.

Another dimension of the building industry, which generally speaking has received very little attention, is known in the UK by the cacophonous term, 'the lump'. In the Soviet Union, these are the private building contractors known as shabashniki. Involving "hundreds and thousands of people in unplanned construction" these workers form part of the clandestine army of labour who take on a second job - car repairs, private taxis, manufacturing goods in short supply

- in addition to working in the state sector. Attempts by the state to curb the black economy, it is now recognised, have been singularly unsuccessful (Aganbegyan 1988a: 26).

These workers are typically male and aged 25-40 who hire themselves out as work brigades composed of carpenters, plasterers and bricklayers to build houses (or roads, garages, farm buildings) on collective and state farms working in teams of 3-9 persons, occasionally rising to 25 people [Murphy 1985: 48-57]). They earn 3-4 times as much as the average industrial wage and up to 2,000 roubles a month. In 1986 the average industrial worker's monthly wage was 216 roubles, the average wage of the leading trades in the construction industry were: carpenters, 288 roubles; plasterers, 263 roubles; painters, 239 roubles; bulldozer operators, 324 roubles (Goskomstat 1987: 431, 433). Some also take their payment in kind.

Many of the 'professional' shabashniki come from Belorussia or the Western Ukraine; some take their families with them and might work for 6-8 months before returning home, or might not return at all. There are also student or post-graduate shabashniki who work for short periods, often during the vacation, and young male intellectuals in Moscow, who take undemanding low-paid state work outside their specialism and simultaneously engage in highly remunerative building work.

Their services are certainly in considerable demand, whether building a private country house (dacha), carrying out a task more quickly than through the 'normal' (state) channels or rectifying poor workmanship. Although shoddy building is universal, sometimes the excesses become virtually intolerable. It is difficult to assess where matters are worst. According to some authors those living in the countryside suffer most. For instance, the chairman of one district soviet in a rural area of the Kirov oblast explained that "in the best cases, builders put up the walls and add the roof. Decoration, pipe-laying and underground systems are left unfinished with the tenants themselves digging trenches and laying pipes without proper heed to sanitary requirements. As a consequence water leaks, underheated homes, cuts in the water flow and so on are common occurrences" (Abramovich 1987: 7, 45). Under such circumstances, it is not surprising that the shabashnik's services are in demand.

This phenomenon, on which the press began reporting in the early 1970s, resembles a type of work association (<u>artel</u>) which dates at least from the mid-nineteenth century. This system, whereby payment is made for work done and workers are remunerated according to their individual contribution to the project, corresponds in certain respects to the brigade system now being introduced by the Government into industry and construction. Despite the functional importance of these work collectives, in the 1970s the government and the courts regarded them - or at least the <u>shabashnik</u> leader in his capacity as commercial <u>middleman</u> - as criminals engaging in private entrepreneurship, bribery (to obtain materials, often from the contractor himself!) or stealing from the state. It is a feature of the Soviet legal system (and of the wider society) that in the majority of cases it is not overly harsh with law-breakers so long as the offence is not committed on a "large scale" or "systematically". These two limiting terms permit a degree of tolerance of direct relevance to housing: the subletting of property (which is permitted by law under certain circumstances) and the charging of illegal rents.

The administrative economy that is currently under attack typically seeks to deal with individual, private activities by creating another bureaucratic formality. In the case of <u>shabashniki</u>, a decree issued in 1978 stipulated that the relationship between state organisations and hired workers should adhere to a model contract: this required each worker to have a labour book confirming that he was free at this particular moment to work away from his normal place of employment (Buzhkevich 1978: 20-27). This seemingly negative response by the state to embrace, codify and regiment 'spontaneous' activity, had its positive aspects. In the first place it was an attempt to protect labour by ensuring that a proper contract existed. This is a difficult area since it involves both the law of contract and the law governing labour relations. Generally speaking, the state was rightly concerned that <u>shabashniki</u> were unprotected by labour legislation and therefore received no paid holidays or guarantees of job security or compensation for industrial injury. Second, the state itself might well already be paying the worker a wage at his formal place of employment and was therefore justified in wanting to have proof that the 'moonlighter' was indeed selling his labour during his free time. Their illegal, or at best quasi-legal, status left shabashniki highly vulnerable to exploitation by their employers who could refuse to pay them the contracted sum on completion of the work - especially if it was large - since the latter would not want to pursue the matter through the courts.

The shabashniki highlight a dilemma faced by many sectors of the economy: A demand for a good or service is unmet by the state. The shabashniki offer to satisfy the demand. However, when individuals and collectives seek to meet the demand they act illegally, because there are virtually no legal ways of obtaining construction materials. The only channels through which the shabashniki can proceed are proscribed: begging, borrowing, bribing or stealing. Yet they are unable to buy these materials; when they do, this is, if not illegal, then on the boundary of legality. Beef, potatoes and honey might be exchanged for a load of cement, gravel or other required building material. A shabashnik leader requiring a bulldozer or excavator for a day can hire one, for a price from a state building agency. In some cases the exchange process attains a far higher degree of sophistication with government organisations in different economic sectors and across the breadth of the country involved.

Yet, just prior to Mr Gorbachev's term as General Secretary, a decree of the USSR Council of Ministers in July 1985 (Postanovlenie 1985: no. 19, art. 89) requested that ministries responsible for the production of building materials should increase their availability to the population. It also required that local soviets should find ways of transporting building materials from warehouses to consumers. These were tentative administrative steps towards dealing with a situation which witnessed even the Minister of the Building Materials industry being dismissed for nefarious and dishonest practices regarding defitsitnyi building materials.

Government legislation in the 1980s corresponded with a change in media representation of the shabashnik. He was no longer prosecuted as an altogether money-grabbing, greedy parasite receiving wholly unwarranted payments for poor quality workmanship. Instead, he came to be depicted as a positive response to inefficiencies, mismanagement and shortages in the local economy by young people who, through demonstrating considerable initiative by travelling long distances to take up contracts, remained unprotected by the law. The fact that these workers now find defenders and their manifestation treated as part of a much more complex problem arising from the organisation of the economy at large, has become part of a much broader emerging debate on 'independent' economic activity and 'unceilinged' wages which has begun to crystallise into the policy programme of perestroika.

Such has been the scale of investment in the research and development of an industrialised building industry, it would be difficult in the short run to initiate a major change in policy away from high-rise construction systems which rely on the prefabrication of components and the production of completed flats in factories for assembly on site. Nevertheless, radical shifts in policy are being expressed. One senior academic, in quoting the last (1986) Party Congress reaffirmed that the most important task facing the government is to raise the quality of housing, which will not be accomplished merely by transferring onto a newer, better designed series of large panelled blocks of flats or even by eliminating waste and defects in construction. It will require formulating an alternative approach to housing, one which takes into account people's needs. This in turn necessitates radically restructuring the 'material-technical base' of the construction industry in the direction of increasing the proportion of low-rise dwellings. Well-constructed small houses must become one of the principal types of structure found in (at least) both small and medium-sized towns and in the suburban areas of large cities (Grebennikov 1987: 195). Such a policy if pursued would see a resuscitation of ideas advanced by urban planners during the 1920s.

Thus, the building labour force will probably increasingly bifurcate into two quite different occupational groups. One will consist largely of 'house assembly workers' lacking in traditional building skills, but who might not be altogether 'deskilled' for they will gain other knowledge and qualifications. The other group, engaged in housing rehabilitation and modernisation, building under contract for individually designed dwellings and carrying out interior decoration for private individuals and collectives, will deploy traditional building trade skills. The latter, already working as shabashniki, will probably expand in number. The problem here is likely to be the need to establish facilities and channels for apprenticeships and re-training since their numbers will be inadequate to cope with the anticipated increase in demand for their services. Certainly in the short term, this will give rise to quite high levels of remuneration being paid to workers with the requisite skills. The significance of this development will be that these necessarily higher charges will be borne by the customer. Experiments have already been conducted in a number of cities to improve the assortment of fabrics and furnishings from which prospective tenants can choose for the interior decor of their new flats and to make it financially worthwhile for building brigades to upgrade their skills and provide a high quality service (Ukolov 1985: 156). As is discussed below, the view of the Government is that the growth in real incomes and accumulated savings offers more individuals the opportunity to become owner-occupiers or members of co-operatives. They can also contribute to solving the qualitative dimension of the housing problem by paying for a variety of services ranging from interior decorating to higher quality fixtures and fittings (Dmitriev 1987:75).

The method of remunerating workers in the 'formal' state-building sector has been subject to revision at different times. One of the primary problems which is only now being radically altered has been that planning the wage fund for builders is based on the volume of building-assembly work taking place. This makes it in the interest of contracting agencies to raise the costs of construction and not to complete construction projects (Vashets 1987). Attempts to establish a bonus system which comes into operation when completed projects have been commissioned have not in the past proved very successful. It remains to be seen whether the far-reaching changes currently being instituted in order to eliminate the high degree of wastage arising from poor organisation and idling time will be effective. Experience

might reveal that the labour shortage - from which
the construction industry is frequently alleged to
suffer - is to a considerable extent artificially
created (Dmitriev 1987: 75).

The already cited government promulgation
(Postanovlenie 1987: no. 27, art. 92) allows
housing department managers - with the agreement
of the local trade union committee - to fix
differential bonuses for different categories of
workers. Wage-rate tariffs and norms for
occupational categories is exceedingly complex;
the central issue is that in future not only will
bonuses not be paid in cases where wastages have
arisen, damage been caused, quality sacrificed and
plans unfulfilled - but that there will be greater
rewards for individual workers and work gangs
(brigades) who perform well. One of the enduring
misconceptions about the Soviet economy and
society - one currently being dispelled - is that
the workforce in the Soviet Union is highly
disciplined and rewarded strictly according to its
performance. The task Mr. Gorbachev and his
advisers have set themselves in their
reconstruction of the economy is to employ
'economic mechanisms' (supply and demand,
marketability of products, reward for work done)
to achieve precisely the disciplined and motivated
workforce that only exists in theory. Another
important component of this endeavour is to raise
the level of qualification in different sectors of
the economy by setting up training programmes.

In order to fill posts in housing management,
administrators in the field of building
construction are to be sent on altogether new
training schemes which will include courses on
repairing and maintaining the housing stock
(Postanovlenie 1987: no. 27, art. 92). This is
particularly important in view of the findings of
one survey that, while in 1986 appeals and
referrals to the soviet had decreased in general,
the reverse was the case as far as the repair and
maintenance of dwellings, central heating and
water supply were concerned, with virtually one in
two people turning to the soviet raising questions
about their accommodation (Smusenok 1987: 44).
The region's poor performance in providing public
amenities - which was considerably below the
republican average - and in carrying out
housing repair work was traced by the author
of the research back to the very way
in which sessions of the city soviets had posed

the problem. Instead of devoting detailed and careful examination to each issue they were "too often dealt with hastily, formally and far too generally with remedies being drowned in a torrent of general phrases". Some deputies to the soviet (local councillors) are more conscientious in the pursuit of their duties. For instance, one, an electrician, strongly rebuked a local factory for the poor quality of the socket and fittings holding up the chandelier in a flat which fell out when the tenant moved in. Decorators and plumbers all left behind them defective work.

The inconvenience to tenants aside, the poor quality of building work is extremely costly to the state: the longevity of the housing stock is considerably reduced and many flats remain empty for years because major repair work has not been carried out, thus diminishing the actual size of the habitable stock. A further negative consequence of this situation is that it contributes to labour turnover and out-migration from areas already suffering from shortages in manpower. This failure to devote sufficient attention to repair work is partly due to the problem of finding accommodation for those whose properties are to undergo major renovation (the so-called manevrennyi fond, akin to the 'decanting stock' used by housing departments of municipal authorities in the UK).

The Government made a major attempt to deal with the problem of repair and maintenance in 1978 (Postanovlenie 1978: no. 22, art. 137). This repeated that resources allocated for the upkeep and maintenance of their housing stocks must be used for that purpose. In the decade which has elapsed since then, dwellings have grown older and some to an advanced stage of dilapidation while the supply of materials remains as bad as ever. But even when the authorities (including enterprise managements) demonstrate good will and are prepared to use funds for the purpose of repair, they may well be thwarted by the absence or unavailability of building contractors and materials. In the already cited survey of 175 enterprises with money to spend on socio-cultural facilities and housing, between 1981 and 1985, on average 43 per cent could not find contractors to undertake the building work. In the Kirov oblast' not a single type of material or product is being provided 100 per cent according to the plan, with the supply of lock fixtures, for

279

example, meeting only 12 per cent of requirements. (The manufacture of lavatories - not to mention lavatory seats - is also far from satisfying demand.) The alleged reason for this state of affairs is that the production of some items requires special substances, or special technological processes or specially trained workers. Under these circumstances, housing departments which have obtained stocks of goods in short supply, behave like other organisations and enterprises finding themselves in this fortunate position: they hoard them - just in case they might be needed at a later date.

The causes of these problems and obstacles to their resolution are not entirely specific to the construction industry. They are tantamount to being permanent features of the existing economic system. In this particular case, those responsible for finance (<u>finansisty</u>) do not regard the expenditure of resources allocated for capital repairs as 'purposeful' or worthwhile. Therefore the tendency for labour and materials to be directed instead towards new construction or used for repairing administrative buildings increases each year. The reason for this practice has already been mentioned: The whole structure of remuneration, bonuses and rewards in general is geared towards undertaking 'new' projects rather than maintaining and improving the existing stock. Yet, those contravening the plan ("which is law") are none other than the central government departments and the executive committees of the local soviets themselves.

In situations such as these, tenants can have recourse to their local elected councillors. Unfortunately, the latter are not always as active as they ought to be. The political reforms inaugurated in 1988 are to a very considerable extent concerned with the revivification of the deputies by increasing their authority and their constituents' confidence in their ability to fulfil their mandates. While officials fail to adhere to the stipulations of the state plan and elected councillors do little to execute their duties, tenants too are culpable for the poor state of the housing stock. Frequently they "behave like barbarians (<u>po-varvarski</u>) towards their flats" (Smusenok 1987: 47). After a few years as tenants such "apology-for-owners" (<u>gore-khozyaieva</u>) reduce their flats to an unsightly condition. The measures recommended to deal with this situation are both prophylactic and penal. In the case of the former, tenants should be given copies of the new regulations on the use of accommodation and its maintenance and upkeep. Where the accommodation is wilfully damaged the powers and penalties laid down in the Housing Code must be invoked. In the case of infringements to the Rules governing a tenant's use of public housing, the tenant, members of his or her family and others living with him "can be subjected to public pressure through the house committee and comrades' courts". If this fails to have the desired effect, then the tenant can be evicted through the courts under Article 98 of the RSFSR Housing Code. The rules stress personal responsibility towards the property and towards other tenants: "It is prohibited to keep animals, birds and bees on balconies" and "the use of televisions, radios, record players and other loud noise-making instruments ... must not disturb the peace of residents. From 23.00 to 7.00 complete silence must be observed" (Postanovlenie 1985: no. 2, art. 10).

The public housing sector is huge; it tends to be organisationally unwieldy, suffers from an inadequately qualified management and is not too well maintained. Legislation, especially over the last few years, has addressed itself to all these issues. Above all it has emphasised that in order to provide each family with its own self-contained flat considerable resources will have to be expended on major capital repair work and the modernisation of the existing stock. Among the new agencies being set up to deal with the whole array of problems are the Institute of the Economics of Housing and Communal Economy, an Institute for the Problems of the Housing and Communal Economy in the North, Siberia and the Far East, and an Institute for Raising the Qualifications of Senior Staff and Specialists working in the field of housing and the communal economy. The same decree requires the Council of Ministers of individual republics to set up in the period 1987-1990 specialised building-repair contracting organisations for the reconstruction of the housing stock (Postanovlenie 1987: no. 27, art. 92). It also stipulated that Gosstroi in conjunction with Gosplan and the Ministry of Finance should have ratified before the end of 1988 regulations on the organisation of the repair, reconstruction and technical servicing of the housing stock.

This increased emphasis on modernisation has as a policy corollary the need to prevent the 'unjustified' demolition of buildings which are still in good physical condition. At present the annual rate of demolition for the whole country is equivalent to about 15 per cent of all new building, ranging from 4.8 per cent in Belorussia and 5.5 per cent in Lithuania to 17.8 per cent in the RSFSR and Uzbekistan (TsSU SSSR 1982: 433). Moreover, in so far as 30 per cent of the urban built area is occupied by 1- and 2-storey houses at very low densities, and often badly provided with amenities, it is anticipated that the demolition rate will rise between now and the end of the century. In view of the shortage in some towns of vacant (or inexpensive to clear) building land these become prime sites for development. As a result demolition and renewal are not restricted to 'low value' wooden houses, but extend to perfectly habitable, post-war 2- and 3-storey, brick-built housing so that housing densities might be increased through the erection of multi-storey blocks.

While demolition will certainly continue, more attention will be accorded to the conservation of older structures, particularly those regarded as comprising part of the national heritage. The modernisation, rehabilitation and renewal of existing utilised space and structures and 'in-filling' (i.e. the building on land within a city's administrative boundaries hitherto regarded as too expensive to develop compared with green-field sites) are attempts to use the social and economic infrastructure more efficiently. The shortage of vacant land, taken in conjunction with the fact that each year 35,000-40,000 hectares are being taken for housing building, provides the rationale for multi-storey building. Energy can be conserved through improved insulation of buildings and the social and economic costs of commuting reduced by higher density living. These cost factors are now being accorded greater consideration (<u>Postanovlenie</u> 1983: no. 11, art. 53). The small economies gained on the initial building outlays by scrimping frequently tend to be outweighed by the higher charges incurred for heating and current and capital repair work, by deteriorating standards of comfort and, in the final analysis, by the premature depreciation of buildings. Hence, on occasions it is cheaper to demolish dwellings long before the expiry of their 'normal life expectancy' rather than incur the expenditure necessary to rectify construction faults and technical defects (Tonskii 1982: 49). The perceived economic benefit of demolition and rehabilitation are accompanied by potential social costs. This is because, as in other countries, rehabilitation and renewal have two important unintended consequences: the ownership and control of the property may change (that is, there is a switch in tenure type) and the social class composition of the district may change. In the Soviet context this can mean the demolition of low-rise buildings in private (departmental or soviet) ownership and their replacement by blocks of flats belonging to the local soviet or house-building co-operative or enterprises. The city centre or sites near water or some historical, original settlement are frequently in demand and therefore command higher values. It might be anticipated that studies by sociologists and human geographers of urban ecology and spatial segregation will reveal tendencies towards class variations within towns and local settlement systems (Vasil'yev and Privalova 1984).

Urban renewal and housing rehabilitation open up new markets for shabashniki and also leads to the cultivation of a DIY ethos. For both these developments to occur legally and on a larger scale, it is crucial that the supply of necessary tools and materials is not only increased but that a proportion of them is actually placed on the 'market'.

Slum clearance - on occasion a not inappropriate descriptive term for the process - has been an issue for residents, newspaper editors, town planners and officials alike for a quarter of a century. Glasnost' has shown a sharpening of language in the polemics over policy. Letters concerned with "unwarranted demolition of good-quality individual dwelling houses" now fill Pravda's postbag. The guilty party, it transpires, is none other than the local soviets which, short of building land and looking for economies, "ignore the protests of the population" and have in the past year demolished a further million square metres of housing, which will cost over 500 million roubles in compensation alone. This raised the populist question of "why should the state, that is the people, pay for this bungling (golovotyapstvo)?" and provoked the populist response that "the building of new neighbourhoods, road networks and generally developing the urban economy must only be carried out after the population has been fully informed and if these policies will be in their interests" (Murphy 1988a: 19, 11).

THE OWNER-OCCUPIED SECTOR

A decree issued on 11 February 1988 (and published in Pravda on the 21 February), entitled "On Measures to Accelerate the Development of Individual Housing Construction", represented at the beginning of the year perhaps the most radical of all post-war government promulgations on housing policy (Postanovlenie 1988: no. 11, art. 28). Stating unreservedly that this sector must play its part in achieving the goal of providing each family with its own flat or house by the year 2000, the decree indicates a major shift in strategy. It reiterated a view frequently expressed over the past decade, that "this type of construction, drawing upon the labour of the population and more fully using the increased spending power deriving from higher incomes and money held in saving banks, is becoming one of the most important ways of 'activating the human factor', increasing the national wealth and raising the standard of living". All agencies of the Party and state, including the local soviets and ministries, and trade unions are exhorted to regard individual housing construction as a "task of paramount importance". Deposits in savings banks have indeed grown. In 1970 they stood at 46.6 milliard roubles. The respective figures for 1980, 1985 and 1988 were: 156.5 milliard, 220.8 milliard and 283 milliard (Goskomstat 1987b: 448; Murphy 1988).

It points out that "in many union republics and regions - mainly because of an 'orientation' to receiving state capital investment - there had been a serious weakening of interest in individual house building. Different types of bans have been imposed and the sale of building materials has been restricted, building plots have not been allocated or connected up to mains supplies". In fact, in some places the attitude towards individual builders has been "improper" and often "spiteful". Enterprises choose not to use their wide-ranging powers to help their employees build their own homes. The latter also experience considerable difficulties in obtaining credit to build, acquiring materials and having them delivered (Pudikov 1980). It was to combat these very problems that legislation had been passed a decade earlier to increase spending on housing for owner-occupiers, initially in the countryside (Postanovlenie 1978: no. 17, art. 102) and later extended to people living in urban areas (Postanovlenie 1981: no. 29, art. 170). The motivating force included a desire to stabilise the rural population and increase the migration of workers to such regions as the Far East or North where labour shortages are acute.

Over the past 25 years this sector has witnessed a threefold decline, so that at present it contributes less than 15 per cent to the total annual amount of new building. To reverse this decline, literally everyone is invited to "look upon individual house building as a social, political and economic task of paramount importance directly related to the activisation of the human factor."

The actual magnitude of this increase is remarkable. Whereas in 1986, this tenure contributed 17.4 million square metres of overall living space to the total stock (i.e. town and country), by 1995 this figure should be no less than 60 million square metres and 200 million square metres during the course of the thirteenth five year plan (1991-1995). This phenomenal increase is to be achieved by: improving the organisation of this type of house building; removing unjustified restrictions on the allocation of building plots; setting up associations of individual builders; developing both contract and direct labour building organisations who can enter into agreements with individual builders or building associations; increasing the production of bricks, tiles and other building materials, while Gosplan, Gossnab and the Council of Ministers of Union Republics are to ensure that more marketable stocks (rynochnye fondy) of building materials, components and equipment are made available to the general public. A number of measures have been taken in the previous few years to make building materials more available to the population. Then in August 1987, enterprises were allowed, first, to make materials and articles used by them available to the population as part of the former's share of 'paid services' and, second, to sell directly at the enterprise cement and concrete articles, timber and decorative materials, plumbing and sanitary fitments, electrical cable etc. The latter are to be paid for through the enterprise's accounts' department at retail prices or, where these do not exist, at wholesale prices with a fixed mark-up (Postanovlenie 1987: no. 42, art. 138).

Banks are also to make credit facilities available to enterprises intending to start or expand production of building materials and other merchandise required to equip and decorate housing. The loans are to be repaid from profits made on the manufacture and sale of these materials and goods within five years of the work being completed at an annual rate of interest of 3 per cent on the sum lent.

The allotment of building plots should be based on a priority system with those on housing waiting lists to improve their living conditions given precedence, especially employees with good work records and those who have retired, invalids and veterans of the Second World War. Preference is also shown to people surrendering their public-sector accommodation to the local soviet. The areas set aside for individual housing development must be provided with access roads, mains water, electricity and (where available) gas supplies. Central heating and sewerage systems should, where conditions permit, be extended to these estates. All these amenities together with social infrastructural facilities are to be paid for by enterprises and organisations, the local soviet budget and the builders themselves.

Furthermore, state savings banks are to be allowed to make advances to individuals living in the countryside of up to 20,000 roubles, repayable over 50 years. (Loans of up to 4,000 roubles repayable over 10 years are also available for rebuilding or undertaking major capital repairs on houses and outbuildings, the erection of new cattle sheds or connecting up to mains supplies.) In towns the size of the loan is to go up from 3 to 20,000 roubles, with the repayment period being extended from 10 to 25 years. The same favourable terms are to be granted to people wishing to purchase individual homes. Payments begin three years after the loan has been made - five years in the case of large families, war veterans, invalids and analogous groups of citizens, who are also freed from paying interest on their loans. The annual rate of interest for other borrowers is either 2 or 3 per cent, depending on whether they live in the countryside or in urban areas.

Additional finance is to come from a number of sources. First, enterprises and institutions are recommended to draw upon their material incentive funds to pay off part of the bank loan made to their employees for housing building. This assistance varies with the length of a persons's service: for those with 5 years service, up to 10 per cent can be paid off; over 10 years, up to 30 per cent, and over 15 years up to 50 per cent. Long service is not the sole criterion for being awarded this privilege. If resources are limited then employees with exemplary work records will be the first to benefit from this scheme. Moreover, with the work collective's consent, such meritorious workers may be granted even larger sums of money (i.e. in excess of 50 per cent) to pay off their loans. The local soviets can make similar payments from their budgets and other sources of finance at their disposal.

In order to make it worthwhile (i.e. enhance their self-interest/zainteresovannost') for enterprises to undertake the building, reconstruction and repair of individual houses, this work and the cost of delivering building materials to individual builders will be regarded as part of their fulfilment of their targets of providing the population with services for which the latter have to pay. Second, funds can be allocated from the local soviet budget to help medical, educational and other service sector personnel, invalids, war veterans and similarly classified local residents.

In addition, enterprises and organisations have been given the right to sell houses to their workers if the latter pay them no less than 40 per cent of the value of the house in equal monthly payments over a period of 50 years in the countryside. The corresponding figures for urban dwellers are 50 per cent and 25 years. This is the nearest the Soviet Union has ever come to the UK policy of selling council housing. Should the purchaser cease to work at the establishment from which the property was purchased before the expiry of 25 years, then the house has either to be handed back to the enterprise - which has to compensate the owner for the sums of money that he has paid (with a deduction being made for depreciation) - or the enterprise has to be recompensed for the financial assistance it has provided. The discrepancy in assistance offered to residents in town and country in both instances is indicative of attempts to stabilise the rural population, for the ageing and declining agricultural population in some regions is now a cause of some considerable concern. At the same time, however, large cities are also suffering from labour shortages, frequently in the building industry and service sector - occupations historically filled by rural migrants. This represents but one of several initiatives made in recent years offering distinctive (housing and other) privileges to those living in the countryside. One in particular merits summarising.

There are, nationally, over 725,000 empty houses, of which 490,000 (68 per cent) are in the RSFSR (mainly in the Black Earth belt) and 110,000 (15) per cent in the Ukraine. The uncultivated ('waste') land attached to these dwellings amounts to 200,000 hectares. Some local party, soviet and economic organisations have resigned themselves to this situation and do nothing to the properties or to bring the land back into use. As a result the Government now requires all local political and economic agencies to maximise the use of the land for agricultural production by removing 'artificially erected barriers' and to encourage the widescale adoption of leasing arrangements. Empty houses in rural areas are first of all to fall to the collective and state farms and consumer co-operatives and service-providing agencies. The recommendation is that they should then be used as accommodation for agricultural workers and other rural employees. They can also be converted into reception centres for consumer co-operatives and used for recreation, shops or eating places. People living in towns are allowed to rent them as 'second homes' on a seasonal or temporary basis, with the rent charged being sufficient to cover the running and maintenance costs of the property. The decree also allows for the sale to urban dwellers of empty houses for their 'sporadic use' on the understanding that the purchaser will cultivate the land and sell any surplus agricultural produce to the state, the collective or state farm or to a consumer co-operative. The sale of such houses in the Moscow, Leningrad and Kiev regions (oblasti) is only permitted to people living in these cities or other urban centres within the region. Whether individuals are renting or buying, the size of the plot including the house itself and associated outbuildings (if any) should not exceed 600 square metres. The actual size is left to the discretion of those disposing of it. Larger land allotments are possible and it has been left to the Ministries of Finance and Justice to define tax rates applicable to these higher allocations. Sales are prohibited to individuals if they already have kitchen gardens or larger plots attached to their privately owned homes. The prohibition also applies if they possess their own country villas (dacha) or even enjoy the use of a state dacha. The cost of purchase of an empty house (and its outbuildings) is determined by a valuation committee composed of representatives of the local soviet ispolkom, the vendor and the purchaser who rely for their assessment on valuation norms drawn up in 1981 (Postanovlenie 1981: no. 28, art. 165). The purchaser on acquiring ownership of the property has the right to sell it at a later date and is allowed to erect other buildings such as a garage or 'bathroom' on the land (Postanovlenie 1987: no. 41, art. 134). The details of this legislation given here are intended to highlight the crystallisation of two particular developments.

The first concerns the changing attitude to property ownership which is occurring. Limits to the acquisition of property remain, but these should be seen in conjunction with recent Government measures allowing people to rent land for up to fifty years and changing attitudes towards the charging of land rent. There may be a shift in philosophy towards property ownership in some sectors. Basically, countries which have retained private ownership have been unable to ignore completely the arguments in favour of communal ownership. On the other hand, countries which have chosen collective ownership have been obliged, after a certain period of time, to recognise the social benefits accruing from a degree of private ownership. Although the two systems clash in their theoretical premises, they are simply granted. The other system derives from Thomas Aquinas who believed that the temporal goods received by men from divine providence belonged to them as to ownership but not as to use. Consequently, in preference to collective ownership and granted use, these countries favour the system of private ownership (moderated by public expropriation) and controlled use. Geography, moral, political and philosophical traditions and culture have tended to be more influential than rational argument in determining the choice of system. Practice has invariably involved, especially in the industrial era, compromise and intermediate solutions. 'Reason' would at present seem to be moving the Soviet Union towards a reappraisal of its scholastism ("narrow or unenlightened insistence on traditional doctrines") in regard to property and ownership. The second central point in this revision of policy is that it presages a formal institutionalisation of a housing market. Even a cursory examination of Tables 6.10 and 6.11 indicating the quite considerable size of the private housing sector in some republics, would suggest that an established and well-organised, albeit largely unofficial, market in domestic property already exists.

Another innovation, creating important precedents, is the decision to permit the formation of designing and building co-operatives which will undertake to draw up design estimates for house building and reconstruction. In the first place it is an organisational novelty: at present, design work is carried out by specialist institutes which come within the overall framework of the State Committee for Construction (<u>Gosstroi</u>), or by design departments within ministries and enterprises. Second, it might well lead to an enhancement of the architectural profession, freeing it to charge 'consultancy' level fees and allow architects to exercise their imaginative talents. (Already cited legislation on the subject of architects can only further this development.) As a consequence, they too will be harnessed to <u>perestroika</u> by building for the Gorbachevian revolution just as they did during the 1920s. The underlying ethos would be a more 'individualistic' style consonant with the stress being placed on the 'human factor' as against bureaucracy and massiveness which expressed the ideology of the impersonal 'masses' and sometimes a vulgar, deterministic marxism (Andrusz 1987).

Fully in the spirit of <u>glasnost</u>', and to ensure that people are made fully aware of their rights in this sphere, the 1988 decree also requires all the agencies of the mass media - the radio, press and T.V. - "to strengthen the propaganda for individual house building and to provide the widest possible dissemination of information on the aims and opportunities that this type of housing offers for solving the country's housing problem". Thus, certainly in its general provisions and final stipulation that the Council of Ministers of republics, All-Union ministries and departments must act quickly to repeal all earlier acts which contravene this decree, it surely is an indicator of the Government's sincerity and resolve to remove barriers to the expansion of this tenure.

An index of the population's confidence in the genuineness of the present administration's intentions and in the likelihood of its remaining in power is the public's eagerness to take advantage of the higher credit limits being offered. In the first ten months of 1988, individuals borrowed 1.8 billion roubles in the form of private home loans, representing a fourfold increase over the previous year (Trehub 1988).

The ramifications, however, could be far greater, especially if the Government does act to "meet the greatest source of popular demand" namely, a substantial increase in the output of private cars. (The goal at present is 35 cars per 1000 families by the year 2000-2005.) There are now many economists and sociologists who consider that the 'automobilisation' of the population is an alternative to drinking as a pastime: "The complex of the 'house and car', particularly when taking into account the necessity of substantially expanding the construction of low-rise housing, must become the framework within which to satisfy the population's demand that the range and availability of paid services should be expanded" (Grebennikov 1987: 190).

While the country's leaders seek to deal with problems of corruption, bribery and nepotism - seemingly endemic to (tsarist and socialist) bureaucratic and administrative systems of distribution - they are overseeing the emergence of another syndrome frequently attendant upon private economic activity, namely tax evasion (Postanovlenie 1988: no. 15, art. 41).

Table 6.10: USSR: Size of urban housing stock in each
republic by tenure, 1980 and 1986

Republic	Urban housing stock			
	Socialised (i)		Private	
	mln. m^2	%	mln. m^2	%
		1980		
USSR	16555.0	75.2	547.0	24.8
RSFSR	1047.3	81.1	243.9	18.9
Ukraine	283.4	64.3	157.3	38.7
Belorussia	54.2	76.1	17.0	23.9
Uzbekistan	38.7	60.5	25.3	39.5
Kazakhstan	69.3	71.2	28.0	28.8
Georgia	21.6	56.8	16.4	43.2
Azerbaidhzan	22.9	62.6	13.7	37.4
Lithuania	23.5	76.8	7.1	23.2
Moldavia	13.2	68.0	6.2	32.0
Latvia	22.6	83.1	4.6	16.9
Kirghizia	8.4	54.5	7.0	45.5
Tadzhikstan	9.6	66.7	4.8	33.3
Armenia	17.1	66.3	8.7	33.7
Turkmenia	9.7	70.3	4.1	29.7
Estonia	13.3	79.6	3.4	20.4

Note: (i) Socialised refers to : state, co-operative
organisations, house-building co-operatives and
collective farms

HOUSE-BUILDING CO-OPERATIVES

Earlier it was noted that the XXVII Party
Congress, Mr. Gorbachev and other prominent
government officials and housing specialists have
remarked on the necessity of further enlisting the
aid of housing co-operatives in the drive to solve
the housing problem. However, these injunctions
notwithstanding, it is by no means certain that
the hopes placed upon co-operatives will be
justified in the foreseeable future (Bessonova
1988:41).

Table 6.10 (continued)

Republic	Urban housing stock			
	Socialised (i)		Private	
	mln. m^2	%	mln. m^2	%
		1986		
USSR	2050.0	77.6	590.0	22.4
RSFSR	1286.4	83.6	252.2	16.4
Ukraine	352.6	67.6	168.9	32.4
Belorussia	70.9	79.0	18.9	21.0
Uzbekistan	52.4	59.3	36.0	40.7
Kazakhstan	87.0	73.9	30.7	26.1
Georgia	26.5	59.0	18.4	41.0
Azerbaidhzan	27.8	64.2	15.5	35.8
Lithuania	29.8	78.6	8.1	21.4
Moldavia	18.0	72.5	6.8	27.5
Latvia	26.4	84.3	4.9	15.7
Kirghizia	10.9	59.2	7.5	40.8
Tadzhikstan	12.1	68.4	5.6	31.6
Armenia	21.3	70.3	9.0	29.7
Turkmenia	11.7	72.7	4.4	27.3
Estonia	16.3	82.3	3.5	17.7

Source: Goskomstat 1987b: 518-519

Contemporaries writing on the earlier experiment from 1924 to 1937, considered that housing co-operatives were a practical way of attracting workers' savings into the state's economic plan (Zhilishchnaya kooperatsiya 1928: 1). Others conceived of their role in much broader terms: workers would be drawn into housing administration, costs of providing accommodation would be reduced, quality would be improved and tenants would take greater care of their accommodation. Overall, the view which prevailed considered that of all the various and valid objectives pursued by housing co-operatives, the

Table 6.11: USSR: Total new building in each republic by
tenure, 1980 and 1986

Republic	Total New Building : Town and Country			
	Socialised		Private	
	thousand m^2	%	thousand m^2	%
		1980		
USSR	88806	84.7	16001	15.3
RSFSR	55389	93.3	3962	6.7
Ukraine	12685	73.2	4641	26.8
Belorussia	3727	86.9	564	13.1
Uzbekistan	3387	59.0	2355	41.0
Kazakhstan	5205	89.8	588	10.2
Georgia	1136	66.9	562	33.1
Azerbaidhzan	874	64.8	474	35.2
Lithuania	1243	73.7	443	26.3
Moldavia	915	60.0	611	40.0
Latvia	987	89.8	112	10.2
Kirghizia	588	55.5	471	44.5
Tadzhikstan	605	57.7	444	42.3
Armenia	785	78.6	214	21.4
Turkmenia	535	52.0	493	48.0
Estonia	745	91.7	67	8.3

most important was to draw upon the population's
savings and earnings.

Between 1958 - when legislation again
permitted the formation of housing co-operatives,
but without state assistance - and 1962, twenty
co-operatives with a combined floor area of
300,000 square metres of overall living space had
been established. Legislation passed in 1962 and
1964 provided a 60 per cent loan repayable over
10-15 years (Postanovlenie 1962: no. 12, art. 93;
Postanovlenie 1964: no. 25, art. 147). Ever
since, the Government has placed great stress on

Table 6.11 (continued)

Republic	Total New Building : Town and Country			
	Socialised		Private	
	thousand m^2	%	thousand m^2	%
		1986		
USSR	102336	85.5	1704	14.5
RSFSR	62273	94.1	3920	5.9
Ukraine	15085	73.8	5362	26.2
Belorussia	4953	91.8	441	8.2
Uzbekistan	3978	62.4	2396	37.6
Kazakhstan	6219	91.6	571	8.4
Georgia	1212	71.5	483	28.5
Azerbaidhzan	1035	49.2	1068	50.8
Lithuania	1607	80.8	382	19.2
Moldavia	1022	56.7	780	43.3
Latvia	1076	84.9	191	15.1
Kirghizia	752	62.8	446	37.2
Tadzhikstan	735	61.1	468	38.9
Armenia	1001	77.6	289	22.4
Turkmenia	656	55.6	524	44.4
Estonia	732	89.8	83	10.2

Source: Goskomstat 1987b: 515-516

the importance it attaches to the growth of this sector (Postanovlenie 1982: no. 23, art. 120). At first this tenure form experienced an expansion. But then during the 10th five year planning period (1976-1980), whereas the state sector erected more than had been planned, the private sector failed to achieve its target by 8 per cent and, more significantly, the house-building co-operative failed to fulfil its plan by 30 per cent (Meerson and Tonskii 1983: 5).

The official favour bestowed upon house-building co-operatives (zhilishchno-stroitel'nyi

<u>kooperativ-ZhSK</u>) in words is not reflected in
deeds. The central government undoubtedly regards
their development as a means of saving on its
housing budget by compelling co-operative members
to bear the 'true cost' of their accommodation.
The underperformance of this sector justifies
speculation that local and institutional barriers
and informal mechanisms are impeding policy
implementation. As one senior official put it, "a
high level of economic development favours the
growth of co-operative construction since an
increase in living standards makes it possible for
the population to participate in financing housing
construction". Many commentators have on numerous
occasions been drawn to remark that, despite
government decrees and resolutions directly
concerned to encourage people to use their incomes
and savings to meet their housing requirements,
formidable bureaucratic obstacles continue to
beset the formation of house building co-
operatives (Alekseev 1987: 195-196).

In 1981 house-building co-operatives owned
2.1 million properties (5 per cent) out of a total
45.2 million flats in towns and urban-type
settlements. Their contribution to house building
over the past twenty years has been small and
erratic: 2.7 per cent in 1961-1965; 6.5 per cent
in 1966-1970; 6.0 per cent in 1971-1975; 5.2 per
cent in 1976-1980; and 5.9 per cent in 1981-1985.
The average annual rate of construction of 130,000
co-operative flats is not even enough to meet the
demand of those for whom there is no other way
than the co-operative to obtain accommodation -
that is, those families working on contract in the
far north, single-person households of 'average
age' and young families living with their parents
and thus formally provided with housing (Bessonova
1988: 43).

The figures should, however, be approached
with some caution. First of all, between 1976 and
1980 only 3.3 per cent of all ZhSK construction
took place in the countryside, the bulk of which
was concentrated in Lithuania and Uzbekistan.
Second, co-operatives are not just an urban
phenomenon, they are also disproportionately over-
represented in the largest cities. For instance,
in 1980 they contributed about 11 per cent of all
new house building in Moscow and 23 per cent in
Leningrad. In fact, in that one year ZhSKs
erected 830,000 square metres of living space in
Moscow and Leningrad, which may be compared with

the 900,000 square metres built during the 10th five year plan in the countryside. Since the co-operatives are essentially an urban phenomenon, the figures are higher if taken as a percentage of building in towns: 1961-1965: 4.6 per cent; 1966-1970; 10.0 per cent; 1971-1975: 8.6 per cent; 1976-1980: 7.2 per cent; 1981-1985: 8.5 per cent. Perhaps one of the most significant features about this set of statistics is that in absolute terms the co-operatives in 1981-1985 failed to reach the level attained in 1966-1970 (32.8 million square metres against 33.6 million) and that proportionately far less was erected in this sector.

Housing policy makers are almost certainly persuaded of the budgetary advantages of encouraging this quasi-private sector, citing with approbation the key role played by this tenure in the majority of European socialist states, such as Czechoslovakia, the GDR and Poland where in recent years the contribution of co-operatives has been much higher. Its attractiveness to them and the loans at extremely low interest rates which they offer to seduce home-seekers, bear resemblance to the British Government's policy of privatising council housing. In the latter's case, in exchange for a 'gift' (to the tenant of an undervalued asset), the Government relieves itself of the burden of maintaining the property. Although there is no direct parallel in the USSR with the British fiscal practice of mortgage interest tax relief - which imposes an enormous burden on the central Government - it might be that the current rate of interest paid by co-operative members (0.5 per cent per annum) also represents a heavy government subsidy which can be measured by comparing it with other interest rates. The co-operative housing form may well carry a similar social and political appeal as does home ownership in the United Kingdom. Some of the consequences may also be similar.

There are possibly five main lines of investigation to find reasons for the 'underperformance' of the co-operative sector. First of all, it might be that the government really intends gradually to run down this type of tenure. But this case is difficult to sustain in the light of the new Party Programme and various central Government and Party decrees and resolutions issued over the last 25 years advocating the promotion of housing co-operatives as a means of solving the housing problem.

Another explanation might be that they do not enjoy popular support among broad sections of the population. Yet, this too fails to persuade: At the beginning of 1980, 1.1 million people were on the waiting list to join co-operatives (Grafova 1980). By the beginning of 1986 the figure had reached 1.3 million (Smusenok 1985) and the chairman of a Moscow co-operative estimated that the actual number of people interested in joining a ZhSK to be 10-20 times greater than this (Vashets 1986). Third, the legal, organisational and economic aspects of this tenure contain a number of contradictions. For example, the Great Soviet Encyclopedia defines the house-building co-operative as "that form of co-operative in which the members combine together to jointly construct and run housing at their own expense" (Bol'shaya Sovetskaya Entsiklopediya 1972: 637). This view is not congruent with reality. Today, it is not the member-shareholders who design, supply the building materials and erect the buildings but state organisations. The latter in the majority of cases also take upon themselves responsibility for the buildings' running and upkeep. In other words, the modern house building co-operative is a "voluntary association of citizens created with the purpose of combining their resources for the joint acquisition of accommodation and future use of communal services at the expense of their own monetary resources" (Matveets 1982; Batishchev 1982). This means that typically the ZhSK is a consumer rather than building co-operative.

Most of the difficulties plaguing housing co-operatives thus appear to derive from a contradiction between the real economics of, and the legislative acts governing, the construction and running of co-operatives. The latter are, for example, relieved of having to meet a number of expenditures. They do not pay the cost of survey work, site clearance and preparation - and in Novosibirsk, 60 per cent of all co-operative building is on already developed land. Neither do they bear the cost of providing social amenities and shops and other daily services which amount to 6 percent of housing construction estimates. The setting aside of 10 per cent of all newly constructed housing for the construction workers themselves and 6 per cent for residents displaced when their houses are demolished to permit new building is not borne by the co-operatives but are included in the state plan

(Bessonova 1988: 45).

For these reasons, Bessonova maintains, building workers oppose the inclusion of building for co-operatives in their construction plans. The building departments attached to the local soviets are themselves reluctant to increase the number of co-operatives in so far as not a single decree makes clear where the resources for site clearance, development and landscaping are to be found. This hostility together with opposition from industrialists, who are supposed to be the champions of the state's policy to "expand", "assist" and "encourage" the house-building co-operative, ensures that they remain of little consequence in the overall volume of housing construction.

Fourth, the system of planning and financing still constantly reproduces all these problems. Each year Stroibank plans for a sum of money to be set aside for providing long-term credit for co-operative house building. These sums are determined by the claims presented by the local soviet. The bank's institutions analyse these requests and establish whether they correspond to the draft economic plan. In order to determine the level of expected use of the loans, they study the use of credit in the past and the actual plan fulfilment for co-operative housing construction in the previous year. Eventually Stroibank finalises the credit plan for the country as a whole. In the course of the year, the Council of Ministers of any republic (its highest executive body) can allow additional housing co-operatives to be built - but only within the credit limits allocated to co-operatives and building for owner occupancy. Furthermore, additional construction may only be carried out if the building industry has the necessary capacity to do so. Thus, the co-operative's underprivileged status is reproduced, first, by the fact that credits are determined by the national economic plan and the co-operative's plan fulfilment of the previous year and, second, by the existing capacity of the construction industry (Bessonova 1988: 46).

Lastly, the explanation for underperformance is at least in part to be found in the recalcitrance of the local soviets who have not always favoured this form of tenure on the grounds that it leads to queue jumping. Annual surveys conducted by Gosstroi indicate that the local soviets are responsible for, among other things, delays in providing the requisite design and estimate documentation and for failing to observe norms stipulating construction times (Pudikov 1980: 63). For instance, against an average construction time of 9-12 months, the buildings frequently take five or more years to complete (Buzhkevich 1985). In one Leningrad co-operative, the chairman of the board of governors complained that they were being treated like 'private traders' (i.e. they were being abused) and that although a report had shown that "it was time to repair the roof, cost estimates had been prepared and the money was available, the plan to carry out the repairs had not even been included in the five year plan" (Arkhangel'skii 1986).

In this sense, the local soviets have acted like some Labour Party-controlled local authorities in the UK over the issue of council house sales between 1972 and 1980, before the Government granted council tenants a statutory right to buy. Only an examination of the minutes of departments under the executive committee of the local soviets would reveal the extent to which social justice was the principal motivating factor behind their obstructive behaviour towards co-operatives. To continue the parallel, one mooted suggestion is that the co-operative sector could be increased by extending an experiment introduced in some cities of selling housing erected by the state in new neighbourhoods to co-operatives (Dmitriev 1987: 75).

There is quite a high probability that the new law on co-operatives in general published in June 1988 and the role to be assigned to local soviets in their development will see a change in policy towards all forms of co-operative, including those in the field of housing. Imaginative local soviets will be able to mobilise the human factor to augment the supply of goods and services within their administrative jurisdictions. Furthermore, if the present change in attitude towards finance, credit and taxation is reflected in changes in the revenue sources of local budgets - for example, the charging of differential rents for properties, depending on their location, a wholly plausible policy development - then the local soviets could gain a double benefit from fostering them. On the one hand, their electorates would benefit from the wider range of goods and services made available and, on the other, the local soviets' revenue could increase from rent, local income tax and forms of corporation/profit tax. That this might become important is indicated by a decree of 14 March 1988 which stipulated that the income tax from co-operatives should be included in the budget of the city or district soviet within whose jurisdiction the co-operative is registered (Postanovlenie 1988: no. 15, art. 41).

It appears that a radical reorganisation of the whole system of housing co-operatives is imminent. One proposal considers it is necessary that four different forms of co-operative be distinguished. Construction co-operatives would erect housing using the labour and resources of its members. These would be one- and two-storey units, in so far as multi-storey, large-panel blocks require highly qualified labour and industrialised building methods. They would receive 80 per cent loans from the state budget. Repair co-operatives are distinguishable from the former only in terms of the nature of the dwelling: they would concentrate on the reconstruction and modernisation of old buildings in large cities for which, as indicated above, there is usually a shortage of labour and resources. Loans for such co-operatives are restricted to a maximum of 50 per cent of the costs involved. The third type, young people's co-operatives, catering for the under-35s, includes housing in both the repair and new construction categories. Provided with 100 per cent loans from enterprise social funds, they would coexist along-side the earlier mentioned, Youth Housing Complexes, and make use of their members' own labour. Fourth, the leasing co-operative - which bears some resemblance to its 1920s forerunner - would cater for those over 35 years old and in higher income categories. The state erects the dwellings which local soviets then rent out to co-operatives. The latter would be formed after the building has been completed. The local soviet would decide what proportion of the building would be leased out to a co-operative. The principal objective of this type of organisation is that it should meet the demand for accommodation by higher income groups who could afford a high initial deposit and a minimal repayment period so that construction costs are quickly recouped (Bessonova 1988: 48).

Housing Policy in The Soviet Union

Some of the complaints, recommendations and prognostications made by commentators on housing co-operatives have been incorporated into legislation enacted in March 1988, "On Measures to Accelerate the Development of Housing Co-operatives" (Postanovlenie 1988: no. 16, art. 43). Indeed, it was perhaps to be expected that the increased benefits accruing to the owner-occupied sector would soon be accompanied by reforms to regulations governing co-operatives. The new law emphasises the role to be played by this tenure: not only will it increase the supply of accommodation but it should encourage labour stability, greater attentiveness towards property and a better balance between the public's growing income and expenditure. It should also present people with an incentive to increase their productivity and earnings.

The poor performance of this tenure was attributed by the above decree to "a serious underestimation on the part of virtually every major institution" - from Gosplan, Gosstroi, Republican Council of Ministers and ministries dealing specifically with construction to local party organisations and local soviets - of the "social and political significance of house building co-operatives". This situation had arisen because these institutions are basically oriented to the state providing funds for building. The low priority usually accorded to co-operatives and the "formal bureaucratic approach adopted towards them, means that the allocation of a building site and the furnishing of designs, specifications and cost estimates is a protracted process". The preamble of complaints then stated that the new provisions - which are to be elaborated in detail in the near future - are designed to make the housing co-operatives "one of the main directions for expanding housing construction ... so that by 1995 it will contribute no less than 2-3 times more than at present to the overall volume of housing construction". In other words, it is intended that this sector should account for 14-20 per cent of all new building. Past precedent suggests that this figure will probably be much higher in large cities and certain regions.

However, the manner in which this expansion is to take place includes a novel element, not unlike one of Bessonova's recommendations. The law defines two types of co-operative. The first

consists of house-building co-operatives created
for the purpose of constructing and running
dwellings by and for the benefit of the member.
These are also allowed to acquire (purchase)
buildings in need of major capital repairs,
renovate them and then occupy them. The second
type of housing co-operative is not concerned with
building at all: it is set up in order to acquire
(buy) existing housing, either newly erected or
renovated, from enterprises, organisations and the
local soviets. Enterprises are free to determine
(in consultation with work collectives) the amount
of co-operative housing to be built and how many
dwellings erected using their own and the state's
resources it should offer for sale. Local soviets
too, when formulating their house-building plans,
define the size of the increase in the number of
both types of co-operative. The procedures for
planning and financing the construction and sale
of housing to co-operatives was to be defined by a
group of ministries and central state committees
by the end of 1988. When purchasing a property
from an enterprise, organisation or the local
soviet, the co-operative must pay not less than
20-25 percent of the value of the property (the
lower price being payable in the Far North, Far
East, Siberia, Kazakhstan, mining settlements and
rural areas). The balance is repayable in equal
monthly instalments over 25 years.

The cost of construction is based on prices
set for state housing. Then, since local soviets
have hitherto tended to act obstructively in
regard to co-operatives, the decree refers to the
need to provide them with an incentive to assist
co-operative formation. This it does by allowing
the soviets to add a supplement of up to 15
percent to the total cost of construction in the
case of the co-operatives being located "in the
most favourable neighbourhoods and places". This
supplementary charge is to be paid into the local
budget. This policy innovation has a twofold
significance: First, this is tantamount to a
fiscal device to strengthen the financial base of
local soviets and thus to enhance their autonomy.
Second, the payment represents a differential rent
and thus is an important step towards making
rental charges for accommodation (and for non-
residential premises) reflect location - a topic
discussed below.

Housing Policy in The Soviet Union

In order to stimulate the construction of housing and subsequently its sale to housing co-operatives, specialist banks are now able to grant credits to enterprises and local soviets up to a maximum of 50 per cent of the estimated cost of construction with repayments due within a six year period after receipt of the loan. At the same time, so that co-operative members are in a position to acquire these properties, they can be advanced loans of up to 5,000 roubles repayable over ten years at an annual interest rate of 2 or 3 per cent (the differential depending on whether the property is located in a rural or urban area respectively). Invalids, war veterans and individuals of equivalent status do not have to pay any interest. People who use the money borrowed for purposes other than house purchase will be surcharged at 12 per cent on the total loan. This may be regarded as one indicator of the intrusion and growing use of credit, finance and interest to influence economic behaviour and regulate economic and social transactions. It is in this fashion that money and finance will gradually come to assume a significance hitherto absent in Soviet society.

Enterprises are recommended to provide those of their employees wishing to join either form of co-operative with financial assistance to repay the cost of the property by drawing upon their social development funds. This help - to be regarded as "an effective means of raising the level of labour activity and securing personnel" - is to be granted freely and related to length of service, ranging from up to 10 per cent for those with more than 5 years' service to 50 per cent for those in the enterprise's employment for over 15 years. These sums may be higher in deserving cases recognised by the labour collective. This facility is also available to local soviets. It is payable to employees in the educational and medical fields and other occupational spheres in the so-called 'non-productive sector' from the local budget and other unspecified financial sources.

Houses sold to co-operatives will, as a rule, be those erected according to standard designs. However, in order to more fully meet the demands of co-operatives, local architects' offices can make changes to the standard designs and also draw up individual plans. If the co-operative member-ship has the necessary financial resources, they can request the construction of additional buildings to meet household needs, including garages and sporting facilities. The flexibility to vary exterior forms and structures is accompanied by other provisions to extend the practice of allowing co-operative members - again at their own expense - to commission building and interior design co-operatives to complete the interior finishing to a higher standard.

In the absence of details on the types of custom-designed housing permissible, cost estimates and credit conditions, it is difficult to assess the importance of these developments. There have already been experiments in the field of individualised housing plans and in allowing people to pay for more costly decoration (for example, parquet floors). But experiments are frequently sanctioned by high-level personnel and therefore do not have to contend with the problem of shortages of resources and obstructionist behaviour on the part of local officials and management. (Experiments succeed for the same reason as did the norm-breaking coalminer, Stakhanov, in 1935.) Nonetheless, bearing in mind the earlier mentioned decree on Soviet architecture and town planning (Postanovlenie 1987: no. 45, art. 149), it does seem that the mood of the Government, the architectural profession and certainly of the population is that the urban environment urgently requires a much more variegated physiognomy.

It cannot be certain that the initiative for change will originate in the private and quasi-private (co-operatives) sectors. However, the fact that residents are being given the opportunity to pay for their own architect's drawings and for the extra costs of more expensive designs and finishes does suggest that they will avail themselves of the chance of living in distinctive dwellings and environments. On the other hand, it would be premature to see the outcome of recent legislation as leading to a polarisation of the 'public' and 'private' sectors. If rents in social housing (understood here to refer to local soviet (municipal) and 'departmental' housing) remain, even after a likely upward revision, substantially below the monthly outlays required of those repaying loans, then the possibility has at least to be countenanced that tenants in the former will use their extra disposable incomes to engage architects and interior designers to ameliorate the uniformity of so many large housing estates. The main point that has to be stressed is that a confluence of pressures, opinions and circumstances (without a royal impetus) will yield in future decades a visually more attractive and differentiated urban landscape.

Finally, the creation of co-operatives to purchase both older and newly constructed properties from the state sector - a policy qualitatively different from that which existed from 1924 to 1937 when local soviets leased their housing to co-operatives - represents a preliminary step towards the privatisation of the housing stock. When all the seemingly disparate strands of current government policy towards housing are examined together, it is not too preposterous to hazard a prediction: When the state housing stock is finally concentrated in the hands of the local soviets, estates and neighbourhoods (mikroraiony) will be transferred on a selective basis to the second type of co-operative - which may bear a resemblance to the newly formed housing action trusts in the United Kingdom. For this to occur on a wide scale will require, however, major conceptual, ideological and institutional changes. It would also herald a major advance for Mr. Gorbachev's perestroika. Evidence for this change is now to be found in proposals made by the USSR Council of Ministers and accepted by the Politburo at the end of 1988. Local soviets and enterprises are permitted by the decree to transfer their flats into private ownership. The decree which comes into force in January 1989 allows sitting tenants to buy their flats and for other citizens to bid for unoccupied apartments in blocks of flats subject to re-construction and major capital repairs (Rozanov 1988; Batalin 1988).

In a question and answer session between a newspaper correspondent and the first deputy chairman of Gosstroi, the latter referred to earlier precedents. For instance, in the post-war period (1956-1960), the private sector contributed 53 per cent of all new housing (25 per cent in towns and 82 per cent in the countryside). By 1986 the respective percentage figures were 20, 9 and 50. Second, this decision to offer state-financed and socially-owned flats for sale was 'defended' as a logical extension of other policies introduced in 1988 and had been mooted by delegates to the XIX Party Conference in June 1988. Third, this move is only extending to flat dwellers the 'rights' and 'advantages' already enjoyed by private house owners.

Priority for purchasing unoccupied accommodation space will be given to people already on waiting lists for rehousing, in order to ensure that "social justice is not infringed by those with large monetary holdings". Freshly commissioned flats will not be for sale. However, naturally, families who are allocated flats will, after a period of (as yet unspecified) time, be eligible to elect to purchase their flats as sitting tenants. All prospective buyers must apply to the owner of the property (the local soviet, enterprise or other public body) who has to value the flat. The purchaser then has to deposit not less than 50 per cent of the total sum, with the remainder being repayable in equal instalments over ten years. In the case of both large and poorer families the repayment period may be extended to 15 years and the deposit reduced to 30 per cent of the value. Although the value of the flat will vary depending on age, general condition, number of storeys and location, given an average cost of 200-250 roubles per square metre, a one-roomed flat with 25 square metres of overall living space will cost the purchaser about 5,000 roubles.

At present the local soviets have been given considerable discretion in deciding on their sales' policy and setting the rules for service and maintenance charges for individuals who buy their flats. The guidelines recommend that the soviets should actively assist in the organisation of private flat-owners' associations in order to facilitate the running of the accommodation on a "collective self-government or mutual aid basis". Since this decree introduces the new concept of a "flat as personal property" an amendment will now have to be made to the Principles of Housing Law.

The Government's decision to remove from both local government and other state-controlled enterprises and agencies some of the responsibility for providing accommodation may be viewed in different ways.

First, it furthers the process of taking away the function of shelter-provider from bodies for whom the ownership of housing is functionally anachronistic. Arguably, self-financing will be made easier if organisations specialise in those activities that they were originally established to perform and are unencumbered by a whole range of loss-making operations.

Second, the legislators remain consistent in believing that the transfer of accommodation into private ownership will "immediately improve the maintenance of the housing stock generally" because individual owners will take care of their property. Moreover, "it will no longer be necessary to expend state funds on repairing and servicing these flats".

Third, if the flats were built using centralised capital investment, the money received from the sale will be paid into the local budget for use in expanding social amenities. In the case of accommodation sold by enterprises (as part of the vedomstvenny stock), the proceeds of the sale remain in the hands of the seller for new construction, repair and maintenance to the existing stock and for expanding the range and quality of social provision.

The wider logic of perestroika has propelled policy makers further along the path of 'privatisation'. It should yield benefits, but also carry some costs. The latter will be exacerbated by the haste of this innovation's introduction, which perhaps even the general urgency of the situation does not warrant.

RENTAL PAYMENTS

After 1925 it was accepted policy that the level of rent charged should cover upkeep. Yet, in his report to Gosplan in March 1926 the celebrated economist, Strumilin, referred to the fact that

> A very modest increase in rents has taken place, tied to wage increases ... . But this supplement is quite inadequate for housing construction to become profitable. Without profitability every incentive to this construction will cease, particularly if it is a matter of attracting private capital. (Strumlin 1926: 47,50)

The publication of the decree "On Rent and Measures to Regulate the use of Housing in Urban Settlements" (Postanovlenie 1926: no. 44, art. 312), later supplemented by decrees of January 1928, "On Housing Policy" (Resheniya 1967: 696-703), and of May 1928, "On payment for Residential Premises in Cities and Workers' Settlements" (Postanovlenie 1928, no. 53, art. 402) established rental rates based on a fixed tariff which varied from 3 to 4.4 kopeks for each square metre of actual dwelling space, with the highest charge occurring in cities with more than 40,000 inhabitants. Small adjustments were made to the charge depending on the distance from the centre of the town and whether or not the accommodation was supplied with a full range of amenities, was damp, below pavement level and so on. The final rate per square metre, the apartment tax, was thus determined solely by the quality and location of the property. A further small supplementary charge was added (or deducted) depending on the wage of the highest paid member of the family, thereby including in the rent a 'personal' charge reflecting the social and economic status of the family. Despite these supplements, the total rental charge could not by law exceed 13.2 kopeks per square metre of dwelling space per month. The Government took a norm of nine square metres per person as the basis for computing rent so that, with the exception of certain specified privileged categories of individuals, space in excess of 4.55 square metres of the (nine square metres) norm per family was paid for at three times the standard rate (Rotkov and Grif 1968: 42-53).

It was anticipated that this level of charge would cover operating and amortisation costs. In other words, housing should be self-financing. In 1937 rental payments accounted for three-quarters of the housing economy's income - the remainder coming from, for example, rent paid for non-residential premises within an apartment block. The standard charges were not, however, sufficient to cover the running and maintenance costs. In order to offset this deficit, the legislators sought to transfer the burden onto so-called "non-working elements" - accounting for 5.1 per cent and 1 per cent of the urban population in 1926 and 1931 respectively (Sosnovy 1954: 153). Needless to say, the differential charge did not cover the deficit. Most important of all, the average rent paid by 85 per cent of the tenants met just 40 per cent of the operating costs and amortisation charges. Although the January 1928 decree was partially successful in reducing the discrepancy between revenue and costs, the onset of the five-year plans after 1928 saw the erosion of any gains made and the gap began to widen again (Walker 1986).

More than sixty years have now passed since legislation defining rental charges was introduced. During this period the rouble has been devalued by a factor of ten (500 pre-1961 roubles became 50 roubles), incomes have risen substantially as have living standards generally. The quality of a modern self-contained flat is altogether different from that of a 1930s (or 1950s) designed communal flat.

Despite these changes and the requirements and demands of today's public, the legislation enacted in the 1920s remains in force. As a result, the rent charged at present lacks any real economic meaning, with virtually all rental payments having reached the legal maximum. This has given rise to situations such as that revealed in a survey of the city of Novosibirsk where, in 1982, although the average per capita living space in local soviet owned accommodation stood at nine square metres, the range varied from two square metres to 43 square metres per person. According to the author of the article reproducing these data, such variations would have been evened out by the economic mechanism of charging treble the rate for 'surplus' space. In the 1930s and 1950s this would have amounted to five roubles per square metre; today, as a result of devaluation, it is 50 kopeks (or 50 pence at the official rate of exchange). The recommendation is that payments for this surplus space have to be set at a level which will either equalise consumption of living space or equalise disposable income levels. The new tariffs should also ensure that the aggregate revenue from these charges cover overall housing costs (Bessonova 1985: 56).

The research conducted by Bessonova set out to devise an index to measure housing quality in order to: (i) demonstrate the injustice of rental charges being determined without regard to the quality of the accommodation; and (ii) construct an index for evaluating housing conditions when letting accommodation.

Using data from the housing exchange office in Novosibirsk it was found that the "exchange equivalent" for a flat is not its size but quality as measured by, for instance, the district in which it is located and the number of storeys. A matrix of coefficients - based on coefficients of preference for a particular district (the level of social amenity provision), distance from the city centre and quality of the accommodation - was constructed for the transfer of space in flats of the same standard of comfort from one district to another. The coefficient showed how much living space extra a person in one district must offer in order to secure a flat elsewhere of comparable standard.

Classes of accommodation were derived from this index of location and an index of "comfort" or "convenience" (based on the number of rooms in the flat - for instance, two small rooms were accorded a higher value than one large room of the same overall size - building materials and number of storeys). This showed that the "exchange equivalent" for a flat of 35 square metres within a particular "housing class" varied between 20 and 35 square metres depending on location and comfort (Bessonova 1985: 58-61).

The analysis and recommendations presented in this study represent an important first step in the direction of a radical revision of the formulae employed in the calculation of rents. The 1989 Population Census will no doubt provide the data necessary for sophisticated computer modelling as a basis for such revision. If housing is indeed the cornerstone of perestroika, as has been suggested, then it might be anticipated that rent reform will be introduced in the near future. The tempo of change will depend in part on the degree of success (or failure) of the already discussed legislation on the co-operative and owner-occupied sectors.

Improvements in living conditions, as a consequence of higher building standards, modernisation and urban renewal, will lead to both the type and location of housing being included in the criteria for assessing housing quality. A classification of these evaluations and categorisations of an increasingly differentiated housing stock could serve as a basis for charging differential rents. A housing index derived from such a classification would take into account, inter alia, the location of accommodation vis-à-

vis the place of work, parks (and recreational
facilities generally), the availability of public
transport, garaging and parking plots and the
general environment (including the level of air
and noise pollution). Calculating different rents
for different properties would seem to be a
daunting task, for this would have to take into
consideration the evaluation made both by
officials and by existing and prospective tenants.
Any attempt to define administratively the
criteria for 'valuing' a differentiated housing
stock in order, perhaps, to devise a system of
'fair rents' would prove extremely difficult.
Nonetheless, the methodological problems
notwithstanding, pressures for change are
mounting.

A view gaining widespread support advocated
that, when all families have been provided with
their own self-contained flats, any further
improvements in living conditions will be rent-
free up to a set standard, but beyond the new
universal norm tenants would be obliged to pay
(Rimashevskaya 1975). A decade later the argument
had advanced slightly. The recommendation put
forward was that everyone should be provided with
guaranteed rent-free accommodation, which would
have to be differentiated by region to take into
account the standard already achieved and climatic
factors. (This is necessitated by the fact that
even in such a 'high profile city' as Novosibirsk
in Western Siberia residents have 22 per cent less
space on average than people living in the
European part of Russia, the Ukraine, Belorussia
and the Baltic Republics.) The disparity is far
greater if these other republics are compared with
most settlements in the rest of Siberia. A
differentiated norm would therefore assist in
equalising conditions between regions and also
serve as a means of stimulating population
migration to areas where labour is required
(Zaslavskaya 1986: 65, Yanowitch 1989). After the
Constitutional right to accommodation has been
met, then housing "becomes an ordinary economic
good, the level of demand for which has to be
treated as reflecting consumer choice". This
'additional' consumption has to be paid for "on an
economic basis, covering both the capital cost of
construction and current outlays on maintenance"
(Grebennikov 1987: 197).

Central to the arguments of most housing reformers is that the rates at which rents are charged do not sufficiently reflect the quality of the accommodation. This gives rise to a contradictory situation where the policy of "equal pay for equal work" co-exists with the principle "equal payments for unequal housing conditions". This anomaly is further aggravated by the fact that only the dwelling area (zhilaya ploshshad'), not the overall living space, is subject to payment. Thus, with the increases in the amount of auxiliary space the inequalities inherent in accommodation charges become still greater since families with the better accommodation receive a higher rent subsidy. Furthermore, research conducted in 1981 revealed that 44.3 per cent of families living in cities were inhabiting accommodation whose dwelling area exceeded the established norm, while at the same time a substantial proportion of all families were still housed below the norm. There is no automatic mechanism to rectify this situation for, although families with 'surplus' space have to pay treble the normal rate, rents are so low that they have no economic incentive to surrender all or part of the surplus. This may be less true for pensioners on low incomes who are in some cases under occupying flats.

The solution would appear to be that rather than relate rents to wages or family income (for to do so would contradict the socialist maxim of "from each according to his ability"), they should reflect the quality and location of the accommodation. Such a policy would make rental payments more equitable, lead to a certain amount of housing and subsidy re-distribution and would also help to reduce the state housing sector's large deficit.

The recommendation mentioned earlier that the state should allocate resources for new housing construction according to a standard norm of so many roubles per person has a number of merits. One of them is its compatibility with the state's welfare culture which derives from a deeper, collectivistic cultural ethos. (In this sense the Soviet state ideology is largely in harmony with underlying Russian societal values.) As such it can undoubtedly lay claim to the support of large sections of the population. From the point of view of 'reformers' it is tainted for attempting to reform the system of housing distribution by 'tightening up' existing institutional structures and practices and adding more administrative functions to the bureaucratic apparatus. Indeed, the charge being levelled by the 'human factor' faction (to which Mr Gorbachev belongs) is that hitherto, when 'shortcomings' or corruption in some sector of the economy or society have reached peaks of unacceptability, the remedy has been to appoint more officials and print more forms, thereby adding to the bureaucratic edifice. It also contradicts a policy element integral to the spirit of perestroika, that: "Paid services must be expanded".

For the reform process to succeed a 'dual track' approach is required: On the one hand, the population must undergo 'psychological reconstruction' which means, among other things, that if people want change then they must perceive that they possess the capacities to effect those changes. On the other hand, the institutional structures must also be subjected to re-evaluation and restructuring. The courage and conviction necessary for radical alterations in norms and mores to occur will be heavily influenced by the extent to which the Government demonstrates to the public's satisfaction that it is genuine in its pursuit of social justice.

The press and television constantly carry articles and documentary portrayals of the manner in which social injustice manifests itself, particularly in the case of housing. The following may be taken as typical examples. A short note on the front page of Pravda of 29 February 1988 referred to "violations of social justice in the distribution of housing space in the Krasnoyarsk district of the Kuibyshev oblast". Following a letter to the newspaper from employees in the local housing and municipal services department accusing the chairman of the local soviet (raiispolkom) of having personally distributed flats, an investigation was held into the affair. This found the chairman guilty of breaking the law and of adopting an "unprincipled attitude towards the distribution of housing and issued a severe reprimand". Further investigation revealed additional "personal indiscretions" and he was relieved of his post.

A seminar held in Sverdlovsk in 1986 for senior Party and local government officials to discuss the decree, "On Measures for Further Strengthening the Responsibility of Soviets of

Local People's Deputies" heard one report on an incident of squatting by a cleaner (a low-paid worker) and her family. The speaker pointed out that this would not have happened "if the heads in charge of soviet departments and the Party had acted on the principle of social justice". As a consequence, the "Party, local soviet and the agents of law and order had to deal with the problem of what to do with the family since, manifestly, the law had been broken". This led to the observation that "in order to put a stop to all manner of rumours, injustices and abuses, it is absolutely necessary to devise a system of greater openness in the distribution of living space".

These two examples provide evidence, documented many times over, of abuse of office, insensitive bureaucratic neglect and, possibly, hidden homelessness (in the form of a single-parent household forced to resort to squatting).

One of the principal protagonists of the whole spirit of <u>perestroika</u> is Tatyana Zaslavaskaya. Her ideas are clearly expounded in the above cited article in the theoretical journal of the Communist Party, <u>Kommunist</u>. She notes that some commodities and services (for example, food products, clothes, furniture) are sold for money, while others (housing, education and health) are provided free of charge or at "extremely advantageous prices". This practice suffers from a number of weaknesses, principally in that it "artificially limits the assortment of goods which people want to acquire with the money that they have earned" and this diminishes their incentive to work efficiently. It is difficult to explain to a well-qualified and well-paid worker why he can buy a suite of furniture, refrigerator and television set, but yet has to wait years for a flat where all these acquisitions can be put. The proposition that the way to resolve this dilemma is to create more house-building co-operatives only "muddles the question still further for it leaves undecided the criteria to be used for determining when goods should and should not be paid for" (Zaslavskaya 1986: 72).

According to Zaslavskaya, the free distribution of those goods and services which are in most short supply - and housing falls into this category - cannot but exacerbate the problem. The distribution of free and semi-free goods becomes the focus for specious and spurious deals and transactions of dubious legality: "speculation occurs over state housing, surplus living space is rented out at high prices and living space is exchanged for large supplementary monetary payments and so on" (Zaslavskaya 1986: 73). Elsewhere, sociological research has demonstrated that the distribution of free goods from the 'social consumption fund' actually benefits the better-paid groups. Taken together these and other "facts seem to us to point to the necessity of broadening the sphere of paid services, which includes raising and justifiably differentiating rents to take into account their quality and location".

CONCLUSION

The decision to maintain rents at the 1928 level might be seen as a politically benign gesture. It offered some compensation for the hardships endured by the population during the industrialisation campaign of the 1930s when housing conditions deteriorated because of both the pressure of vast migrations of people into the cities, and the policy decision to concentrate investment on providing the industrial infrastructure at the expense of housing, public utilities and social amenities. It remained a moral imperative in view of the suffering and destruction of the meagre housing stock during the Second World War. A blend of political expediency, ideological consistency and popular expectations all combined to ensure that the notion of low rents remained hermetic, sacrosanct, a symbol of the building of socialism and an important source of legitimation.

The persisting housing shortage in the Soviet Union is acknowledged by the government and is a perpetual source of complaint by the population. Over the past quarter of a century huge resources have been invested in the housing sector and as a consequence living conditions have improved. Over-crowding has not been eradicated but it is less severe. Quantity is still an issue, but quality ever more so: contemporary satires on various aspects of housing and domestic life abound but they are qualitatively different from those of the 1920s.

As the stock has grown so has the cost of providing and maintaining it. This has provoked the government to adopt a number of measures to reduce the disparity between revenue and expenditure. Three of the five different strategies examined here for reducing expenditure which operate on the 'supply side' - modernisation in order to use the built environment more efficiently, the application of science and technology to reduce manufacturing and on-site assembly costs, and improvements in organisational structures and methods - though undoubtedly important, are likely to bring relatively limited economies.

Any major assault on the deficit will eventually have to come from the 'demand side'. There is no shortage of evidence that "making use of the population's own resources" to provide accommodation is a popular option in policy-making circles. The house-building co-operative, although officially favoured, suffers from the fact that it has tended mainly to cater for "well-provided citizens, specialists and highly qualified workers". While hitherto ideologically more acceptable than the individual owner-occupier, there have almost certainly been strong grass-root pressures for an expansion of the latter. Official data representing government prognostications on housing output indicate that these two sectors are anticipated to account for over 40 per cent of housing construction in 1991-1995, which is double the equivalent percentage figure for 1986. The Chairman of the State Committee for Construction put the figure as high as 66 per cent by 1995 (Trehub 1988). This huge expansion will be linked, to an unspecified (and probably uncalculated) degree, with the workplace.

Those in favour of extending the role of the workplace as provider of accommodation argue - not entirely accurately - that whereas previously social planning was on a nation-wide scale, today "planning of social development takes place at the enterprise and regional level". Enterprises allegedly have greater knowledge of the actual needs of the workforce - as a collective and of its individual members - than central authorities and, therefore, are better placed to decide how resources should be allocated. The decentralisation of decision-making implied in this view has its parallel in the trend towards granting local soviets more responsibility (and financial resources) to respond to the demands of their constituents. These parallel trends also represent the underlying tension between the 'productive' and 'territorial' spheres of organisation (i.e. between enterprise managements and local soviets) over the control of resources for housing and other objectives of collective consumption. The former treat accommodation, for example, as an element in their reward structure, whereas the latter can use 'need' criteria as the basis for distributing housing which it controls. As far as enterprises are concerned, since the public ownership of the means of production does not automatically make labour for the good of society an individual's primary life need, people must be motivated to work in a specific place and at specific tasks. In this context Lenin's observation that "attracting people to work remains the most important and difficult problem for socialism" has lost none of its pertinence (Lenin 1962: 285).

As a central component in the reward structure housing has long been used by enterprises and organisations as a means of attracting, motivating and retaining workers. A number of enterprises and organisations in "the most important sectors of the national economy" are permitted under Article 62 of the Principles of Civil Law of the USSR and Union Republics to evict from 'company homes' workers who have terminated their employment contracts. In the long run labour may be 'bonded' to an enterprise more effectively through the offer of loans to join house-building co-operatives or to erect their own dwellings than through 'tied accommodation'. The advantage of such a policy is that it would help to overcome one of the contradictory consequences of housing policy: between, for instance, the reproduction and stabilisation of the labour force by offering cheap, but 'tied' accommodation on the one hand, and the objective of reducing the subsidy to housing by transferring a larger proportion of costs onto 'owner-occupiers' on the other.

Numerous ideas have been in circulation over the past decade recommending solutions to the financing and distribution of housing, with the question of rent being one of the foremost among them. East European countries have been cited and commended for the improvements which they have introduced in the financing of house building and for the setting of rents in accordance with the standards of comfort provided and tenants' income. Most prominent politicians, economists and other commentators still consider that as far as the foreseeable future is concerned, housing needs must continue to be met principally through the state sector, albeit of reduced size. Others regarded the house-building co-operative as possibly becoming the main channel for allocating accommodation, with some even proposing that state housing could be sold to co-operatives (Rutgaizer 1981). Yet others made the goal of "eliminating differences between town and country" into the premise for their argument that the responsibility for financing housing in towns should be transferred on to the population who could use their own resources. This would be the "most rational way of removing the existing social difference", for after all, 71 per cent of the housing stock in the countryside is in the hands of owner-occupiers - a figure which ranges from 83 to 96 per cent in nine republics (Goskomstat 1987: 520-521). The implementation of this proposal would, it was averred, entail the "transfer of housing belonging to state and co-operative organisations into the property of the population" (Sarkisyan 1983: 186). All the above mentioned suggestions voiced at the very beginning of the decade combined to form the theoretical basis of policies enacted in the dramatically short period since 1986.

One author writing in the "reform period" and forcefully advocating the right of individuals to use their "justly earned income" to improve their living conditions, prescribed that savings should be paid into the state budget and then used to advance the depositor in the queue for a self-contained flat or one with a better layout or more living space (Alekseev 1987: 196). If an individual's savings are 'requisitioned' in this way, then they should be guaranteed specific privileges when offered accommodation. But, according to one variant, this privilege should not be confined to individuals living in overcrowded, poor and unpleasant conditions, but also to people inhabiting 'average' or 'normal' accommodation, but who, because they have saved the money, want to live in greater comfort and enjoy more living space without having to endure a long wait.

Housing Policy in The Soviet Union

There seems to be evidence of a Rawlsian pragmatism in current discussions: if it were true that a policy relying on the use of individual resources for house building would encourage an increase in the scale of housing construction, then such a move would be justified. The case against this alternative rests on the fact that in a socially stratified society those with lower incomes would be disadvantaged: "Research has shown that families in need of housing are usually those who have relatively low incomes and who, therefore, lack the requisite resources to join a housing co-operative or build their own house" (Sarkisyan 1983: 187; Shatalin 1982; Rutgaizer 1987). Legislation enacted over the past year has attempted to synthesise the 'thesis' and 'antithesis' by acknowledging the virtue of the thesis - the justification of requiring individuals to use their own resources - and the antithesis - that far from everyone is in a position to act in a 'Smilesean' way - by encouraging the owner-occupied and co-operative sectors through the application of a system of differentiated subsidies to different socio-economic groups and types of property. Nevertheless, as the general standard of living rises and the population's demands and expectations grow, the Government has made it quite clear that it envisages the co-operative and private sectors will make a greater contribution to house building.

The only way of reducing the size of the overall housing subsidy, however, is to raise revenue through higher rental charges. Anything beyond a minimum rent increase probably remains politically unpalatable and ideologically suspect: low rents comprise one of the Soviet Union's "essential myths" - and these are hard to dispel. Yet there are sound social as well as economic reasons for introducing differential rents: a crucial flaw in the housing system is that people are "paying equal rent for unequal accommodation". Indeed, an appeal for 'fair rents' on moral grounds may be more acceptable than a proposal for a more 'realistic rent' whose aim is to reduce the size of the Government subsidy. The introduction of differential rents would mark not just the end of an era of low rents but would signal the departure from 'universalism' in housing policy. A change in policy towards rents and other fixed charges for services would create a much more sophisticated, albeit still restricted, housing market in which consumers could reveal their preferences by paying the higher tariffs set by planners.

The debate is not yet over. Many writers who acknowledge the shortage of housing and the need for new initiatives adhere to the belief that the state should remain the principal provider. Moreover, a strong strain of egalitarianism runs through the society. The current leadership's aim would appear to be a Soviet version of a liberal meritocracy which is, for historical reasons, alien to a society whose polity for the past seventy years has subscribed to an ideology of positive discrimination for the underprivileged. However, it is not privilege as such that the leadership necessarily finds unpalatable, but rather its 'unacceptable face' - that is, officials "occupying cosy little niches" who benefit disproportionately from the society yet offer far less in return, and workers who, protected by labour legislation and tradition, are content to draw a wage from jobs to which they have a low commitment and on which they expend little energy. In this lies the kernel of the problem of perestroika and thus why housing is the keystone of the restructuration process.

The economic reforms, democratisation, glasnost' and the cause of social justice must provide each household with the incentive to strive to become an 'owner-occupier' or a tenant in a modern state flat of which the owner or tenant feels the 'proud' master or mistress. To invoke and arouse the 'human factor' is a monumental task akin to the Reformation in the primacy that it accords to the (autonomous) individual: The abolition of serfdom in 1861 did not herald the immediate demise of feudal structures. Paternalistic attitudes and relationships were not only maintained in the countryside but were transferred to the towns. The latter were most visible in the form of 'rent-free' accommodation provided by employers for their employees, representing an attempt by the emerging industrial capitalist class to adapt pre-capitalist relations to the emergent wage-labour economy. The close relationship that existed before 1917 between manufacturing industry and the ownership of dwellings continued after the October 1917 revolution when housing belonging to factory owners was nationalised at the same time as the manufacturing plant. Thus, after the revolution, one set of reciprocal social relationships within a fairly closely defined system was replaced by another. The whole array of economic reforms that have been introduced, which include allowing the creation of producer, consumer and banking co-operatives, and the steps taken towards laying the foundations for the formation of a 'property-owning democracy', potentially represent a highly significant rupture with that past - but it is a past to which the present is attached by many threads.

REFERENCES

Abramovich, A. (1987) Zhilishchnoe stroitel'stvo v sel'skom raione, Sovety narodnykh deputatov, 10

Aganbegyan, A. (1988a) The Challenge: Economics of Perestroika, London: Hutchinson

_____ (1988b) USSR May Open Zones to Western Countries, The Independent, 10 November

Alekseev, N. (1987) Ekonomicheskii eksperiment. Sotsial'nye aspekty, Moscow: 'Mysl'

Andrusz, G.D. (1984) Housing and Urban Development in the USSR, London: Macmillan

_____ (1986) Sowjetische Wohnungsbaupolitik: Das Ende einer Epoche, Osteuropa, Zeitschrift fur Gegenwartsfragen des Ostens, 4

_____ (1987) The Soviet Built Environment in Theory and Practice, International Journal of Urban and Regional Research, volume 11, number 4

Arkhangel'skii, V. (1986) Kooperativnyi dom, Sotsialistich-eskaya Industriya, 11 March

Artemov, V., and Khakhulina, L. (1984) A Typology of Urban Places in West Siberia, unpub. paper presented at the Meeting of Research Committee 21 on Sociology of Regional and Urban Development of the International Sociological Association, Bratsk-Irkutsk, September

Batalin, Y. (1988) Kvartiry na prodazhu, Pravda, 12 December

Batishchev, V. (1982) Ekonomicheskaya sushchnost' zhilishchnoi kooperatsii, Zhilishchnoie i kommunal'noe khozyaistcvo, no. 5

Bessonova, O. (1985) K voprosu o kvartirnoi plate v SSSR, Izvestiya Sibirskogo Otdeleniya Akademii Nauk SSR, seriya Ekonomika i prikladnaya sotsiologiya, no. 1

_____ (1988) Problemy zhilischchnoi kooperatsii, Izvestiya Sibirskogo Otdeleniya Akademii Naul SSSR, seriya Ekonomika i prikladnaya sotsiologiya, no. 1

Housing Policy in The Soviet Union

Blekh, E. (1972) Bolshaya Sovetskaya Entsiklopediya, 3-izd.,
    Moscow
_____ (1987) Povyshenie effektivnosti ekspluatatsii zhilykh
    zdanii, Moscow: Stroiizdat
Borovkov, I. (1984) Sotsial'no-ekonomicheskaya effektivnost'
    zhilishchnogo stroitel'stva, Moscow: Stroiizdat
Broner, D. (1980) Zhilishchnoe stroitel'stvo i
    demograficheskie protsessy, Moscow: 'Statistika'
Buzhkevich, M. (1978) Byulleten' Goskomtruda (Moscow), 8
_____ (1980) Constitution of the Union of Soviet Socialist
    Republics, Moscow: Novosti Press
_____ (1985) Stroitel'nyi kooperativ, Pravda, 30 September
_____ (1988) Byulleten' Normativnykh Aktov Ministerstv i
    Vedomstv SSR, Moscow: Izd. 'Yuridicheskaya Literatura'
Dmitriev, S. (1985) Estafeta sotsialisticheskikh traditsii,
    Pravda, 22 September
_____ (1987) Zhilishchnoe stroitel'stvo na novum etape,
    Planovoe khozyaistvo, no. 2
Fesenko, D. (1986) Problemy etazhnosti i sotsial'no-kul'
    turnyi faktor, Arkhitektura SSSR, no. 2
Fetisov, T. (1980) Zaboty ne tol'ko Saratovskie, Sovety
    Narodnykh Deputatov, 1
Gorbachev, M. (1987) Political Report of the CPSU Central
    Committee to the XXVII Party Congress, Moscow: Novosti
    Press
Gordon, L. (1988) Social Policy on the Remuneration of Labour,
    Sociology. A Journal of Translations, vol. 27, no. 2
Goskomstat (1985) Narodnoe khozyaistvo SSSR v 1984g, Finansy,
    Moscow 1985
_____ (1987a) Vsesoyuznaya perepis' naseleniya 1989 goda.
    Moscow: 'Finansy i Statistika'
_____ (1987b) Narodnoe khozyaistvo SSSR za 70 let, Yubileinyi
    statisticheskii ezhegodnik, Moscow: 'Finansy i
    Statistika' soobshchaet (1988) Argumenty i fakty, no. 32
Grafova, L. (1980) Strasti po ustavu, Literaturnaya Gazeta,
    no. 51
Grebennikov, V. (1987) Intensifikatsiya obshchestvennogo
    proizvodstva: sotsial'no-ekonomicheskie problemy.
    Moscow: Izd. politicheskoi literatury
Gribanov, V.P. (1983) Osnovy sovetskogo zhilishchnogo
    zakonodatel'stva, Moscow: 'Znanie'
Kalmantaev, B. (1988) Otlozhennyi shtraf, Sovetskaya Rossiya,
    2 March
Khakhulina, L. and Trofimov, V. (1985) Uroven' zhizhni
    naseleniya Sibiri v sravnenii s drugimi regionami RSFSR,
    Izvestiya Sibirskogo Otdeleniya Akademii Nauk SSSR,
    seriy Ekonomika i prikladnaya sotsiologiya, no. 7
Kutsev, G. (1982) Novye goroda, Sotsiologicheskii ocherk na
    materialakh Sibiri, Moscow: 'Mysl'

Housing Policy in The Soviet Union

Lebedev, P. (1980) Sistema organov goroedskogo upravleniya, Leningrad: Izd. Leningradskogo Universiteta

Lenin, V.I. (1962) Polnoe sobranie sochenenii, Moscow: 5th ed., vol. 23

_____ (1976) Materialy XXV S"ezda KPSS, Moscow: Izd. Polit. Lit.

_____ (1981) Materialy XXVI S"ezda KPSS, Moscow: Izd. Polit. Lit.

_____ (1982) Materialy XVII S"ezda professional'nykh soyuzov SSSR, Moscow: Profizdat

_____ (1986) Materialy XXVII S"ezda KPSS, Moscow: Izd. Polit. Lit.

Matveets, G. (1982) ZhSK: Skol'ko obyazannostei, stol'ko i prav, Literaturnaya Gazeta, no. 21

Meerson, D. & Tonskii, D. (1983) Zhilischchnoe stroitel'stvo v SSR v XI Pyatiletke, Moscow: Stroiizdat

Murphy, P. (1985) Soviet Shabashniki: Material Incentives at Work, Problems of Communism, vol. XXIV. no. 6

_____ (1988a) Nesostoyavshchiesya novosel'ya - upor stroitel'am, ne vypolnyayushchim plan, Pravda, 19 November

_____ (1988b) Nesostoyavshiesya novosel'ya - upor stroitel'yam, ne vypolnyyushchim plan, Pravda, 19 November

Osnovy zhilishchnogo zakonodatel'stva Soynza SSR i Soynznykh respublik. Zhilishchnyi kodeks RSFSR (1986). Moscow: "Yuridicheskaya litertura"

Politicheskoe obrazovanie (1987):10

Poltorygin, V. (1983) Razvitie material'no-teknicheskoi bazy neproizvodstvennoi sfery (na materialakh zhilischchnogo khozyaistva), Moscow: 'Ekonomika'

Postanovlenie (1926) O kvartirnoi plate i merakh k uregulirovaniyu pol'zovaniya zhilishchami v gorodskikh poseleniyakh (Postanovlenie TSIK i SNK. SZ SSSR 1926, no. 44, art. 312)

_____ (1928a) Ozhilishchnoli politike, (Postanovlenie YsIK i SNK of 4 January. Resheniya partii i pravitel'stva po khozyaistvennym voprosam, vol. 1. Moscow: 1967)

_____ (1928b) Ob oplate zhilykh pomeshchenii v gorodakh i rabochikh poselkakh (Postanovlenie VTsIK i SNK RSFSR. SU RSFSR 1928, no. 53, art. 402)

_____ (1962) Ob individual'nom i kooperativnom zhilishchnom stroitel'stve (Postanovlenie TsK KPSS i Soveta Ministrov SSSR. SP SSSR 1962, no. 12, art. 93)

_____ (1964) O dal'neishem razvitii kooperativnogo zhilishchnogo stroitel'stva (Postanovlenie Soveta Ministrov SSSR. SP SSSR 1964, no. 25, art. 147)

_____ (1969) O merakh po uluchsheniyu kachestva zhilishchnogo stroitel'stva

(Postanovlenie TsK KPSS i Soveta Ministrov SSSR.
SP SSSR 1969, no. 15, art. 84
_____ (1978a) O dal'neishem razvitii stroitel'stva individual'
nykh zhilykh domov i zakreplenii kadrov na sele
(Postanovlenie TsK KPSS i Soveta Ministrov SSSR.
SP SSSR 1978, no. 17, art. 102)
_____ (1978b) O merakh po povysheniyu effektiavnosti
nauchno-issledovatel'skikh rabot v oblasti stroitel'stva,
arkhitektury, stroitel'nykh materialov...
(Postanovlenie TsK KPSS i Soveta Ministrov SSSR.
SP SSSR 1978, no. 20, art. 123)
_____ (1978c) O merakh po dal'neishemu uluchsheniyu
ekspluatatsii i remonta zhilishchnogo fonda
(Postanovlenie Soveta Ministrov SSSR.
SP SSSR 1978, no. 22, art. 137)
_____ (1980) O dal'neishem ukreplenii trudovoi distsipliny i
sokrashchenii tekuchesti kadrov v narodnom khozyaistve
(Postanovlenie TsK KPSS i Soveta Ministrov SSSR.
SP SSSR 1980, no. 3 art. 17)
_____ (1981a) O merakh po povysheniyu effektivnosti
kapital'nykh slozhenii, vydelyaemykh na zhilishchnoe
stroitel'stva
(Postanovlenie Soveta Ministrov SSSR.
SP SSSR 1981, no. 29, art. 169)
_____ (1981b) Ob individual'nom zhilishchnom stroitel'stve
(Postanovlenie Soveta Ministrov SSSR.
SP SSSR 1981, no. 29, arat. 170)
_____ (1981c) O zhilishchno-stroitel'no kooperatsii
(Postanovlenie Soveta Ministrov SSSR.
SP SSSR 1982, no. 23, art. 120)
_____ (1982) O srokakh vvedeniya besprotsentnoi ssudy na
uluchshenie zhilishchnykh uslovii ili obzavedenie
domashnim khozyaistvom molodym sem'yam, imeyushchim detei
(Postanovlenie Soveta Ministrov SSSR.
SP SSSR 1982, no. 9, art. 47)
_____ (1983) "O merakh po obespecheniya vypolneniya planor
stroitel'stva zhilykh domor i sotsial'no-bytovykh
ob"ektor ot 25 fevralya 1983", Pravda, 26 February 1983
_____ (1985a) Ob utverzhdenii Pravil pol'zovaniva zhilymi
pomeshcheniyami, soderzhaniya zhilogo doma i
pridomovoi territorii v RSFSR i Tipovogo dogovora
naima zhilogo pomeshcheniya v doma gosudarstvennogo
fonda v RSFSR
(Postanovlenie Soveta Ministrov RSFSR.
SP RSFSR 1985, no. 2, art. 10
_____ (1985b) O dal'neishem uvelichenii proizvodstva
stroitel'nykh materialov, izdelii i konstruktsii
dlya prodazhi naseleniyu
(Postanovlenie Soveta Ministrov SSSR.
SP SSSR 1985, no. 19, art. 89)

_____ (1985c) O dopolnitel'nykh merakh po stroitel'stvu
molodexhnykh zhilykh kompleksov i kooperativnykh
zhilykh domov dlya molodezhi
(Postanovlenie Soveta Ministrov SSSR.
SP SSSR 1985, no. 22, art. 111)

_____ (1985d) O dal'neishem razvitii industrializatsii i
povyshenii proizvoditel'nosti truda v kapital'nom
stroitel'stve
(Postanovlenie TsK KPSS i Soveta Ministrov SSSR.
SP SSSR 1985, no. 26, art. 133)

_____ (1986) Ob osnovnykh napravleniyakh uskoreniya resheniya
zhilishchnoi problemy v strane (ot 17 aprelya 1986)
(Postanovlenie TsK KPSS.
SP SSSR 1986, no. 19, art. 100)

_____ (1987a) Ob utverzhdenii Pravil o dogovorakh podryada na
kapital'noe stroitel'stvo
(Postanovlenie Soveta Ministrov SSSR.
SP SSSR 1987, no. 4, art. 19)

_____ (1987b) Ob utverzhdenii Pravil finansirovaniya i
kreditovaniya stroitel'stva
(Postanovlenie Soveta Ministrov SSSR.
SP SSSR 1987, no. 7, art. 31)

_____ (1987c) O merakh po dal'neishemu sovershenstvovaniy
raboty zhilishchno-kommunal'nogo khozyaistva v strane
(Postanovlenie TsK KPSS i Soveta Ministrov SSSR.
SP SSSR 1987, no. 27, art. 92)

_____ (1987d) O merakh po obespecheniyu vypolneniya
utverzhdennykh na dvenadtsatuyu pyaatiletku zadanii po
razvitiyu material'no-tekhnicheskoi bazy sotsial'no-
kul'turnoisfery
(Postanovlenie TsK KPSS i Soveta Ministrov SSSR.
SP SSSR 1987, no. 28, art. 96)

_____ (1987e) O sovershenstvovanii sistemy bankov v strane i
usilenii ikh vozdeistviya na povysheniya effektivnosti
ekonomiki
(Postanovlenie TsK KPSS i Soveta Ministrov SSSR.
SP SSSR 1987, no. 37, art. 121)

_____ (1987f) Ob ispol'zovanii pustuyushchikh zhilykh domov i
priusadebnykh uchastkov, nakhodyashchikhsya v sel'skoi
mestnoisti
(Postanovlenie TsK KPSS i Soveta Ministrov SSSR.
SP SSSR 1987, no. 41, art. 134)

_____ (1987g) O dopolnitel'nykh merakh po obespecheniyu
naseleniya stroitel'nymi materialami i izdeliyami i
okazaniyu emu platnykh uslug
(Postanovlenie Soveta Ministrov SSSR.
SP SSSR 1987, no. 42, art. 138)

_____ (1987h) O dal'neishem razvitii sovetskoi arkhitektury i
gradostroitel'stva
(Postanovlenie TsK KPSS i Soveta Ministrov SSSR.
SP SSSR 1987, no. 45, art. 149)

_____ (1988a) O merakh po dal'neishemu razvitiyu dizaina i rasshireniyu ego ispol'zovaniya dlya povysheniya kachestva promyshlennoi produktsii i sovershenstvovaniya ob"ektov zhiloi, proizvodstvennoi i sotsial'no-kul'turnoli sfery (Postanovlenie Soveta Ministrov SSSR. SP SSSR 1988, no. 1, art. 1)

_____ (1988b) O dal'neishem razvitii arkhitektury i gradostroitel'stva v RSFSR (Postanovlenie Soveta Ministrov RSFSR. SP RSFSR 1988, no. 5, art. 15)

_____ (1988c) O dal'neishem razvitii stroitel'stva molodezhnykh zhilykh kompleksov v RSFSR (Postanovlenie Soveta Ministrov RSFSR i VLKSM. SP RSFSR 1988, no. 6, art. 20)

_____ (1988d) O merakh po uskoreniyu razvitiya individual'nogo zhilishchnogo stroitel'stva (postanovlenie TsK KPSS i Soveta Ministrov SSSR. SP SSSR 1988, no. 11, art. 28)

_____ (1988e) O nekotorykh voprosakh kooperataivnoi i individual'noli deyatel'nosti (Postanovlenie Soveta Ministrov SSSR. SP SSSR 1988, no. 15, art. 41)

_____ (1988f) O merakh po uskoreniyu razvitiya zhilishchnoi kooperatsii (Postanovlenie TsK KPSS i Soveta Ministrova SSSR. SP SSSR 1988, no. 16, art. 43)

_____ (1988g) O sovmeshchenii proektirovaniya i stroitel'stva zhilykh domov i drugikh ob"ektov sotsial'nogo naznacheniya (Postanovlenie Soveta Ministrov SSSR. SP SSSR 1988, no. 18, art. 53)

Prokopchenko, I. (1986) Zhilishchnoe i zhilishchno-stroitel'noe zakonodatel'stvo, Moscow: Stroiizdat

Proshkin, B. & Povarich, I. (1987) On the issue of material non-monetary work incentives, Soviet Sociology. A Journal of Translations, vol. XXV, no. 4

Pudikov, D. (1980) Finansirovanie i kreditovanie kooperativnogo i individual'nogo zhilishchnogo stroitel'stva, Moscow: 'Finansy'

Rimashevskaya, N. (1975) Struktura lichnogo i obshchestvennogo potrebleniya v sotsialisticheskikh stranakh, Voprosy ekonomiki, no. 12

Rotkov, V. and Grif, Yu. (1968) Planirovanie zhilishchno-kommunal'nogo khozyaistva stroitel'no-montazhnykh organizatsii, Moscow: Stroiizdat

Rozanov, E. (1988) Kvartira stanet lichnoi sobstvennost'yu, Izvestiya, 9 December

Rumer, B. (1984) Investment and Reindustrialisation in the Soviet Economy, London: Westview Press

Rusakovskii, A. (1983) Rabochee obshchezhitie, Moscow: Profizdat

Rutgaizer, V. (1981) Chelovek v sfere raspredelenya i potrebleniya, EKO, no. 9

_____ (1987) Raspredelenie po trudu, EKO, no. 3

Ryvkina, R. (1988) Overcoming the Braking Mechanism, Soviet Sociology. A Journal of Translations, vol. 27, no. 2

Sarkisyan G. (1983) Narodnoe blagosostoyanie v SSSR, Moscow: 'Mysl'

Sedugin, P. (1983) Pravo na zhilishche v SSSR, Moscow: Yuridicheskaya Literatura

Shatalin, S. (1982) Sovershentsvovanie raspredelitel'nykh otnoshenii, EKO, no. 1

Smusenok, S. (1985) Stroit' kooperativ, Pravda, 30 September

_____ (1986) The Programme of the Communist Party of the Soviet Union, Moscow: Movosti Press Agency Publish House

_____ (1987) Vremya obeshchanii proshlo, Sovety narodnykh deputatov, 7

Sosnovy, Y. (1954) The Housing Problem in the Soviet Union, New York: Research Program on the USSR

Strumlin, B. (1926) Problemy planirovaniya. Itogi i perspektivy, Moscow

Trehub, A. (1988) Soviet Construction Chief on the Housing Programme, Radio Liberty Research, no. 49 (3514) 25 November

Tolstoi, Yu. (1984) Novoe sovetskoe zhilishchnogo zakonodatel'stva, Leningrad: 'Znanie'

Tonskii, D. (1982) Ekonomicheskii prognozirovanie gorodskogo zhilishchnogo stroitel'stva, Moscow: Stroiizdat

TsSU SSSR (1982) Narodnoe khozyaistvo SSSR 1922-1982, Moscow: Tsentral'noe statisticheskoe Upravlenie SSSR 'Finansy i Statistika'

TsSU RSFSR (1985) Moskva v tsifrakh v 1985g, Moscow: Statistika

Ukolov, V. (1985) Sotsial'no-ekonomicheskaya effektivnost' bytovogo obsluzhivaniya, Rostov: Izd. Rostovskogo Universiteta

Vashets, S. (1981) Vedomosti Verkhnovogo Soveta, no. 26, arts. 834, 835

_____ (1986) Vokrug kooperativnoi kvartiry, Izvestiya, 23 March

_____ (1987) Podryad nabiraet sily, Zhilishchnoe stroitel'stvo, no. 9

Vasil'yev, G. & Privalova, O. (1984) A social-geographic evaluation of differences within a city, Soviet Geography, 25

Volkonskii, V.A. and Koryagina, T.I. (1985) O teoreticheskikh obosnovaniyakh narodno-khozyaistvennykh prioritetov, Izvestiya Sibirskogo Otdeleniya Akademii Manka SSSR, seriya Ekonomika i prikladnaya sotsiologiya, no. 2

Walker, D. (1986) Housing in the Soviet Union, unpublished
    M.Phil thesis in 3 vols (Middlesex Polytechnic library at
    Enfield)
Yanowitch, M., ed. (1989) A Voice of Reform. Essays by
    Tatiana I. Zaslavskaya, N.Y.: M.E. Sharpe
Zagorul'kin, V. & Kolesnikov, V. (1983) Fond sotsial'no-
    kul'turnykh meropriyatii i zhilishchnogo stroitel'stva,
    Moscow: Profizdat
Zapal'skii, L. (1988) Zhil'e: zhguchie voprosy, Politicheskoe
    obrazovanie, no. 3
Zaslavskaya, T. (1986) Chelovecheskii faktor razvitiya
    ekonomiki i sotsialnaya spravedlivost', Kommunist, 13
Zhelezko, S. (1982) Zhilaya yacheika v budushchem, Moscow:
    Stroiizdat
_____ (1986a) Naselenie krupnogo goroda, Moscow: 'Mysl'
_____ (1986b) Zhilishchnyi Kodeks RSFSR, Moscow: Yuridicheskii
    Literatura
Zhilaya yacheika (1982), Moscow: Stroiizdat
Zhishchnaya kooperatsiya (1928), no. 1

GLOSSARY

local soviet	the functional equivalent of the British "local authority" or municipal government
gorispolkom	executive committee of the city (gorod) soviet
ispolkom	executive committee
postanovlenie	decree (normally by the Council of Ministers and/or the Central Committee of the Communist Party (TsK KPSS)
RSFSR	Russian Soviet Federal Socialist Republic
rouble	on par roughly with the pound (officially, 1R = c. £1)

Chapter Seven

HOUSING POLICY IN THE GERMAN DEMOCRATIC REPUBLIC

Hanns Buchholz

THE PRESENT SITUATION

Housing in the German Democratic Republic (GDR) is the concern of the state. According to its ideological self-understanding as well as by law the state has to take the responsibility for housing needs and the housing standard of its subjects. Seen quantitatively, the supply of dwellings for the population of the GDR does not look bad. On 31 December 1981, the time of the last census, there were 6,569,000 dwellings (in dwelling-houses) occupied by subjects of the GDR for dwelling purposes. At this time the total population of the GDR numbered 16,705,635 persons, living in 6,509,932 households, with 1,728,691 one-person households. Accordingly, the ratio was 393 dwellings per 1,000 persons. An intermediate census took place in 1985. The results are not available in detail. But on this basis all dwellings used for non-dwelling purposes were excluded from the official figures which read for 1986 as follows (see Statistical Pocket Book, 1987, p. 63): 6,910,720 dwellings for 16,639,877 persons, resulting in a ratio of 415 dwellings per 1,000 persons.

Every dwelling has an average living-space of 58 square metres; the average living-space quota per person amounted to 22.7 square metres per person. (Data for 1981; data not available for the more recent years.)

Most of the dwellings are composed of two or three rooms, i.e. 35.2 per cent and 34.8 per cent respectively (Melzer, 1983 (C), p. 98). The percentage of one-room flats - 10.9 per cent - is diminishing, whereas the proportion of dwellings with four or more rooms has slightly increased.

330

Housing Policy in the German Democratic Republic

The equipment of the dwellings (all data from 1981, after Melzer, 1983 (c)) leaves much to be desired. Admittedly, 90.8 per cent of all dwellings are fitted with piped water; but just 56.4 per cent are equipped with a toilet inside their dwelling, and only 54.3 per cent with bath or shower, 43.2 per cent with a heated water supply, and only 25.7 per cent with any type of central heating.

Likewise, the condition of the buildings shows considerable deficiencies. The data are available for 1971 only. At that time the dwellings were divided into four categories: I good condition; II minor damage; III serious damage; IV uninhabitable. According to this classification just 20.3 per cent of all dwellings in the GDR were in good condition (I), 63.0 per cent showed minor damage (II), and 15.7 per cent of all dwellings had serious damage or were uninhabitable (1.0 per cent).

The remarkable lack of quality by Central European standards results partly from the old age of many dwelling houses and partly from the careless treatment of the buildings resulting from a lack of repair and maintenance over a period of decades. In 1981 about 46 per cent of the 6,569,000 dwellings had been built before 1919, 19 per cent between 1919 and 1945, and 35 per cent since 1945 (Melzer, 1983 (c), p. 196). It is quite clear that most of the older buildings lack up-to-date technical equipment, and that they are susceptible to damage. Nevertheless, this cannot explain the relatively poor building conditions mentioned above. Additionally, we have to consider that until the 1970s the dwellings had not been restored or modernized systematically, neither by the state nor by the private owners. Until that time the state established other priorities for political reasons, and the private landlords, the owners of apartment houses in particular, were not able to maintain or to modernise their buildings in a proper way: the rents were and are extremely low, and construction material as well as craftsmen have always been difficult to obtain. Therefore in most of the towns and cities of the GDR we can see on the one hand large new residential estates, and on the other hand many degraded old building areas, partly with ruins from the Second World War.

Housing Policy in the German Democratic Republic

THE PROPERTY SITUATION

As was already mentioned, housing in the GDR is not a private matter; it is directed and controlled by the state. The construction of tenement blocks is exclusively done by the state or by co-operatives. The building of single-family houses is state regulated. There is no construction of freehold flats. Apartments in private old buildings from before 1945 are subject to public rent control; the maintenance of these blocks is generally not supported by the government. Therefore, many old rented blocks were given to the state as a gift, because the owners could not afford to create reasonable conditions; other blocks fell into decay and were demolished.

There are four different forms of dwelling properties:

I. Socialist dwelling properties:

    (a) Nationally-owned dwelling property (volkseigenes Wohnungseigentum). These dwellings are managed by the so-called "Nationally Owned Enterprise Communal Housing Administration" (volkseigener Betrieb/VEB Kommunale Wohnungsverwaltung). This category comprises the following dwellings: every flat erected by the state since 1945; every dwelling belonging to the public even before 1945; every dwelling confiscated by order of the Soviet Military Administration in Germany; and every dwelling given to the state by its former proprietor. Additionally, the Communal Housing Administration administers as a trustee all the flats and property of persons who, before 1945, resided in foreign countries or in what became later the Federal Republic of Germany, or of persons who left the GDR illegally after 1953.

    (b) Co-operative dwelling property. This means, on the one hand, housing and dwelling property of non-profit or other co-operative societies, which were transferred completely to non-profit housing co-operative societies (Gemeinnützige Wohnungsbaugenossenschaften) in 1957, and, on the other hand, of Workers Housing Co-operative Societies (<u>Arbeiter-wohnungsbaugenossenschaften</u>), founded in 1954.

II. Private Dwelling Property:

(a) Capitalist private dwelling property. This means dwelling in private rented blocks, which were almost all erected before 1945.

(b) Personal private dwelling property. These are owner-occupied private dwellings, mainly single-family or two-family houses. It is not permitted to make a profit from these private dwellings.

As is mentioned below, private dwelling property is limited generally to owner-occupied dwellings and the new erection of private rented flats used to be strictly forbidden. There seems to appear a more tolerant policy in recent years: in several towns, in urban renewal areas in particular, the existence of new private two-family houses with one rented unit has been observed.

Table 7.1: GDR: Dwellings by type of property 1971 and 1981

		Socialist dwelling property		Private dwelling property		Other dwelling property
	dwellings in residential houses	nationally owned	co-operative societies	"capitalistic" private property	"personal" private property	
1 Jan. 1971	5971000 (100%)	1642000 (27.5%)	591000 (9.9%)	1373000 (23 %)	2341000 (39.2%)	24000 (0.4%)
31 Dec. 1981	6562000 (100%)	2067000 (31.5%)	912000 (13.9%)	1096000 (16.7%)	2467000 (37.6%)	20000 (0.3%)

Note: There may be some slight inaccuracies in the figures because of missing public statistical data in the GDR, but the tendency of the figures will be correct.

Source: Melzer, 1983(c), p. 156; Statistisches Jahrbuch der DDR 1987, p. 162; author's calculations

Housing Policy in the German Democratic Republic

SOME IDEOLOGICAL EXPLANATIONS FOR THE DEVELOPMENT OF HOUSING

For a better understanding of the present housing situation in the GDR some politico-ideological facts have to be explained. After the end of the Second World War housing was declared the responsibility of the state, first by the Soviet Military Administration in the Soviet Occupation Zone, and later by the government of the GDR. Regarding the measures during the Soviet Occupation, problems of the severe housing need were of great importance. As a result of the war the housing stock in the area, which later became the GDR, was reduced by 530,000 units, from 4,591 million to 4,061 million (without East Berlin). Of the 4,061 million dwellings about 757,000 were slightly or more seriously damaged. In East Berlin 184,000 dwellings were completely destroyed during the war. At the same time the population of the Soviet Occupation Zone increased remarkably, resulting from the war and from the separation of the eastern provinces of the former German Empire and their transfer to Poland and the Soviet Union. The population grew from 16.7 million (1939), 18.4 million (1946) to 19 million (1948). Through Order No. 9 (21 July 1945) and some subsequent orders the Soviet Military Government enforced rent pegging, thus continuing rent pegging that had already been introduced by the German government in 1936. Later this regulation was adopted by the government of the GDR.

Apart from this extremely low rent level dating from 1936 for flats in pre-war buildings, the flats in new apartment blocks are also very cheap. At present the rent for a flat in a newly constructed block in East Berlin amounts to Mark 1.00 to Mark 1.25 per square metre. Outside of East Berlin the rent is Mark 0.80 to Mark 0.90 per square metre per month (without costs for heating, water etc.). Generally these rents of about 2.7 to 4.0 per cent of the average monthly household income (see Deutsches Institut für Wirtschaftsforschung, 1984, p. 180) are kept stable for socio-political reasons. But this rent level has nothing to do with the real costs of the houses. Just for the running of the residential estates the state has to pay costs amounting to Mark 3.00 per square metre per month on average (Staemmler, 1982, p. 1412).

Housing Policy in the German Democratic Republic

With this rent policy the government of the GDR offered not only low rents for social reasons, but, as far as the rent control of the private pre-war buildings is concerned it also neutralised the contradiction between the demand for the socialisation of all capital goods (i.e. including apartment houses for rent) and the continuing existence of privately owned dwellings for rent. Because of the extremely low rents, the private ownership of apartment houses is not economically attractive, so that it is almost impossible to sell such buildings, despite the legal possibility to do so.

Additionally, the state has the right to intervene decisively in housing matters by law. Based on an act regarding the right of the state to direct all housing matters (Verordnung über die Lenkung des Wohnraumes, 14 September 1967) each letting of a flat and each exchange of a flat has to be authorised by the responsible department of the respective municipality, or partly, too, by the chief office of principal enterprises or of public authorities, as well as of the appropriate bodies of co-operative societies. The same act regulates the criteria for the occupation of dwellings, which generally means social aspects (family size, health situation, change of employment) and political aspects (extraordinary performances for the GDR). Family-independent residential types (private collective households) are practically not able to exist (Staemmler, 1982, p. 1417).

CONSTRUCTION OF TENEMENT HOUSES

Basic ideological principles prevent any private person from constructing tenement houses. The constitution of the GDR lays down that private property may only be accumulated by personal work or by social grants (e.g. pension, family allowance) (see Hoffmann, 1972, p. 58). There must be no private tenement ownership, because this would lead to profits from capital. The toleration of the private ownership of pre-war tenement buildings is regarded as a temporary arrangement. Since 1945 nearly all the tenement houses have been constructed and managed by nationally owned dwelling construction companies (Melzer, 1983 (c), p. 73) or by non-profit co-operative societies. The possible derestriction of this policy has been mentioned already.

The development of the state's responsibility for all housing matters started after the Second World War by transferring all publicly owned buildings and plots as well as all confiscated houses and real property to nationally owned property administrations (volkseigene Grundstücks-verwaltungen). In 1958 they were reshaped to nationally owned dwelling administrations, as already mentioned. These enterprises were given the additional function of managing the nationally owned housing construction operation and of arranging the repair and maintenance work. The dwelling administration is subordinate to the municipality (city council or local authority). But substantial decisions cannot be taken at this administrative level. This is done by the National Planning Commission (Staatliche Plankommission), the central body for planning and controlling the political economy of the GDR, which is directly subordinated to the central government (Ministerrat), and ranks as a ministry. The National Planning Commission, together with the Ministry of Construction, establishes a spatial and quantitative housing plan, under which the local authorities have to produce an annual sub-plan, called "complex housing construction". This sub-plan does not only include the details for the construction of the tenement houses but also those for the so-called "social establishments" (retail trade, services); in addition the sub-plan defines tasks for repair, building alteration and modernisation. Finally, the plan gives the exact location of the planned construction work.

From 1945 to 1986 about 1,562,000 nationally owned flats were constructed (see Table 7.2). During this period of more than forty years several phases can be distinguished.

(a)　The first period after the war is character-ised by clearing and repair work and by the reconstruction of damaged and partly destroyed buildings. The Soviet Military Government could not be encouraged to take an interest in comprehensive actions for new buildings. The first German government only appeared in 1949. Between 1945 and 1949 about 36,500 nationally owned dwellings were constructed (see Table 7.2).

Housing Policy in the German Democratic Republic

(b) The first systematic housing programme covers the time from 1949/1950 to 1955. Based on the "principles of town planning", passed by the government (<u>Ministerrat</u>) of the GDR on 27 July 1950, representative multi-storey residential buildings arose, made from brick and in their style following Soviet architecture. The housing projects were located - with the exception of East Berlin - according to the housing demand of the employees of large industrial plants. On a number of occasions housing construction was connected with the reconstruction of city centres, guided by a policy of also placing a considerable proportion in larger city centres. The number of newly constructed nationally owned flats increased to about 115,000 in the five-year period 1950-1955.

(c) The middle of the 1950s saw the beginning of a process of rationalisation and the systematisation of building techniques as well as of housing estate planning. Building types and tenement patterns became more and more standardised everywhere. Whole housing estates, generally at this time composed of 3- to 4-storey blocks, resembled each other more and more, because the centralised planning authorities tended towards uniform settlement patterns. It became the main task of housing construction to erect the highest possible number of dwellings in the shortest possible time by saving as much construction material as possible.

Industrialised housing construction was created for this purpose in the second half of the 1950s, based on a decision of the GDR government of 21 April 1955. A system of prefabricated construction parts was developed, out of which standardised tenement blocks were erected in a highly sophisticated cycle technique. The position of the tenement blocks followed the requirements of the crane tracks and were grouped together into so-called "socialist residential complexes".

The size of a "socialist residential complex"
(see e.g. Junghans, 1954), a development
from the Soviet "housing-cell", was based on
a primary school unit. It consisted of a
group of multi-storey, densely located tenement blocks with about 5,000 inhabitants,
surrounding as a shopping centre, including
supermarket, restaurant, a combined receiving
office for maintenance and other services,
kindergarten, school etc. Several "residential complexes" form a "residential
district" with its respective district centre
of higher order (compare principle 10 of the
"16 principles of town planning" 16 Grundsätze des Städtebaus). The size of the
socialist residential complexes increased
later to 20,000-30,000 inhabitants. This was
not attained by an enlargement of the
building area, but chiefly by concentration
and raising the height of the buildings, so
that the population density increased from
200 persons per hectare to about 600-800
persons per hectare.

From the technical point of view there was a
long time of experimentation with many
different construction parts, which generally
became larger and more complicated as time
went on. At present, in almost every
location of housing construction the so-
called "housing construction series 70"
(Wohnungsbauserie 70) is to be found.

(d) The speed of this new development, backed by
the full support of the state (with the
consequence that it is hard to find brick
layers and bricks in the GDR) is reflected in
the fact that in 1960 32 per cent of all
newly erected flats were already constructed
using the assembly technique.

During the first half of the 1960s the share
of dwellings constructed from pre-fabricated
parts rose to 90% of all new units. The
number of nationally owned flats erected from
1961-1965 was 141,000. During the second
half of the 1960s this figure increased to
210,000.

These figures seem to be very high, but they never matched the published aims of the respective 5-year or 7-year plans. And there was another growing problem: each year an increasing number of pre-war tenements had to be subtracted from the available stock of dwellings, because they became uninhabitable. The housing census of 1 January 1971 demonstrated the dangerous situation: out of the given 6.057 million dwellings at the census date 56.7 per cent were constructed before 1919; 22.4 per cent between 1919 and 1945; and only 20.8 per cent since 1945.

The poor state of the buildings has already been mentioned: 14.7 percent of all dwellings belonged to category III (serious damage) and 1.0 per cent was uninhabitable. If we relate the technical building conditions to the type of property, it becomes clear that the share of inadequate dwellings was distributed among all the different types of ownership: 17.5 per cent of all nationally owned dwellings; 12.9 per cent of all co-operatively owned dwellings; and 16.1 per cent of all privately owned dwellings belonged to the low quality classes III and IV.

In this situation the government of the GDR made the decision to intensify the programme of housing construction, the aim of which was "termination of the housing problem by 1990". In 1973 the GDR government announced a housing construction programme for the period 1976-1990. During these 15 years 2.8-3.0 million residential units were to be newly constructed or modernised to improve the housing conditions for nearly 60 per cent of of the population of the GDR.

To meet this ambitious aim, the housing construction programme was given a very high priority in comparison to all the other political responsibilities and its instruments were diversified: besides increased planning figures for nationally owned housing construction, the housing co-operatives - the Workers Housing Co-operative Societies in particular - had to take over a higher share of the programme, and even the private con- struction of single-family houses was stimulated. In addition the government started a campaign to renovate and to modernise the older and inadequately equipped tenements.

MODERNISATION OF INADEQUATE HOUSING STOCK

Whereas until 1970 the reconstruction of old flats and houses was more or less neglected, the development in the 1970s and the 1980s demonstrates the changing policy. The over-ageing process of buildings and the growing number of dwellings which had to be demolished or to be taken from the housing stock for other reasons (between 30,000 and 56,000 in every year since 1971!) called for drastic measures. Therefore it was decided· to modernise about 900,000 tenements from 1976 to 1990. After about 11,000 modernisations in 1971, the yearly quota rose from the next year to more than 54,000 modernised dwellings in 1981 and to more than 90,000 in 1986.

In this context modernisation means the upgrading of the housing stock in three categories: (1) installation of piped water, drainage, and a toilet inside the tenement; (2) additionally to (1) the installation of bath or shower and heated water; (3) additionally to (1) and (2) the installation of a central heating system.

In 1986 13.5 per cent of all modernised units belonged to category (1), 49.8 per cent to category (2), and 36.7 per cent to category (3).

Despite these efforts there will still be huge problems in the future as regards the older dwelling stock. From 1971 to 1990 a total of about 1.05 million dwellings will be modernised. This means only about one third of all the dwellings built before 1919. Additionally, we may estimate that about one quarter (or about 300,000 units) of dwellings constructed between 1919 and 1945 are also in need of modernisation.

Housing Policy in the German Democratic Republic

HOUSING PROVIDED BY CO-OPERATIVE SOCIETIES

Besides nationally owned housing construction there used to be and there still are housing stocks and construction activities of co-operative societies. The non-profit and other housing co-operative societies were seen as "capitalist" during the first years after the war. They were discriminated against; their housing plots were confiscated. After the uprising by the population in 17 June 1953 the GDR government tried to improve living conditions - including housing conditions - as quickly as possible. So-called Workers Housing Construction Co-operative Societies were therefore founded in 1954 (Anordnung über die Zulassung and Registrierung der Arbeiterwohnungs-baugenossenschaften vom 14, Mai 1954, in Zentralblatt der DDR, p. 213). These societies were always connected with a specific enterprise, and only employees of this enterprise could become members of the co-operative society. In 1957 the concept of workers housing construction co-operative societies was extended, so that they could also originate from universities or public authorities etc. From 1972 onward, employees of other firms or bodies which do not have a co-operative society could also become members.

The co-operatives received nationally owned building plots free of charge, cheap loans and tax exemptions. By means of these co-operatives it became possible to direct private capital into housing construction without creating private housing property. In principle, the members of the co-operative were also permitted to substitute cash investment by personal work; but this very seldom became significant, because industrialized housing construction offered hardly any opportunity for active work by laymen. It was essential that the enterprises who were backing the respective society should very often provide building plots, labour, construction material, and transport facilities.

In 1957 the long neglected non-profit housing construction co-operative societies (about 420 of them) were asked to adopt the same articles as the workers' co-operatives, and then they received more or less similar public support.

Both types of housing construction co-operatives were directly included in the public housing construction programmes. They were given their planning figures by the National Planning Commission just as for nationally owned housing construction. After slow progress at the beginning (see Table 7.2) the number of newly constructed dwellings after 1958 grew to more than 50,000 units for the two-year period in 1961-1962. After 1962 the construction activities of the non-profit co-operative societies were reduced more and more, and since 1980 the societies have no longer been engaged in housing construction, but only in the maintenance and modernisation of their housing stock.

The workers' housing construction co-operative societies still have important functions. Their output was only reduced in the 1960s, because at that time housing construction was very much concentrated on very large housing estates at important industrial locations and at sites for the redevelopment and reconstruction of large city centres. These dimensions were beyond the capacity of the co-operatives. So the number of completed housing units went down from about 42,000 in 1962 to 10,000 in 1971. Probably this decline was also connected with the decentralization of planning responsibility, following the introduction of the "new economic system of planning and management" in 1963. Whereas housing construction became part of the responsibility of the cities, the supply of labour and construction material was not liberalized in the same way, so that work became ineffective as a result of competition and of failures in planning construction. In 1970 the old system of central planning and management was reintroduced. In 1986 nearly 30,000 flats were erected by the workers' housing construction co-operative societies, which had a total output of about 790,000 residential units from 1954 to 1986.

PRIVATE HOUSING CONSTRUCTION

Private housing construction is restricted to single-family houses, which should be owner-occupied. Generally the building of owner-occupied dwellings was never forbidden, but for a long time it was unwanted for ideological reasons. During the first years after the Second World War private housing construction was more or less limited to separate houses for displaced persons and for former farm workers, who became

343

smallholders as a result of the land reform. In all other cases it was easy to limit private building activities by refusing to grant the licence required. Therefore only 19,300 private dwellings were constructed between 1945 and 1950.

As a reaction to the incidents of 17 June 1953 (when the population of the GDR made an uprising against the government, which was kept in power only by military action of the Soviet army) private housing construction was made easier: building loan contracts could now be signed; nationally owned single-family houses were sold to private persons (to the present tenant where possible); workers received public building plots free of charge and free of time limitations; the employers were asked to assist their employees by drawing up the building plans, by arranging the finance contracts and by providing building plots, construction material and transport facilities. Consequently, the number of new private dwellings rose from 4,000 in 1953 to nearly 9,000 in 1958. But then - after the construction of "the wall" in 1961 - the private construction activities diminished to about 2,000 units per year between 1965 and 1971. This was less than 4 per cent of all new dwellings.

It was not until the new housing policy was announced in 1973 that private housing construction experienced a boom. Since 1979 the output has amounted to about 18,000 private dwellings yearly, i.e. about 10 per cent of the total new dwellings in the GDR per year. There are several public instruments of assistance like cheap loans and probably building plots free of charge. But, on the other hand the state makes licencing dependent on family size and other social and political conditions. In addition, the construction of a private house is often hindered by the lack of building plots, construction material and construction workers etc., because the builder of a private house has to work mainly by himself, assisted by friends, neighbours and by the enterprise that employs him.

Housing Policy in the German Democratic Republic

REGIONAL AND LOCATIONAL DIFFERENTIATION

Population development inside the GDR is becoming increasingly concentrated. This is caused by the decisions taken on the location of investment in industrial enterprises and their dependent housing construction. Apart from the private construction of single-family houses, housing construction is closely connected with industrial enterprises, with the main administrations of so-called "combined works" and with other large establishments. The larger these factories and institutions are, and the more they are agglomerated, the larger and more agglomerated is the housing construction that takes place. Additionally, it is important that for ideological reasons a few locations are specifically promoted. These are:
- the leading cities as regards the administrative hierarchy of the GDR, i.e. East Berlin and the capitals of the Bezirke (or districts) [1] and
- the most important places of industrial production, which are often identical with the capitals of the Bezirke.

The centralistic system of policy also governs the spatial system.

The spatial mobility of the population is highly dependent on the housing supply. In the GDR the housing market is limited mainly to private pre-war tenements, and even these are controlled by the state. All other parts of the housing supply as regards new constructions, conversion or modernisation have been decided in principle by state authorities. Private initiatives of persons willing to move to other places are very restricted. Cities, in particular, offer very few opportunities for building a single-family house, if we consider the limited number of building sites. Sometimes even these limited opportunities are taken up by single-family houses constructed by a factory or another institution, which thus try to attract special employees. But generally, private housing construction is unimportant at present for larger cities. Most of these activities take place in small towns and villages because these places offer much better opportunities.

With nationally owned and co-operatively owned new housing construction the state governs spatial population mobility.

Table 7.2: GDR: Newly constructed dwellings by type of property

	New dwellings in total	Of which:				
		Nationally owned	Co-operatively owned	Workers' co-operatives	Other non-profit co-operatives	Private property
1945–						
1947	10,110					
1948	10,890	36,500				19,300
1949	15,320					
1950	19,470					
1951	28,950	20,000				9,000
1952	22,640	11,800				10,800
1953	29,340	25,000				4,300
1954	31,409	27,000	500	400	100	3,900
1955	29,736	22,000	3,200	2,500	700	4,500
1956	31,294	21,500	4,900	4,400	500	4,900
1957	45,702	29,500	7,800	5,900	1,900	8,400
1958	49,561	26,000	14,700	7,800	6,900	8,900
1959	67,314	27,500	31,400	12,000	19,400	8,400
1960	71,857	25,500	40,600	19,700	20,900	5,800
1961	77,680	24,000	54,100	29,600	24,500	7,500
1962	80,139	23,300	50,700	42,000	9,000	6,100
1963	69,321	23,200	40,300	33,000	7,000	5,800
1964	69,345	35,300	30,500	25,000	5,000	3,500
1965	58,303	35,200	21,000	18,000	3,000	2,100

Table 7.2 (continued)

	New dwell- ings in total	Nation- ally owned	Co-ope- ratively owned	Workers' co-ope- ratives	Other non- profit co-op- ratives	Private property
				Of which		
1966	53,366	35,100	15,500	12,000	3,000	2,800
1967	59,107	39,700	17,000	14,000	3,000	2,400
1968	61,863	44,800	15,000	13,000	2,000	2,100
1969	56,547	40,600	13,900	12,000	2,000	2,000
1970	65,786	49,900	13,600	13,000	1,000	2,300
1971	65,021	51,585	11,238	10,000	1,000	2,198
1972	69,552	47,557	19,558	17,000	2,000	2,437
1973	80,725	49,364	26,172	24,000	2,000	5,189
1974	88,312	47,186	31,577	30,000	2,000	9,549
1975	95,976	48,113	36,656	35,000	2,000	11,207
1976	103,091	54,615	37,366	36,000	1,500	11,110
1977	106,826	54,544	40,498	39,000	1,500	11,784
1978	111,909	56,894	43,102	42,000	1,000	11,913
1979	117,355	62,200	37,327	37,000	500	17,828
1980	120,206	62,020	39,188	39,188	-	18,998
1981	125,731	64,872	39,396			21,463
1982	122,417	64,558	39,514			18,345
1983	122,636	64,523	39,652			18,461
1984	121,654	65,923	37,878			17,853
1985	120,728	72,390	30,151			18,187
1986	119,335	72,489	29,551			17,295

Sources: Melzer, 1983 (c); Statistisches Jahrbuch 1987

Housing Policy in the German Democratic Republic

Where there is no state-planned housing construction there is no area of immigration or vice-versa: The centres of housing construction are at the same time the target areas of migrants. The number of significant housing construction locations is limited. Soon after the Second World War the government of the GDR concentrated its housing measures on (i) some large cities; (ii) some harbour towns (because the GDR was separated from all the harbours it traditionally used); and on (iii) industrial locations, basic industries in particular. Until 1950 these were mainly:

(i)   (East) Berlin, Leipzig, Dresden, Magdeburg;
(ii) Rostock, Wismar, Wolgast;
(iii) Hennigsdorf (steel works and mills) Bitterfeld (chemical works), Riesa (steel works and mills), Unterwellenborn (iron and steel), Sangerhausen (copper ore mining), Leuna (chemical works), Böhlen and Berna (lignite, energy production, chemical works), Annaberg (uranium ore mining).

As time went on, more and more towns and cities were added. But this was always based on governmental decisions. So the implementation regulations of the "law on the construction of the cities of the German Democratic Republic and of the capital of Germany, Berlin" of 6 September 1950 named a total of 53 cities as centres of reconstruction. Until the end of the 1950s about 40 per cent of all newly constructed dwellings was erected in 15 main building cities (<u>Hauptaufbaustadte</u>) and Berlin:

(East) Berlin	Potsdam	Halle
Rostock	Frankfurt/Oder	Leipzig
Schwerin	Erfurt	Dresden
Neubrandenburg	Gera	Eisenhüttenstadt
Magdeburg	Karl-Marx-Stadt	Hoyerswerda
		Dessau

These main centres of building activities were almost exclusively the capitals of the 14 districts (<u>Bezirke</u>), created in 1952, and (East) Berlin. The accentuation of the district centres is underlined by the following figures. Of all newly constructed dwellings in the respective district the following quotas were reserved for the district centre: 48 per cent in Dresden, 40 per cent at Gera, and about 38 per cent for Rostock, Magdeburg and Leipzig.

Housing Policy in the German Democratic Republic

Housing construction was not concentrated on those locations where the housing need was greatest, but rather to places selected for political reasons.

During the eleven years from 1971 to 1981 the greater regional disparities have not changed, apart from the decline in the advantage of (East) Berlin, which still surpasses the GDR average by about 14 per cent.

If we look at the distribution of dwellings in smaller regions, then the spatial disparities have increased. Manfred Melzer (1983 (a) (b)) has analysed the housing situation in a highly labour-intensive way on the level of the Landkreise and the Stadtkreise. These figures (Melzer, 1983 (a), p. 113) show a growing polarization in the supply: Whereas in 1971 the difference regarding the number of dwellings per 1,000 persons oscillated between 287 (Landkreis Greifswald) and 421 (East Berlin), in 1981 the range had increased from between 314 (Landkreis Heiligenstadt) and 493 (Stadtkreis Jena).

Table 7.3: GDR: Newly erected and converted dwellings and
their ratio per 1000 persons by districts
(Bezirke) 1953-1986

	1953 - 1960 Dwelling units	Ratio (1) per 1000	1961 - 1970 Dwelling units	Ratio (2) per 1000
(East) Berlin	55,767	50,2	68,986	63,8
northern districts:				
Neubrandenburg	14,570	21,8	36,493	57,3
Rostock	25,724	31,0	51,413	60,8
Schwerin	11,086	17,4	29,145	48,8
middle districts:				
Cottbus	27,104	33,9	61,303	73,0
Frankfurt/Oder	22,068	33,4	41,456	62,2
Magdeburg	28,467	20,3	48,082	36,2
Potsdam	23,524	19,8	44,236	39,1
more urbanised districts:				
Dresden	43,059	22,6	68,822	36,4
Halle	38,931	19,5	87,597	45,3
Karl-Marx-Stadt	44,542	20,8	66,274	31,9
Leipzig	34,742	22,5	52,459	34,7
south western districts:				
Erfurt	20,255	16,0	47,765	38,1
Gera	18,569	25,4	32,949	44,8
Suhl	9,390	17,3	19,104	34,6

1) Referred to population figure of 1957
2) Referred to population figure of 1966
3) Referred to population figure of 1976
4) Referred to population figure of 1983

Table 7.3 GDR (continued)

	1971 - 1980 Dwelling units	Ratio (3) per 1000	1981 - 1986 Dwelling units	Ratio (4) per 1000
(East) Berlin	100,225	91,5	152,014	128,2
northern districts:				
Neubrandenburg	44,552	71,1	39,530	63,6
Rostock	69,651	80,2	59,390	66,3
Schwerin	39,353	66,7	37,346	63,0
middle districts:				
Cottbus	63,175	72,4	54,422	61,6
Frankfurt/Oder	51,862	75,2	49,358	69,6
Magdeburg	84,035	65,2	94,136	74,8
Potsdam	62,940	56,2	66,649	59,4
more urbanised districts:				
Dresden	101,791	55,5	119,081	66,3
Halle	109,846	58,5	119,509	66,0
Karl-Marx-Stadt	104,682	52,9	141,391	74,3
Leipzig	79,205	54,8	109,635	78,7
south western districts:				
Erfurt	78,152	62,9	79,839	64,5
Gera	53,219	72,1	60,827	81,9
Suhl	30,780	56,0	30,976	56,3

Source: Statistisches Jahrbuch, several issues

Note: All the data in Table 7.3 demonstrate the considerable concentration process on the regional level resulting in a spatial disparity of housing supply in the GDR.

Housing Policy in the German Democratic Republic

Table 7.4: GDR: The regional housing supply in 1971 and 1981

	1st January 1971		31st December 1981	
	Dwelling units per 1000 persons	Deviation from average of GDR (in %)	Dwelling units per 1000 persons	Deviation from average of GDR (in %)
(East) Berlin	427	+ 20.3	447	+ 13.7
northern districts	316	− 11.0	357	− 9.2
central districts	347	− 2.3	380	− 3.3
more urbanized districts	365	+ 2.8	407	+ 3.6
southwestern districts	338	− 4.7	378	− 3.8
GDR total	355		393	

Source: Melzer, 1983 (c), p. 196

The disparities become even more clear by analysing the quality of dwellings regarding the equipment. In 1981, of 192 Landkreise and 27 Stadtkreise (including East Berlin) the following data appeared:

Table 7.5: GDR: Distribution of facilities by Stadtkreise
and Landkreise

Facility	More than 50 per cent	33.3 - 50 per cent	Less than 33.3 per cent
bath/shower	64 Landkreise 27 Stadtkreise	113 Landkreise	15 Landkreise
toilet inside	60 Landkreise 26 Stadtkreise	108 Landkreise 1 Stadtkreise	24 Landkreise

	more than 25 per cent	10-25 per cent	less than 10 per cent
central heating	27 Landkreise 24 Stadtkreise	142 Landkreise 3 Stadtkreise	13 Landkreise

Source: Melzer, 1983 (a) (b)

Generally dwellings in district capitals
(Bezirksstädte) are much better equipped than the
surrounding Landkreise because the Landkreise had
less construction activities and less
modernisation. The same differentiation is given
between other bigger towns and their surrounding
Landkreise.

Since the beginning of the 1970s the GDR has
executed a policy of "intensification": i.e. all
systems existing already in a location shall be
used more intensively to get an increase in
production (for instance). In the same way
factories and population shall be concentrated
into relatively few locations to get a more
rational use of the given infrastructure.

In this way housing becomes part of the
national spatial policy, and this means in the
case of the GDR that it becomes part of the
national policy of intensification and
concentration.

Housing Policy in the German Democratic Republic

CONCLUSION

The main characteristic of housing in the GDR is the shut-down of the formerly existing and predominant private and individual responsibility for housing construction of rented blocks on the one side, and the private and individual choice regarding the housing market on the other side. Because of ideological reasons there is no individual construction of apartment buildings, and the building-up of single-family houses is extremely limited by bureaucratic obstructions and by the lack of building material and builders' labourers. And further, the situation results in a considerable limitation of the individual options regarding size, type and location of a dwelling.

The principal conversion of housing matters to the use of the state results in a bipartition of the housing stock: the widely degraded private stock of the pre-war buildings and the state-owned or co-operatively-owned stock of new buildings. The low rents of the older flats, limited to the rent level of 1936, hamper any measure to maintain or - more than ever - to modernise the dwellings. And the state-owned or co-operatively-owned new buildings are less attractive because of their boring uniform pattern and because of their often unattractive location at the fringe of the towns or cities. The realization of individual requests or needs is possible very often only to those who are provided with the capacity for manual or financial (free convertible currency in particular) actions. Symptomatic is the enormous desire of people for the building-up of private single-family houses. People without practical abilities or useful relations are in a less favourable situation: e.g. many elderly people cannot provide building material, they are not able to improve their flats by their own manual work, and because most of the newly built-up dwellings are given away via enterprises or other employers they have little chance to get such a flat. Of course, generally there are no homeless people in the GDR; but the quality of housing is very limited for large parts of the population.

Despite the remarkable quantitative performance of the state regarding housing construction, there are principal deficiencies in the system of central planning and guidance: that is, the lack of flexibility and adaptability to the specific and changing personal and regional needs of the people. The ideological doctrine limits individual initiatives also there, where the state prevents an individual's acting in his own capacity, e.g. because of the small size of a housing project or because of individually motivated deviations from the given system of prefabricated building material (e.g. in urban renewal areas) or regarding the needs for repair and modernisation of old buildings. It seems that in this way the political goal of a quantitatively and qualitatively balanced dwelling supply cannot be reached - the extremely low rents result in a degrading stock of old buildings, the state monopoly of the construction of rented blocks leads to strategies of concentration and rationalisation, and this leads to regional imbalances and to large and uniform housing estates. Many supply difficulties in the country-side have been solved with the opening for private single-family house construction since the 1970s; a certain increase of rents for dwellings in pre-war buildings would limit the degradation process of the older residential areas, and more toleration of private rent housing construction would assist enormously the renewal and modernisation process of towns and cities.

NOTE

[1] The GDR is divided administratively into Bezirke (i.e. districts), Landkreise and Stadtkreise. Landkreis means an administrative unit which combines several communes or local authorities with a central administration responsible (i) for many matters which exceed the local authority's boundary and (ii) for supervising the local authority. Bigger towns or cities do not belong to a Landkreis; they form a Stadtkreis of their own which acts on the same hierarchical level as a Landkreis. They are also divided into Gemeinden (i.e. local authorities).

Housing Policy in the German Democratic Republic

REFERENCES

Arndt, K.D.: Wohnverhältnisse und Wohnungsbedarf in
   der sowjetischen Besatzungszone. Berlin 1960
Autorenkollektiv (Ed.): Organisation und Gestaltung von
   Wohngebieten. Wissenschaftliche Beiträge zu den
   Grundsätzen. (East) Berlin 1972. (Schriftenreihen
   der Bauforschung. Reihe Städtebau und Architektur, 33)
Bartholmai, B. and M. Melzer: Wohnungsbau in beiden
   deutschen Staaten. In: Vierteljahreshefte zur
   Wirtschaftsforschung (DIW), 4, 1986, pp. 293-310.
   (as well as in Archiv für Kommunalwissenschaften,
   1987, p. 1)
Bönisch, R., G. Mohs and W. Ostwald (Eds.): Territorial-
   planung, 3rd edition, Berlin 1982
Buchholz, H.J.: Die DDR und ihre Städte. In: Geographie
   heute, 6, 1985, no. 30, pp. 30-35
Bundesministerium für innerdeutsche Beziehungen (Ed.):
   DDR Handbuch. 2 vols, 3rd edition. Köln 1985
Chronik Bauwesen. Deutsche Demokratische Republik.
   Ed. by Ministerium für Bauwesen, Bauakademie der
   Deutschen Demokratischen Republik. 1945-1971,
   Berlin 1974; 1971-1976, Berlin 1979
Deutsches Institut für Wirtschaftsforschung (DIW) (Ed.):
   Handbuch DDR-Wirtschaft. 4th edition, Reinbek bei
   Hamburg 1984
Friedrich-Ebert-Stiftung (Ed.): Wohnungs- und Städtebau
   in der DDR. Zur Wohnungsfrage. Bonn 1981
Gesetz über den Aufbau der Städte in der Deutschen
   Demokratischen Republik und der Hauptstadt
   Deutschland, Berlin, (Aufbaugesetz) vom 6.
   September 1950. In: Gesetzblatt der Deutschen
   Demokratischen Republik, no. 104, 1950, pp. 965-967
Gesetz über den Fünfjahrplan für die Entwicklung der
   Volkswirtschaft der DDR 1986-1990 vom 27. November
   1986. In: Gesetzblatt der Deutschen Demokratischen
   Republik, Teil I, Nr. 36, 1986, pp. 449-465
Grundsätze des Städtebaus. In: Ministerialblatt der
   Deutschen Demokratischen Republik, no. 25, 1950,
   pp. 153-154
Hagemann, F.: Ergebnisse der Wohnraum- und Gebäudezählung
   1971. In: Statistische Praxis, 9, 1972, pp. 371-376
Hoffmann, M.: Genossenschaftlicher Wohnungsbau in der DDR.
   In: Jahrbücher für Nationalökonomie und Statistik.,
   vol. 187, 1972/1973, pp. 522-542
_____ : Wohnungspolitik der DDR. Das Leistungs- und
   Interessenproblem. Düsseldorf 1972
_____ : Die Wohnungsbaugenossenschaften in der DDR in der
   "sozialistischen Entwicklungsperspektive". In:
   Gemeinnütziges Wohnungswesen, 27, 1974, no. 11,
   pp. 583-586

_____ : Wohnungsbau Privater in der Deutschen Demokratischen
Republik. In: ORDO, vol, 25, 1974, pp. 125-187
_____ : "Lösung der Wohnungsfrage" bis 1990. Zur Wohnungs-
und Bodenpolitik der DDR. In: Bauwelt, 1975, 5,
pp. 118-135
Homann, W. : Die Arbeiterwohnungsbaugenossenschaften im
Rahmen der Wohnungsbaupolitik der DDR. Berlin 1981.
(Forschungsstelle für gesamtdeutsche wirtschaftliche
und soziale Fragen. FS-Analysen 5)
_____ : Veränderungen beim sozialistischen Eigentum im
Bereich dere Wohnungswirtschaft der DDR. Berlin 1983.
(Forschungsstelle für gesamtdeutsche wirtschaftliche
und soziale Fragen. FS-Analysen, 3)
Jenkins, H.W.: Wohnungswirtschaft und Wohnungspolitik in
beiden deutschen Staaten. Versuch eines Vergleichs.
Hamburg 1976
Junghans, K., F. Boesler and R. Günther: Der Wohnkomplex
als Planungselement im Städtebau. (East) Berlin 1954
Junker, W.: Bauen für das Wohl des Volkes. In: Einheit, 6,
1979, pp. 607-615
_____ : Das Wohnungsbauprogramm der Deutschen Demokratischen
Republik für die Jahre 1976-1990. Berlin 1973.
Krenz, G.: Stand und Perspektiven der städtebaulichen
Engwicklung in der DDR. In: Stadt, 30, 1983, 1,
pp. 12-17
Langhof, M.: Zum Bedeutungswandel der Wohnungspolitik in der
DDR. In: Deutschlandarchiv, 12, 1979, pp. 390-405
Lembke, K.: Standpunkte und Auffassungen zum inner-
städtischen Wohnungsbau. In: Architektur der DDR, 31,
1982, no. 5, pp. 286-289
Melzer, M. with W. Steinbeck: Qualitative Aspekte der
regionalen Wohnungsversorgung in der DDR. In: Die DDR
im Entspannungsprozeß. Lebensweise im realen
Sozialismus. Edition Deutschlandarchiv, 1980,
pp. 148-162
_____ : Probleme und bisherige Erfolge des
Wohnungsbauprogramms. Nord- und Mittelregionen der
DDR. In: Deutschlandarchiv 16, 1983 (a) no. 1,
pp. 76-92
_____ Wohnraumlenkung und Mietverhältnisse. Zur
Wohnungspolitik in der DDR. In: Gemeinnütziges
Wohnungswesen, 4, 1983 (b), pp. 191-194, 199
_____ : Wohnungsbau und Wohnungsver-
sorgung in beiden deutschen Staaten - ein Vergleich.
Berlin 1983 (c). (Deutsches Institut für Wirtschafts-
sforschung. Beiträge zur Strukturforschung, 74)
_____ : Zwischenbilanz des Wohnungsbauprogramms. Ballungs-
und Südwestregionen der DDR. In: Deutschlandarchiv 16,
1983 (d), no. 12, pp. 1289-1302.
Minowsky, B.: Im Wohnungsbau auf das Neue orientieren!
In: Statistische Praxis, 1960, no. 8

Housing Policy in the German Democratic Republic

Ostwald, W.: Territoriale Bedingungen für die weitere
     Ausprägung der sozialistischen Lebensweise in der DDR.
     In: Probleme der sozialistischen Lebensweise.
     (East) Berlin 1977, pp. 125-129. (Abhandlungen der
     Akademie der Wissenschaften der DDR, 1977, no. W5)
Schöller, P.: Wideraufbau und Umgestaltung mittel- und
     norddeutscher Städte. In: Informationen. Institut für
     Raumforschung, 21, 1961, pp. 557-583
     _____ : Städtepolitik, Stadtumbau und Stadterhaltung in der
     DDR. Stuttgart 1986. (Erdkundliches Wissen, 81)
Sozialistische Wohnungspolitik. Ed. by Akademie für Statts-
     und Rechtswissenschaften der DDR. Berlin (East) 1979
Staemmler, G.: Wohnungswunsche von DDR-Bürgern -
     exemplifiziert an einer Untersuchung der
     Wohnungstauschwünsche Ostberliner Bürger in bezug zum
     Wohnungsbauprogramm der DDR bis 1990. In: Die DDR im
     Entspannungsprozeß. Lebensweise im realen Sozialismus.
     Edition Deutschlandarchiv, 1980, pp. 163-172
     _____ : Wohnungsbauplanung und Wohnungspolitik in den
     Städten der DDR. In: Bauwelt, 73, 1982, no. 35,
     pp. 1412-1421
Statistical Pocket Book of the German Democratic Republic
     1987. Central Statistical Board. Berlin 1987
Statistisches Jahrbuch der Deutschen Demokratischen
     Republik. Several editions
Storbeck, D.: Die Wohnungswirtschaft in Mitteldeutschland.
     Regionale Aspekte und Tendenzen. Bonn/Berlin 1973
Volks-, Berufs-, Wohnraum- und Gebäudezählung am 31.12.1981
     in der Deutschen Demokratischen Republik. Ausgewählte
     Ergebnisse. (East) Berlin 1983
Werner, F.: Stadt, Städtebau, Architektur in der DDR.
     Aspekte der Stadtgeographie, Stadtplanung und
     Forschungspolitik. Erlangen (Institut für
     Gesellschaft und Wissenschaft an der Universität
     Erlangen-Nürnberg) 1981
Winkler, G.: Okonomische und soziale Probleme der weiteren
     Ausprägung der sozialistischen Lebensweise. In:
     Probleme der sozialistischen Lebensweise. (East)
     Berlin 1977, pp. 7-40. (Abhandlungen der Akademie
     der Wissenschaften der DDR, 1977, no. W5.)
Wohnungsbaugenossenschaften. Eine Zusammenstellung der
     wichtigsten gesetzlichen Bestimmungen über
     Arbeiterwohnungsbaugenossenschaften und gemeinnützige
     Wohnungsbaugenossenschaften. (East) Berlin 1960

Chapter Eight

HOUSING POLICY IN ALBANIA

Derek Hall

CONTEXT

> Improvement of the housing conditions of the
> population has always been one of the
> important questions of the Party's policy in
> the field of well-being. (Hoxha, 1981, 54)

DATA SOURCES

There is as yet no abundance of available
hard data on Albanian housing. Although a wide
range of sources have been employed in the
compilation of this chapter, the overall picture
necessarily remains incomplete. For example,
annual (sic) statistical yearbooks were only
published, irregularly, in the 1960s; from the
1970s qinquennial compendia have been the main
general Albanian statistical source. Reports on
five-year plans include no more than a short
paragraph on housing, within a sub-section on
social welfare provision, at best merely outlining
the overall numbers of dwellings built, with
little indication of the type, quality,
distribution, supply and demand equation,
allocation mechanisms or indeed any qualitative
assessment of the housing situation. Regional
monographs are irregularly produced, and while
often useful, appear only to be available within
the appropriate areas of the country, and are, of
course, produced only in Albanian.

The last national census was held in 1979,
but no detailed housing data have been released
from it. A potentially useful new tome on the
Albanian population (Bollano, 1988) had just been
published in Tiranë as this chapter was going to

press. None of the major western texts on post-
war Albanian development (Marmullaku, 1975;
Prifti, 1978; Schnytzer, 1982) address the housing
question.

While movement by foreigners within Albania
is less restricted than it once was (but still
almost exclusively escorted - e.g. see Hall,
1984a, 1984b), whole regions of the country are
still inaccessible unless visited by personal
invitation. Within local areas, access to housing
is also problematic. Most Albanians are
discouraged from communicating with foreigners
such that everyday residential experiences and
first-hand internal observations of Albanian
housing are theoretically impossible to secure.

Data shortcomings therefore unfortunately
preclude the present author from pursuing detailed
debates such as those on housing allocation and
socio-spatial differentiation undertaken for other
Eastern European societies.

THE CONTEXT OF SETTLEMENT PLANNING AND DEMOGRAPHIC
DEVELOPMENT

Albanian state housing management is undertaken through elected people's councils which
carry out internal territorial administrative
responsibilities at village, city, and district
levels. They are charged with economic and
cultural matters and direct affairs of the various
bodies under their aegis, while being responsible
to the higher organs of state power. They meet
twice a year, between which time their work is
undertaken by executive committees elected from
their membership. These supervise and administer
a number of permanent departments. Twenty-six
districts - rrethi - act as the largest internal
administrative units. These are sub-divided into
localities, made up of a number of villages
constituting a territorial unit. At the same
level is the 'regrouped village', comprising
several villages forming a distinct territorial
and economic unit, on the basis of which regrouped
co-operatives have been set up. In the countryside the village is the basic unit of
administrative division, whilst sixty-two cities
and major inhabited centres share equal urban
functions. Larger cities are administratively
divided into quarters.

Housing Policy in Albania

SETTLEMENT PLANNING AND DEMOGRAPHIC TRENDS

In 1944, the settlement system inherited by
the communists exhibited the centuries of neglect
suffered under Turkish rule. Poor construction
methods, inadequate sanitation and hygiene, and
little urban development characterised this
inheritance. Two inter-war decades of
independence had briefly witnessed some suburban
development along 'European' lines to serve a
small administrative and middle-class elite, with
central Tiranë laid out, along Italian architectural lines, as a modern capital. Villa
development at recreational spots such as Himarë
on the Ionian Coast, and on Lake Ohrid near
Pogradec, remained relatively isolated oases of
bourgeois decadence for the very few. Wartime
devastation and impoverishment curtailed further
such developments.

As a consequence, and with post-war socialist
policies emphasising rural development, today
Albania retains by far the lowest level of
urbanisation in Europe (34 per cent), and is
distinctive in having a rural population which is
still increasing in absolute if not in relative
terms (32,000 per year). Unlike some eastern and
many western European countries, such rural
increases are not statistical quirks resulting
from suburbanisation or the growth of dormitory
commuter settlements. Indeed, strong control is
placed upon any form of migration, while centrally
directed ('voluntary') labour mobility provides
strong echoes of Eastern Europe's Stalinist past.

Nevertheless, during the 1951-1955 period of
early industrialisation, the country's urban
population increased by 52 per cent, about 80 per
cent of which was attributable to in-migration
(Geco, 1970). High rural emigration rates in
those early post-war days began to create serious
knock-on effects for rural development, such that,
for example, the increase in agricultural output
for 1956-1960 was only five percent instead of a
planned target of 79 per cent (Borchert, 1975).
Between 1950 and 1960 rural-to-urban migrants
numbered about 130,000, representing 40 per cent
of rural areas' natural increase during that
period (Vejsiu, 1987, 36), or an overall annual
national migration rate of ten per thousand. But
by the end of the 1950s, rural-to-urban migration
had been sufficiently contained to be of
secondary importance to natural increase in

contributing to urban growth (Geco, 1973): Misja
& Vejsiu (1982, 13) report average annual rural-
to-urban migration rates of 3.5 per thousand total
population for the 1960s and 2.7 per thousand for
the 1970s. Skenderi and Vejsiu (1983, 161; 1984,
34) point out that since the mid-1970s rural-to-
urban migration has been substantially slowed down
to a level of six to seven thousand persons per
year (a migratory rate of about two per thousand
population, although sources fail to say
explicitly whether such migration is centrally
directed or 'spontaneous'; nor are levels of
urban-to-rural migration available).

Policies discriminating positively in favour
of rural development, and particularly of upland
areas, have subsequently seen high rural
population growth rates: between 1970 and 1978
representing 72 per cent of the national increase
(Hall, 1987a). However, some Albanian authors
argue that there continues to be an over-
concentration of activity in the Durrës-Tiranë-
Elbasan 'core area'. Here, 26 per cent of the
country's urban population, over 38 per cent of
total industrial output, more than 50 per cent of
national light industrial production and over 58
per cent of the country's engineering can be found
(Qemo & Luci, 1983, 238). They argue that
migration to these areas should be further
restricted.

Although at a much lower level than in the
rest of eastern Europe, some rural-to-urban
commuting takes place: about 120,000 people (about
6.3 per cent of the rural population) are said to
work in industry and other urban employment
sectors but live in rural homes (Skenderi &
Vejsiu, 1983, 162). However, with no private
ownership of cars and a somewhat rudimentary
national public transport system, the physical
nature of such commuting is somewhat less
sophisticated than elsewhere.

Every village is said to possess a doctor,
midwife, nurses and a medical establishment.
Begeja (1984) tells us that all urban births and
98 per cent of rural deliveries take place in
medical institutions (hospitals or maternity
homes). However, in recent years reproduction has
been considered to be in-sufficient - from around
thirty-five per thousand in the late 1960s and
early 1970s, by 1978 the national birth rate
had dropped to 27.5 per thousand (Anon. 1981),

and by 1980 it was down to 26.5 (Anon. 1984). In 1981, therefore, the party leadership declared that the country's birth rate should always remain higher than that of the 1980 level, and an improvement in maternity provision soon began to show results. Whereas the average number of births for the 1978-1981 period had been 71,400, for 1982 the figure was 77,300 (Skenderi & Vejsiu, 1983), representing a rate of 27.7 per thousand (Anon. 1984). Further, 1982 saw an eighteen percent decrease in infant mortality rates, particularly marked in the two-to-five month age group, corresponding to the very period of extended maternity leave granted in 1981 (Skenderi & Vejsiu, 1983, 158-159).

As a consequence, the death rate has continued to decline: from 10.4 per thousand in 1960 to a 1986 figure of 5.7. However, by 1986 the birth rate had also fallen further to 25.3. Significantly, however, the tenth Congress of the Women's Union of Albania, held in June 1988, failed, for the first time in the congress' history, to make an appeal for large families, and unusually, spent little time discussing demographic policy. It has been argued therefore, that rather than attempting to return to previously high levels of natural increase, the Albanian authorities are now more concerned with preventing further decline to levels comparable with other parts of Europe (Zanga, 1988a).

Average family sizes have substantially diminished, a process seen more readily in urban areas than in the countryside. Urban families with more than ten members represented only 0.8 per cent of the total in 1980, compared to 2.2 per cent in 1960. In rural areas, however, the respective figures were 6.3 and 9.1 per cent. Indeed, in 1978 there were still 13,000 mothers with more than seven children, and the Stalinist practice of presenting medals and awards to successfully fecund women continues: on the first birthday of the eighth surviving offspring, the accolade of 'Mother heroine' is bestowed, while third, second and first class awards of 'Mother's glory' are made to women with between four and seven children (Begeja, 1984, 37).

The average family now contains three to five members, in 1979 these representing 79.4 per cent of all urban families and 64.4 per cent of those in the countryside. Most flats appear to be built to meet the need of families of this size. In this closely-knit socialist society, single-person families (households) had been reduced to just 3.3 per cent of all households by 1979 (compared to a 1930 figure of 9.2 per cent, and a current UK figure of some 52 per cent), representing 0.6 per cent of the total population (Begeja, 1984, 22-23).

Four not necessarily compatible goals appear to have been significant in the formulation of Albanian settlement planning policies, all of which possess relevance for housing policy: (a) a reduction of socio-economic differentials within society in terms of: rural and urban, agricultural and industrial, manual and non-manual labour, northern and southern, Albanian majority and ethnic minority dimensions; (b) the provision of guaranteed living standards and facilities for the well-being of the people commensurate with socialist ideals; (c) the preservation and enhancement of agricultural land; (d) the promotion of monuments to the past glories of the Albanian people, in architecture, archaeology, customs and motifs.

As in other socialist societies, the establishment of post-war new towns has been one of the more explicit dimensions of state-directed spatial development. In Albania these settlements are often little more than villages. Their residential areas are invariably made up of apartment blocks, which provide planned residential and service facilities for the workers and their families employed in several categories of economic activity, including mineral extraction centres, mineral processing centres, new heavy manufacturing plants and new centres of food processing.

New town development has added an impetus to the growth of the country's urban population and has aided its wider regional distribution (Hall, 1986). Overall, however, the actual proportion of the country's total population living within urban areas will rise only slowly, even though in absolute terms it is significant. Additionally, the largest urban settlements - those which most benefited from the rapid if brief early post-war rural-to-urban migratory flows - will continue to retain their pre-eminence within Albania's settlement system. But, with increasing population pressure and very little remaining reclaimable agricultural land, the pressing need in Albania is for the establishment of more dynamic rural growth centres and better communications with the more remote regions of the country (Hall, 1987a).

TRENDS IN HOUSING CONDITIONS AND CONSTRUCTION 1945-1985

Overall trends

By the end of World War Two, over 62,000 houses - possibly a quarter of the total stock - had been damaged or destroyed, and with generations of neglect, the Albanians' post-war housing problem was arguably the most formidable in Europe in relative terms. Early efforts to improve housing conditions and keep pace with population growth, saw 185,000 new flats and houses building between 1945 and 1970. But at an average of 74,00 dwellings per year, this effort substantially fell short of the country's requirements at a time when the annual population increase was between 40,000 and 50,000 (Table 8.1). Further aggravated by the ravages of earthquakes (the effects of the November 1967 earthquake, for example, required over six thousand buildings to be rebuilt or repaired), the housing question was subject to rare public debate in 1967.

In this year of the inauguration of Albania's 'cultural revolution', the country's leader Enver Hoxha spoke of a national housing crisis which was particularly critical in urban areas. While citing rapid industrialisation and a high birth rate as contributory factors, the Party Secretary appeared unable to offer a solution to the problem, and in 1971 the matter was again raised by a leading party official in the press, who argued that if such social problems as housing were not frankly discussed in public, forcefully expressed social discontent could result (Logoreci, 1977, 147). One approach, which gained particular significance in the 1970s, was to further emphasise rural development to the extent of rusticating urban youths and cadres - on a basis not dissimilar to Chinese practice - to ease the pressure on urban areas and enhance food production. In the wake of such policies, an emphasis was laid upon the extension of rural house building programmes to meet the needs of the urban-to-rural influxes:

> Of new houses to be built by the state ...
> (during the 1976-1980 plan period) one
> third will be built in the villages, to
> provide housing for the families of
> cadres who have gone to work in the
> countryside but have their families still
> in town, as well as for the town youth who
> have begun to go as volunteers to work and
> live in the countryside. (Shehu, 1976,
> 71)

> The improvement of the housing conditions
> of the people constitutes a question of
> major importance for the Party and the
> state ... great efforts will be necessary
> in this field, especially in the villages,
> because apart from the need to improve
> housing conditions for the co-operativists,
> many young men and women from the towns
> will be going to work and live there
> permanently. (Hoxha, 1977, 61)

In the 1970s, mechanised and pre-fabricated
building systems were introduced to speed the
development of new urban residential areas. In
practice, however, only limited application of
pre-fabrication methods appears to have taken
place. In new flat developments, floors are often
made in pre-fabricated sections, but walls are
usually brick with cement facing, if any. A plant
('Josif Pashko') for producing pre-fabricated
sections was developed on the outskirts of the
capital, but only isolated examples of fully pre-
fabricated housing areas are known: five- and six-
storey blocks on the western fringes of Tiranë, in
the town of Kavajë, and at the Kamzë Red Star
state farm in the centre of the country.

It is claimed that over 80 per cent of the
total population now lives in accommodation built
since 1944 (Alia, 1984, 5). Since at least the
mid-1970s, this has meant invariably blocks of
flats in urban areas. In the countryside,
dwellings normally comprise one- or two-storey
family houses for the co-operative peasantry, with
a greater likelihood of small flat blocks for
state farm workers, although exceptions to both
patterns are not uncommon. However, no official
data are available on density/occupation rates,

nor on dwellings deemed to be sub-standard or inadequate.

Quantitative achievements

Table 8.1 attempts to chart dwelling completions and state investment in housing by five-year periods (five-year plan periods did not actually commence until 1951) in the post-war period, from a diversity of sometimes conflicting sources. A number of general trends are evident:

(a)   a continued increase in the number of dwellings constructed up to the 1966-1970 period, and then a falling off, to some extent relating to the relative decline of the country's birth rate. Further, despite an apparent resurgence in the 1980s, plan targets have continued not to be fully realised;

(b)   through formulating a 'housing quotient' - dividing the annual average population increase by the annual average number of new dwellings, to produce a figure representing the annual increase in population per new dwelling built for each five year period - an overall trend can be seen, reflecting both generally declining population growth and increasing numbers of new dwellings constructed, suggesting that the country may be beginning to catch up with its quantitative housing problem. However, such a quotient ignores the additional need for new housing to replace existing stock. Given that 20 per cent of existing dwellings were built before the socialist period, and that much will still be considered inadequate, there still remains a substantial task of replacement. This is now constrained by the physical limits being placed upon the expansion of built-up areas, given the need to conserve as much agricultural land as possible. Sulko (1988, 10) for example, points to the fact that although two thousand new apartments were to be built in Tiranë during 1988, the capital had no vacant building sites, such that old, lower-density dwellings must be demolished and their residents decanted elsewhere, before new higher-density developments can be constructed. Sulko (1988, 10) also points out that within the 1986-2005 twenty-year development plan for Tiranë, in addition to raising overall construction rates, 'special attention is being paid now (sic) to the quality of buildings': certainly the exterior facings and overall appearances of new apartments in the capital are more appealing than previous constructions. It is still being claimed that greater use of pre-fabrication methods will be made in the future, and that floor space and facilities in apartments will be increased. It is also still admitted, for the capital at least, that because of rapid industrial and demographic growth, the housing problem has not yet been finally solved (Sulko, 1988);

(c) with the state enrolling an urban voluntary construction programme to boost its own urban efforts from the late 1960s, urban self-build housing has been squeezed, both through shortages in individual time for such construction and the availability of building materials. By contrast, self-build rural housing has maintained its pre-eminent role in the countryside, even though numbers have not returned to the 1966-1970 peak;

(d) a relatively steady, if declining proportion of state investment appears to have picked up in the 1980s with greater all-round state spending;

(e) the state's share of new housing, compared with other east European societies, has tended to increase, attaining a level of just over half of all new dwellings, the majority of which are urban;

(f) average annual new house completions have fluctuated, although in the post-war period there has been an overall upward trend.

Internal domestic provision
─────────────────────────

Apartments built in urban areas have between one and three rooms in addition to a kitchen and 'necessary annexes', which normally include a bathroom with piped water. As the deputy building minister has argued:

> Whereas we used to build one- or two-storeyed houses in small quarters, now we have gone over to big blocks of tall buildings, with better economic indices. ... They are very different internally too: the small kitchens have given way to living rooms. These are big enough to fulfill several functions in family life... cooking is done in specially built annexes ... external appearances ... are ... simple without unnecessary ornamentation. (Kuke, 1983, 14)

Even built up to elevations of six storeys, the blocks within which such apartments are located are invariably completed without lifts (which would require imported technology). Given Albanian hardiness, this is perhaps less of a problem than in more pampered societies. One must assume that floor allocation of apartments takes into consideration the age and physical circumstances of household members and their ability to negotiate stairs.

Table 8.1: Albania: Dwelling completions and state housing
investment by five-year period

FYP Period	Dwellings Built	Of which:			
		State		Self-build	
		No.	%	Urban	Rural
1945-1950	12,114	1,114	9.2	-	11,000
1951-1955	26,110	7,596	29.1	1,619	16,895
1956-1960	47,413	11,734	24.7	3,514	32,165
1961-1965	44,693	15,808	35.4	2,091	26,794
1966-1970	73,213	29,045	39.7	2,382	41,786
1971-1975	61,908*	32,038⁺̄	51.8	1,476	28,394
1976-1980	56,390+	26,326	46.7	1,240	28,824
(1981-1983	44,051	23,623	53.6	700	19,728)
1981-1985	>75,000	50,000			31,000
1986-1990	85,000◊				

/ Million leks at 1981 prices
+ Percentage of total state investment for five year period
§ Annual increase in population per new dwelling
* 80,000 new dwellings had been the plan target
‾ State flats now include those constructed by voluntary
  labour
+ 65,000 new dwellings - 42,000 rural, 23,000 urban - had
  been planned for this period, the significant shortfalls
  suggesting disruptions in the wake of the break with
  China
◊ Projected plan figures

Table 8.1 (continued)

FYP Period	State sector investment mn. leks /	% ‡	Housing quotient §
1945-1950	[six years]		8.0
1951-1955	165	6.7	6.6
1956-1960	292	6.2	5.0
1961-1965	429	6.6	5.3
1966-1970	579	5.7	3.7
1971-1975	862	5.6	4.8
1976-1980	790	4.3	4.3
(1981-1983	750	6.1	3.9)

Sources: Alia (1986), 60; Anon. (1977), 41, (1984), 29, 106,
114, (1985a); Çarçani (1986) 15, 57, 58;
DeS (1962), 53, (1968), 27; DPS (1974), 21;
Hoxha (1977), 61; Mihaili (1984); Misja & Vejsiu
(1982), 4; Shehu (1971), 36, 101, (1976), 30.
Author's additional calculations

Heating systems of dwellings are essentially of three types:

(a) central heating systems are provided in new developments in upland areas, to combat winter cold, the most important settlement falling within this category being the south eastern town of Korce, situated at an elevation of about a thousand metres; such systems are usually oil-fuelled;

(b) solid fuel comprises coal, often relatively poor quality lignite, which is mined in a number of locations within the country, and whose burning in domestic fires exacerbates considerable atmospheric pollution in winter; and firewood, which is widely collected and burned in substantial quantities. The extent of domestic solid fuel burning is reflected in the fact that appropriately equipped cooking ovens manufactured in Tiranë are prominently exhibited amongst the 'white goods' section of the capital's economic achievements exhibition (see Gurashi & Zuri, 1982). The outward signs of their use are the stacks of firewood often to be seen on apartment balconies, even in mid-summer, and the flue pipes emerging from the roofs and occasionally sides of apartment blocks;

(c) gas is employed particularly in those areas adjacent to the country's major oilfields, such as in Berat and Qytet Stalin (Stalin City); some conversion is taking place, although its extent is unclear. The draft plan for the 1971-1975 period talked of liquified gas to be used towards the end of the plan period as a domestic fuel in the main cities (Shehu, 1971, 101). However, little reference to this ambition has been seen subsequently.

Communal infrastructure

Efforts to upgrade water supplies have taken place at least since the 1960s. For the 1966-1970 plan, new systems were to be installed into several towns and over 400 villages. Draft plans for the 1971-1975 period talked about new water supply systems to be built and old ones expanded

in seventeen centres.   By the mid-1980s, however,
only just over half of the country's villages were
considered to have adequate access to a water
supply: Shkodra and Ganiu (1984, 36) could point
out that 56 per cent of the country's villages had
been supplied with drinking water through systems
built since 1944, while Meksi (1987, 13) suggested
that this applied to about 1500 of 2719 villages
(55 per cent).   Of the remainder, 1100 villages
were considered without water because of their
distance from the nearest springs, or because
water supplies were seasonal.   The eighth five-
year plan 1986-90, intended to bring drinking
water to all remaining villages and inhabited
centres through the con-struction of pipelines to
connect 1200 previously unserved villages, and to
improve the supply to 300 others.   Parts of the
systems in Tiranë, Durrës and Vlorë were also to
be upgraded.   Under the plan, each village would
have a tap for every sixty to seventy residents,
or group of ten to twelve dwellings, or the
equivalent of sixty litres of water per day per
inhabitant (Meksi, 1987).

Great publicity was given to the fact that
every village in the country was in receipt of
electricity by October 1970.   The subsequent 1971-
1975 plan period saw the completion of telephone
links to all villages, and a striving (not then
completed) to make all villages accessible for
motor transport through an extension of the rural
road network.

For the 1976-1980 period, in the wake of the
break with China, exhortations for quality
improvement in services continued to issue from
the party leadership.   When introducing the five-
year plan, the then prime minister implied
substantial shortcomings, particularly in overall
infrastructural provision, when he argued:

> Workers of communal services should
> extend their network, especially in
> the countryside, and radically
> improve the quality of their
> services. (Shehu, 1976, 71)

Retailing facilities are now built into the
ground floor of new multi-storey developments,
although the proportion of floorspace devoted to
such functions in any development appears to vary
considerably: central place accessibility plays an

important role. For example, in the centrally
located blocks of the upland new towns of Burrel
and Bajram Curri, ground floor shops characterise
most of the five-storey square blocks constructed
since the early 1970s. At the other end of the
country, on the 'pioneer' state farm of Ksamil,
close to the country's south western border, of
the fifteen four-storey blocks only one has shops
on its ground floor, and another a communal
buffet, while separate, single-storey buildings
provide a communal laundry, buffet and food shop.
Skarço (1984, 3-4) speaks of there being over
3,750 rural shops in 1973 to meet 'the ever
growing needs of the peasantry'. This figure
represented one shop for more than every four
hundred rural dwellers.

Other communal facilities found in relatively
recent developments include an underground
defensive shelter, the entrance to which may be
found in courtyards or open space between flat
blocks.

More traditional rural settlements appear to
have had some degree of central place patterning
grafted onto them, with certain villages being
singled out for the location of a post office and
telephone, polyclinic, school, creche,
kindergarten, library and house of culture, public
baths, public service office, MA-PO (People's
Store) 'supermarket', other shops and at least one
buffet, this list appearing to comprise the top
end of the rural service provision hierarchy (e.g.
see Basha, 1986, 30).

TENURE STATUS

Although the term 'housing class' is not one
which the Albanians would wish to use or indeed
readily recognise, one can distinguish two basic
housing sectors: (a) state, involving rented
accommodation; and, (b) private, representing the
most important element of 'personal property'
ownership available to Albanian citizens.

The state rented sector

The state sector has dominated the urban
housing market for much of the post-war period,
but appears only to have made an impression in
rural areas in the last two decades (Table 8.1).

Construction of blocks of flats in urban and rural areas represents and symbolises both a raising of living and hygiene standards compared to the past, and emphasises aspirations of egalitarianism by producing similar internal and external residential environments for the majority of the state's citizens. By building apartment blocks in the countryside, particularly at state farm centres, differentials between town and country, both tangibly and symbolically, are seen to be reduced.

About 60 per cent of the total population now live in flats, most of which will fall into the state sector, whether urban or rural. In Tiranë, some 25,000 new apartments were built between 1945 and 1974, during which time the city's population trebled from 60,000 to 180,000 (Lleshi, 1977); Basha (1986, 16) quotes 36,770 flats having been built in the capital between 1944 and 1983. Today, approximately 80 per cent of Tiranë's population lives in state-built apartments (Carter, 1986).

In terms of allocation procedures, a household's flat size should be matched to the size of family. However, given the country's relatively high birth rate, and thus continuously changing individual household sizes, a strict adherence to this principle would demand constant residential relocation, contributing to high levels of internal circulation. This appears not to be the case. In practice, new flats appear to be limited to two or three size variations at most.

Couples about to marry need to register with their People's District Committee for a flat. Once registered, newlyweds have an automatic right of access to an apartment, which is theoretically allocated within a few months of the initial application. However, one might expect considerable local and regional variations in both housing availability and ease of access, although data on this and other significant variables, such as regional marriage rates and household sizes, are not accessible.

In addition to apartments built directly by state building organisations are those constructed by 'voluntary labour'. The collective construction of housing through mass mobilisation emerged in 1968 when a pioneer programme was launched in Tirane during the country's 'cultural revolution'. The definition of 'voluntary' donation of labour and services appears to be a rather loose one in the Albanian context, and owes its pedigree both to the pre-war corvee system and to Stalinist methods of mass mobilisation, whereby both 'volunteers' and other groups are employed on major construction works. In response to recent earthquakes (1979, 1983, 1988) (e.g. see Dede, 1983; Popovic & Milic, 1981; Qerimi, 1981; Smith, 1979), the whole of the working population of Albania has been required on each occasion to donate one day's pay towards reconstruction costs, articulating the dogma that such work is undertaken relying solely on Albania's 'own efforts'.

The principle involved in voluntary construction efforts is similar to the Soviet Gorkii method (Stretton, 1978), and the later (from 1971) more successful Cuban micro-brigade system (see Hall, 1981, 1989), but does differ in some significant respects. Groups of volunteers are organised within the framework of their residential area or place of employment. Unlike the Cuban system, which sees voluntary workers undertake such construction during their normal working time, whilst their colleagues back at the workplace attempt to compensate for their absence, the Albanian voluntary principle depends upon construction being undertaken by comrades in addition to their normal work patterns - i.e. entirely in their 'free' time. Otherwise, the principle is similar: the state, through the executive committee of the local people's district council, provides the building materials, and technical assistance is made available by state building organisations (Pajcini, 1983).

Final allocation of such housing is said to be a matter of full meetings of all persons at the workplace, given that co-operation based upon the place of employment is the usual basis for activity. However, it is the trade union's role, perhaps in conjunction with the enterprise director, to designate final dwelling allocation, according to 'the existing living conditions of each worker and his contribution to the construction of these buildings' (Anon. 1983, 8). Given the relative dearth of empirical data on this topic, one can only deduce the effectiveness of such effort. As other socialist societies have found, while mass mobilisation may be ideologically important and sustaining, resultant poor work quality and low productivity have tended to see such methods phased out, with the exception of a recent revival of the Cuban micro-brigade system (Hall, 1989). For example, voluntary construction undertaken on behalf of the Korçë Knitted Goods Combine has seen, over the past two decades, just two blocks - housing some thirty families - provided, for a total work force of some 4,000. In Durrës the number of homes built with voluntary labour is said to equal that of those built by state building enterprises (Anon. 1983).

Sixty per cent of rural dwellers live in state-owned dwellings, many of which are not flats. ´ The building. of new one- or two-storey state-owned rural housing has at least three functions: (a) to provide a minimum standard of domestic facilities, such as internal plumbing, where none existed previously; (b) to establish new state farm centres; (c) to aid the nucleation of rural settlement generally. Such dwellings are also built by the state in response to particular needs, such as replacement dwellings for those lost in earthquakes (for example, in response to that of April 1979 in the north of the country, 2,441 new houses and apartments were built). But usually in such latter cases, the replacement housing is provided rent-free.

A number of essential requirements are laid down in the planning of rural developments: (a) they must occupy the least possible productive agricultural land; (b) building foundations must be on solid ground to withstand earthquakes (all of Albania falls within a zone of crustal instability); (c) the buildings must receive ample sunshine and be protected from prevailing winds; (d) the dwellings should harmonise with each other; (e) each home must have convenient road access to all the facilities of the village (Qerimi, 1981, 70-71).

Two standard types of dwelling houses are provided by the Tiranë Design Institute: (a) the 'Elbasan' style: single-storey dwellings with three rooms and a verandah, with internal facilities and provision for solid fuel; and (b) the 'Shkodra' style: two-storey houses for two families, with separate stairway access, again providing three rooms plus internal facilities and solid fuel provision.

Housing Policy in Albania

The private owner-occupied sector

This sector can be divided into two distinct groups:

(a) <u>Surviving pre-socialist housing</u>. Particularly in urban areas, such dwellings have avoided both compulsory purchase and demolition for urban reconstruction, on the grounds of their architectural or historic merit. In particular, areas of traditional housing in the towns of Berat and Gjirokastër have been put under state protection, but not ownership, because of their intrinsic collective importance, and both settlements have been declared 'museum cities'. Areas of pre-socialist construction with merit in other major settlements - such as at Shkodër and Korçë - are also protected by the state. In most cases, the dwellings in these areas are either family houses of one or two storeys, or larger buildings perhaps divided between two or three families. Owner-occupation appears to be the predominant form of tenure. The Albanian literature on these dwellings, emphasising their architecture and historical importance, is relatively abundant and reasonably accessible (e.g. Anon., 1973a; Muka, 1985; Riza, 1971, 1973, 1978, 1981a, 1981b; Strazimiri, 1987; Suli, 1982).

A typical surviving pre-socialist urban house in a town such as Korçë is a nineteenth century two-storey dwelling occupied by a single three-generation extended family. There are five rooms. An outside squat loo has running water, either to a mains sewer or septic tank. Usually there is an interior kitchen possessing mains water. Often there is a courtyard and/or garden with plentiful shade.

In rural areas, where approximately 40 per cent of the population live in privately-owned dwellings, pre-socialist dwellings of an adequate standard for continued habitation have also remained in private hands throughout the socialist period.

(b) <u>Self-built dwellings</u>. Although, theoretically, private self-building can take place in both urban and rural contexts, the primacy of state-sector activities and planning regulations, and the problem of the availability of appropriate materials, would appear to have presented increasing obstacles to the urban private self-build sector, which appears to have consequently declined particularly since 'voluntary construction' was enlisted by the state (Table 8.1).

The state provides credits for the construction of new (private) self-built dwellings, but makes a point of denying such funds for the purpose of outright purchase of existing houses. Urban dwellers within this category are permitted a building plot area of 200-300 square metres, the actual size depending upon local circumstances and the wishes of the local people's council. Such a plot usually includes provision for the personal garden.

The collectivisation of agriculture was completed in March 1967. In the majority of agricultural co-operatives, most dwellings are single-family, privately-owned houses built with family funds. Each co-operative has its own physical development plan, which is formulated jointly by the co-operative's management and the local people's district council. This structure plan contains a 'yellow line' to indicate the planned limit of expansion for the built-up area of the co-operative. Land is set aside for new housing within this yellow line, and each co-operative family requiring a new dwelling receives a plot of about 500 square metres. Usually there is no maximum size of house prescribed, as long, of course, as it can be contained within the pre-determined plot area: it may simply depend upon a family's ability to afford the cost of the building materials. Except along the coastal plain, building stone is readily available locally, so that such a cost should be minimal.

In addition to the dwelling house, every family has its personal garden, which may be typically devoted to grapes, apples, potatoes and tomatoes. Co-operative members are allowed to keep chickens, rabbits, bees and limited numbers of other livestock, and anything is permitted to be grown in the plot, for family consumption or to be sold either to the co-operative or in the nearest town. Personal physical access to the latter may, however, be difficult, and most households appear to sell straight to the co-operative, despite the lower prices usually received. The average size of garden is around 1,000 square metres per family, although between the mid-1960s and mid-1970s co-operativists' plots were reduced

in size by between a half and two-thirds, and their livestock reduced in number also by half (Skarço, 1984, 58). The size of each garden should be determined by a general meeting of the members of the co-operative. Most plots are located adjacent to the homes of their owners, and although there is still talk of phasing the gardens out altogether within ten years, it is being emphasised that if this takes place it will not affect co-operativists' private housing rights.
Overall, dwellers in private owner-occupied housing retain rights of inheritance, at least for their offspring.

> The houses which the citizens have built themselves or have inherited from their parents or other close relatives, along with their furniture and other facilities, irrespective of their value, are personal property. (Haxhiu, 1986, 20)

Private houses can be sold but not for speculation. They can also be rented, if, for legitimate reasons, the owner or his family do not live in part or all of the dwelling. The rent cannot exceed that paid for state-owned houses, and no more than one dwelling can be owned. A private rented sector of sorts therefore exists, but it is assumed to be relatively small; no published data are available.

HOUSING TYPES AND THEIR DISTRIBUTION

Albanian dwellings can be divided into three very general physical types:

(a) Detached dwellings

These are occupied by one or more households, and may be rural or urban, pre-socialist or post-war:

i. Rural one- or two-storey. Many traditional and modern local and regional variations of domestic dwelling types exist, e.g.: (a) the traditional northern upland semi-nomadic pastoralists' fortified kula dwellings, made

380

from local stone and with few windows at ground
level; (b) large zadruga dwellings bordering the
Kosova region (e.g. see Muka, 1985; (c) lowland
settled agriculturalists' outwardly simple single-
storey dwellings of mud, wattle and thatch (e.g.
see Cavolli, 1987a); (d) 'Greek' village
dwellings of two storeys, made from local stone
(usually limestone), and perched on hillsides
overlooking the valleys of the Dropull region
(known to Greeks as northern Epirus); a number of
these have now been abandoned for state-sponsored
settlement developments further downslope (see
Suli, 1982, for an assessment of the fortified
houses of the Konispol district); (e) early post-
war housing developed with particular local
architectural motifs and built en masse in newly
reclaimed agricultural areas: e.g. brick-built
one-storey houses of the Maliq Plain with
characteristic overhanging central roof pieces
(this style perhaps echoes pre-revolution days
when in the then swampy and malarial Maliq region
only beys and other land-owners were permitted to
have chimneys on their dwellings); (f) housing
often built by 'voluntary' means to rapidly
replace that destroyed in earthquakes or other
natural disasters: e.g. the orderly rebuilding of
villages south of Shkodër following the earthquake
of April 15th 1979.

ii. Old urban one or two-storey. These are
rapidly disappearing as most town centres are
being re-developed, such as in Tiranë, Elbasan,
Shkodër, Vlorë and Durrës. In the latter case, a
substantial area of both nineteenth and early
twentieth century family dwellings are being
cleared as excavation of the major Roman
amphitheatre gradually proceeds. In former Moslem
strongholds such as Elbasan, old town walls still
circumscribe, and bazaar-like urban forms still
exemplify, formerly important areas of mixed
residential, retailing, wholesaling and
manufacturing activities and land uses (Cavolli,
1987c).
    Well-built pre-socialist dwellings
(implicitly on 'European' lines) in favourable
locations such as the two-storey villas at Himare
on the Ionian coast, or those adjacent to Lake
Ohrid, to the south east of Pogradec, appear to be
used for recreational purposes by high party and
governmental functionaries. In the city of
Durrës, the country's major port and seaside

resort, a number of quite substantial Italian influenced dwellings exist: Berxholi (1986, 30).

Preserved urban two-storey dwellings are notable in a number of towns. Early in the country's post-war development, legislation on the protection of the monuments of culture and the environment was enacted, which was particularly aimed at ending the plunder and degradation to which the country's ancient monuments had been subjected. As noted earlier, the towns of Gjirokastër and Berat, together with the subsoil (sic) of Durres (e.g. see Toci, 1971; Karaiskaj & Bace, 1975; Riza, 1975; Adhami & Zheku, 1981; Strazimiri, 1987), received the official accolade of 'museum city' on account of their rich architectural and archaeological heritage. Gjirokaster and Berat received this status in 1961 by a 'special decision' of Albania's Council of Ministers. This meant that all buildings and activities within a prescribed area were under the protection of the state. The conservation of existing buildings, particularly in groups, and avoidance of new development, are central to this philosophy. Krujë is a 'heroic city' on account of its historic role against Turkish incursions: somewhat ironically, a Turkish bazaar has been the subject of considerable restoration work here.

All four cities are focal points on the prescribed itineraries of foreign tourists (Hall, 1984a, 1984b; Hall & Howlett, 1976). As national monuments, these towns attract considerable attention from the state, representing an indirect investment in the tourist industry and a more direct investment in housing. One of the embracing arms of such state activity is represented by the Institute for the Restoration of Monuments of Culture. This body has seven major branches, located in the key centres of Tiranë, Berat, Gjirokastër, Korçë, Shkodër, Vlorë and Peshkopia. Of the Institute's five major sections, it is that for Traditional Buildings which is chiefly concerned with dwelling houses in traditional urban and rural styles.

In Berat, three of the city's districts are regarded as conservation areas - 'museum quarters', and receive special attention from the Institute. The Mangalem district, for example, adjacent to the city's fortress, contains many typical two-storey dwellings with slightly protruding wooden balconies on their upper floors, and many finely decorated arched windows and

pantiled roofs. Such dwellings have been built on
terraces, lining the lower hill slopes leading up
to the fortress. With their appearance of many
windows facing outwards, they have contributed to
Berat's characterisation as the 'city of a
thousand windows'. Open landings are also typical
of such dwellings, presenting broad views out for
their occupants, and are often partly or
completely encircled with skilfully carved stones.
The city was hit by an earthquake in 1851, and
although much subsequent rebuilding has taken
place in the old quarters, the distinctive
character of these essentially privately-owned
family dwellings is still much in evidence
(Strazimiri, 1987: Cavolli, 1987b).

The maintenance and enhancement of such pre-
socialist relics has required ideological justification:

> By preserving and developing all the
> healthy and progressive elements of
> the national culture of the past and
> of the popular tradition ... (we)
> assist the Party in the communist
> education of the masses, in preparing
> them to cope with the imperialist-
> revisionist ideological aggression.
> (Xholi, 1985, 29)

(b) Low rise flats

Usually of up to three storeys, but
occasionally four, brick-built, with little
evidence of mechanised methods of construction,
these typify early socialist urban dwellings, such
as those along the main streets of major towns -
notably Durrës and Elbasan - as well as in lesser
urban centres such as Peqin, and in rural developments associated with state farms, such as at
Levan and Lukovë.

(c) Apartment blocks

Rising to five, and occasionally six storeys
(notably in Tiranë, Durrës, Fier and Ballsh), and
constructed with some degree of mechanisation and
pre-fabrication, these are usually urban
dwellings, although they are also represented in
more recent developments at state farm centres,

such as the five-storey blocks at Kamzë and Malic
Muci. As noted earlier, only very limited
comprehensive pre-fabrication has taken place, at
developments in Tiranë, Kavajë and Kamzë. Most
apartment blocks are otherwise brick-built, often
with local stone foundations, with or without
concrete or similar types of facing. None of the
blocks appear to contain lifts, these remaining
the preserve of such public buildings as hotels
and palaces of culture.

FINANCE

Incomes

Up to the early 1960s, Albanian pay scales
were similar to those in the rest of socialist
Eastern Europe. After the break with the Soviet
bloc in 1961, a shortage of suitably qualified
experts saw the emergence of income differentials
as a result of the need to encourage acquisition
of better qualifications. Physically difficult
jobs also offered higher remuneration.
Consequently, by the mid-1960s, workers in the
highest paid sector, transport, received an
average income of 71 per cent more than those in
food, the lowest paid sector (Lika, 1964). In
response to this, attempts began in 1967 at
equalisation, by reducing the highest salaries.
All incomes above 1200 leks per month were cut
(the minimum salary would have been about 500 at
this time), although this was to some extent
compensated for by the abolition of all forms of
income tax. However, 'supplementary' payments
were still significant in making up wage packets,
and official claims that the ratio between highest
and lowest pay ranged from 1:2.5 to 1:3 appear a
little unrealistic.

In April 1976 income equalisation was pursued
further with particular regard to urban-rural and
upland-lowland inequalities. This entailed
measures to reduce: (a) all salaries above 900
leks by between 4 and 25 per cent; (b) incomes of
teachers and scientists by 14 to 22 per cent; and,
(c) bonuses for scientific titles, degrees and
publications by up to 50 percent, together with
the abolition of other bonuses and payments. On
the other hand, wages of state farm workers were
raised, albeit linked to plan fulfilment. All

specialists were to be paid according to their field of specialisation and not in relation to the district or institution in which they worked. Finally, certain unspecified 'disproportions' observed in the wages of workers in the fishing and maritime transport sectors were to be adjusted.

The reported consequences of these measures were to see the ratio between the average pay of workers (i.e. urban and state farm employees) and the salary of a ministry director, brought down from 1:2.5 to 1:2. This ratio has often been subsequently wrongly quoted as representing the whole range of Albanian incomes (e.g. see Bollano, 1984): but it is only a comparison of sector averages, and excludes the co-operativist peasantry.

The 1976 pronouncement on the need to reduce rural-urban differentials was not new. Indeed, between 1971 and 1975 while national average per capita real income increased by 14.5 per cent, that for rural workers increased by 20.5 per cent compared to only 8.7 per cent for urban workers. (Also, in 1972, the co-operativist peasantry were provided with state pensions for the first time.) Neither was attention to improving upland areas new - the major slogan for the 1966-1970 plan period - 'let us take to the hills and mountains and make them as fertile and as beautiful as the plains' - had symbolised that. But by explicitly referring to the need for priority to be given to the improvement of living standards in rural and upland areas, the 1976 signals were suggesting that all was not as well as it should have been. Various subsequent rural development polices have aimed at positive discrimination in favour of rural rather than urban areas and uplands rather than lowlands (Hall, 1987a).

One consequence of this policy has been a faster rise in rural incomes than urban: while for 1960-70 the rate was 140 per cent greater, for 1971-80 it was up to three times greater. From 1980 to 1982 per capita income in the countryside rose by 28 per cent compared to 4.1 per cent for the population as a whole. The 1980 rural population, at 1.75 million, was greater than the country's total population in 1960 (1.61 million), and is officially viewed as reflecting the success of a policy which has attempted to prevent unplanned rural-to-urban migration and to ensure the country's upland areas are not depopulated

(Papajorgji, 1982). Even so, Skenderi & Vejsiu (1983, 161) talk about the demographic imbalance in some of the country's north-eastern uplands resulting from out-migration of young females.

> In some instances the physical movements of the population are expressions of lack of proper co-ordination in the plan of territorial distribution between the material elements of productive forces and the sources of labour. (Qemo & Luci, 1983, 236)

The state rented sector

If Albania has proceeded to sell off state flats to sitting tenants, as other socialist societies have done, then no evidence exists. The state sector within housing is therefore essentially concerned, through the district people's councils, with the renting and maintenance of flats, either urban or attached to state farms.

'Our rents remain symbolic' (Mullaj, 1984, 5), such that

> In practical terms, the daily rent is not more than the price of a cup of coffee in a second-class cafe. (Anon., 1985b, 22)

> A month's rent is equal to two days' pay of an ordinary worker. (Shkodra & Ganiu, 1984, 36)

With average Albanian incomes at 600-650 leks per month, and an overall range (of averages) said to be 530-1000, typical rent levels cited by official sources (Anon., 1985b, 22) for urban apartments are twenty, thirty-three and forty-two leks per months for a one-, two- or three-room dwelling respectively, each with kitchen and 'necessary annexes'. These represent about 3.5, 5.5 and 7 per cent respectively of monthly incomes. However, given that most households will compromise at least two wage earners, as the right to work is forcefully pursued, the proportion of family income spent on rent is effectively a half of these figures. While official sources claim to have abolished income tax, no mention is made in relation to housing costs of payments for such services as water and electricity.

The private owner-occupied sector

State credits are offered for the self-building of new (private) dwellings, but are explicitly not made available for the purchase of existing ones. The state requires credit repayments to be made - with apparently no interest levied - at rates of about fifty to sixty-five leks per month, extending over a repayment period of up to twenty years.

Private houses in urban areas are said to fetch around 20,000-30,000 leks, but the methods by which sale prices are arrived at and monitored by the state are little discussed. The concept of the open market is not one that Albanians would readily recognise.

Both urban and rural dwellers in owner-occupied property are able to retain rights of inheritance at least for their offspring. On marriage, pre-existing property and ownership rights are retained by the individual.

THE ROLE OF HOUSING INSTITUTIONS AND ORGANISATIONS

Design Institutes

In 1947, the country's first institute of studies and designs, including town planning designs, was established in Tiranë - the Projekti state enterprise. Albania had little indigenous experience of such matters at the time, and the enterprise was established with Soviet aid. Today, all new buildings in Albania are supposedly constructed in accordance with standard designs provided by the Study and Design Section of the Institute of Town Planning and Architecture based in Tiranë, while Tiranë University's Department of Architecture and Construction Engineering includes a Town Planning and Architecture section to train cadres and undertake research work (Mosko, 1984).

Housing Policy in Albania

The Institute has town planning and design
branch offices in each of the country's
administrative districts, to assist the people's
district council, which incorporate such a section
within their executive committee. It is this body
which plans the layout and modifies the external
designs of the standard plans, according to local
circumstance (to harmonise with existing buildings
and relate to topographic requirements). Such
modifications do appear, however, to be limited to
such elements as changing balcony positions, the
distribution of windows and external colour
schemes.

> The socialist character of
> architecture is expressed in its
> simplicity and beauty which cannot
> put up with extravagance and
> formalism. (Mullaj, 1984, 5)

Within the framework of five-year plan
targets for construction and services, each town
has its own regulating (structure) plans, which
are continuously updated through consultations
between the rrethi administrations, branch offices
of the Institute of Town Planning and
Architecture, and central ministries. For
example, the regulating plan for Vlorë city,
published in 1972, covered a period of up to
twenty-five years and envisaged a doubling of the
city's population to 120,000 by the end of the
century. Both a major expansion of the urban area
and a comprehensive redevelopment of the city's
central area were planned, the latter encompassing
a new city hall, palace of culture, theatre,
tourist hotel and market place, interspersed with
pools, fountains and greenery. New residential
districts would incorporate public gardens,
schools, kindergarten and creches, shops and playgrounds (Anon., 1972a).

Building and physical development

Pointing to the country's limited arable area
and extremely rugged relief, an
uncharacteristically critical public assessment of
land-use management during the 1971-1975 plan
period (Shehu, 1976) declared that the Party,
state and economic organs had been very negligent
and careless. At a time when great efforts were

being put into opening up new agricultural land, often under very difficult circumstances, misuse of the existing fund of land was considered impermissible. Of particular concern was the loss of some 5,000 ha of the most fertile arable land to various construction projects. Shehu emphasised that land-use planning controls had not been sufficient, either in their theoretical basis or in their implementation. In one of the most explicit urban physical planning directives ever to be voiced by an Albanian leader, Shehu, then Prime Minister, argued that all new urban buildings should henceforth be of several storeys, and implied that all existing single- or two-storey urban dwellings and institutions should be replaced (Shehu, 1976, 78).

In urban areas the state takes an overseeing responsibility for apartment construction and overall administration. Almost without exception, all new dwellings are flats; these are, however, built in two ways: either by state building enterprises - construction arms of the central ministry, or by voluntary labour. It is claimed that the number built by both methods is roughly equal, although detailed figures are not available (see Table 8.1).

> In every district there is at least one enterprise that builds housing, along with other buildings. In the capital there is one specialised enterprise engaged solely in house building. It includes the plant that produces pre-fabricated apartments. Its products go to other districts too. (Kuke, 1983, 14)

The 'voluntary' route has been developed since 1968 when a pioneer programme was inaugurated in Tiranë during the country's 'cultural revolution'.

In the countryside, the co-operative peasantry is responsible for most rural housing which is not collectivised (Table 8.1). The state administers all flats built in association with state farms.

Housing Policy in Albania

Allocation mechanisms

The executive committees of the district people's councils (and of the people's councils of city quarters in the larger urban centres) not only guide the planning and construction of new dwellings, but determine the allocation of those built by state building enterprises. For the urban apartments built by voluntary labour, allocation decisions are undertaken at the workplace - usually a service or production enterprise - through which the volunteers are organised. The existing living conditions of each worker, his/her contribution to the construction effort, and other, ideologically related factors, are taken into consideration at meetings of the workers' collective for this purpose.

PROBLEMS AND POLICY ORIENTATIONS

Housing quantity and demographic trends

A major practical problem appears to lie in shortcomings in the absolute provision of a sufficient housing stock. Given continued high birth and marriage rates, evidence suggests that in Albania, as elsewhere, housing supply cannot keep pace with demand, despite the infusion of mass mobilisation concepts into housing construction.

Emphasising the shortage of urban apartments, new flats are often inhabited before services and outer facades have been completed. Problems of quality control appear to be abundant, while maintenance shortcomings reveal bottlenecks in a number of areas, and at several stages in the housing programme. These questions are now being seriously addressed, although with what long-term consequences remains to be seen.

By the year 2000, Albania plans to have a population of four million, largely by maintaining present birth rates and lowering infant mortality. Some 375,000 new agricultural workers will be available as the rural population increases by 42 per cent between 1982 and the end of the century. The urban population is planned to increase by 44 per cent. Currently, the rural population is annually increasing by 32,000, while the towns are witnessing a national increase of 22,000 per year

(Skenderi & Vejsiu, 1984, 33-35). Up to 1979 about 300,000 urban and rural flats - mostly of two rooms and a kitchen - had been built in the post-war period.

With current annual construction rates of at most 6,000 rural and 9,000 urban dwellings, sustained population increases together with rising marriage rates - up to 8.9 per thousand for 1982 compared to 8.1 for 1980 and 6.8 for 1970 (Skenderi & Vejsiu, 1984, 34-35) - are continuing to place a significant strain on housing provision, given the need for new housing to replace existing inadequate stock as well as to keep pace with demographic demands. Certainly Biber (1980, 551) has reported a shortage of flats, particularly for newly married couples.

Major demographic appraisals of Eastern Europe by 'western' authors have been noticeably reticent in dealing with Albania (e.g. Besemeres, 1980; Kosinski, 1977; Kostanick, 1977; McIntyre, 1975). This in part reflects the unique and distinctive position of Albanian demographic patterns and their role within a spatial planning framework. These are expressed in a number of ways:

(a) the relatively high levels of natural increase partly reflect Islamic and Roman Catholic backgrounds, but are also a consequence of: the almost total absence of public availability and discussion of birth control measures, reflecting the state's desire to maintain high manpower levels for the country's future military and economic strength, and for youthful revolutionary fervour to submerge old beliefs. Some thirty-seven percent of the population is under fifteen years of age, while three-quarters of all Albanian citizens have been born into and educated within the framework of Albanian socialism. The average age of the population is just twenty-six years (Skenderi & Vejsiu, 1983, 156);

(b) birth and death rates take on a greater demographic significance given that international migratory movement into and out of Albania is virtually nil, and internal migration is now almost wholly sponsored by the state in relation to labour construction requirements for major projects;

(c) regional variations, which, in such a small country, have reflected a picture of diversity the authorities would prefer to be not quite so explicit (Hall, 1986, 1987a). Detailed demographic data are only available for up to 1973. Nevertheless, for the period 1960-1973 there were significant differences in the population growth rates between the northern and southern upland areas. Commenting on this for the 1965-1971 period, Borchert (1975, 179) suggests that the contrasts of high birth rates in the northern alps compared to the relatively low rates of the southern uplands, is probably partly accounted for by religious differences, with the north traditionally strongly Roman Catholic as opposed to the Orthodoxy of the south. The official abolition of all religion in 1967 apparently had little effect on relative birth rates in the short term. Although the data are incomplete, for 1960-1973 the two northern districts with by far the country's highest growth rates - Mirdite and Krujë - were also the only two to contain three major new towns each, the construction and subsequent development of which would have significantly contributed to population growth, initially through in-migration and later by natural increase. In the southern districts by contrast, only one or two major new towns were established before the mid-1970s (Hall, 1986).

While the population of Albania is young, particularly in the cities, an increase in numbers of citizens of pensionable age is taking place at a higher rate than for population growth as a whole. Indeed, it appears that although the current birth rate, at around twenty-seven per thousand, coupled with the death rate, at about six per thousand, produced an annual average population growth rate between 1981 and 1985 some four times the European average, measures are still being undertaken to stimulate birth rates in the face of a gradual long-term decline (from a peak of around thirty-five per thousand in the 1960s and early 1970s). In 1981 the period of paid pregnancy leave from the workplace was extended from one and a half to six months, and it was noticeable that 22,000 more births were recorded for the four-year period after this measure than for the four years preceding it. As a corollary, infant mortality was said to have been halved between 1979 and 1985 as improved health and welfare facilities were developed,

particularly in the countryside. Shop prices of
children's articles have been deliberately
lowered, while the need to emphasise pre-school
education has been acknowledged particularly for a
previously unrecognised category - 'low income
families' (Vejsiu, 1987, 32). The unprecedented
use of this relative term by an official Albanian
spokesman says a great deal about changing
attitudes within the country's leadership in the
post-Hoxha era. Emphasis is also now being placed
on the 'protection and strengthening of the health
of women' and of their domestic facilities.

PROSPECTIVE DEVELOPMENTS

The seventh five-year plan, for 1981-1985,
aimed at building 80,000 new dwellings, but a
report to the Ninth Party Congress held in 1986
referred simply to 'over 75,000' being built -
only about 2,000 more than for the 1966-1970 plan
period (Table 8.1). In the late 1970s, Hoxha was
emphasising the need to exert great efforts in the
rural housing sector, not only to improve
conditions for the co-operativists and state farm
workers, but to provide for the permanent transfer
of young urban dwellers. Since Hoxha's death,
rather as in China after Mao, such 'revolutionary'
policies as youth rustication have gradually, and
quietly, been de-emphasised, such that a long-term
reorientation of housing priorities may be
required. The recent apparent shortfall in
construction levels may reflect the uncertainties
of this interim period. As noted earlier, high
priority is, however, being given to improving
infrastructure and services, particularly rural
water supply.
Some 85,000 dwellings were planned to be
built for the 1986-1990 period, and the aim to
provide drinking water for all villages and
inhabited centres by the end of the period was
receiving a high priority. An upgrading of the
quality of new dwellings - initially focusing on
the capital - appeared to have assumed an
implicit, if not explicit emphasis. Outwardly,
the quality of necessary mass-produced housing has
been generally very poor. But to now explicitly
declare a policy shift emphasising quality would
be to admit to past shortcomings, shortcomings
which, in the form of a now substantial state-

owned housing stock, will be with the Albanian people for some considerable time to come. Whether the Albanians ultimately take the road of their neighbours and shift the burden of the upkeep of this stock from the public to the private domain remains to be seen. It will be intriguing to watch Albanian decision-makers grappling with such equity/efficiency trade-off questions in an era of potentially rapid change.

Although the country's leaders continue to denounce Soviet perestroika (Zanga, 1988e), adjustments to remove 'outmoded' mechanisms and ways of thinking (Zanga, 1987d), especially since the party's fifth plenum in March 1988 (Zanga, 1988i), have seen, for example, public debates criticising conservative attitudes towards young people (Zanga, 1988h, 1988n) - a debate which had been rumbling since at least 1984 when the party theoretical paper Rruga e Partise and the daily Zeri i Popullit highlighted slackening discipline amongst the country's youth (Biberaj, 1985, 38). Attacks on 'leftists' for opposing innovation (Zanga, 1988l), and a resolution calling for the replacement of all party bureaucrats after five years in office (Zanga, 1988g, 1988m) have also focused attention on potential policy changes. Such debates and modifications, while in no way representing an overhaul of the system, will, in the medium term, impact on such social policies as housing.

However, at the time of writing, there was no outward indications that collective forms of ownership of the means of production were about to be rolled back, nor was there much slackening in the country's emphasis on the need to maintain its demographic strength, despite a discernible drift downwards in the birth rate. A continued growth rate of 2.2 per cent per annum was projected to achieve a population of four million by the end of the century: between 1982 and 2000 the rural population being planned to increase by 830,000 and the urban by more than 380,000 (Skenderi & Vejsiu, 1983, 161). This period would see an increase of about 65 per cent of the population of pensionable age - an issue of no little concern to the authorities - and a rise in the average age of the population to twenty-eight. By the late 1980s almost 75 per cent of the total population will be under forty years of age (Begeja, 1984, 36).

The pace of improving foreign relations has noticeably increased since Hoxha's death in April 1985 (Biberaj, 1985; Zanga, 1988a, 1988c). These have included the establishment of diplomatic relations with West Germany (Zanga, 1987c; Wachsmuth, 1987), promotion of foreign language teaching and literature (Zanga, 1988j) and an improvement in relations with the country's Balkan neighbours (e.g. Zanga 1987a, 1987b, 1988b, 1988d, 1988f, 1988p), despite the ineffectiveness of the rail link with Yugoslavia (Hall, 1985, 1987b) and the continued problems of ethnic Albanian - Serb relations in Yugoslav Kosovo (Zanga, 1987e; 1988k). One longer-term consequence of such improvements could see the taking on of foreign technology for building construction, and consequent higher elevations for apartment blocks, with the greater use of pre-fabrication, installation of lifts and improved service provision. However, the country's current (1976) constitution forbids any loans or credits from foreign countries and institutions, and precludes any notion of joint venture developments employing foreign capital. One potential future irony of note here is the fact that by the early decades of the twenty-first century there could well be more ethnic Albanians outside of the country than within, irrespective of any changes in thinking on personal mobility. The influence that such a situation may have on the country's relations with its neighbours, particularly Yugoslavia, where the bulk of externally-located ethnic Albanians live, can only be conjectured.

Whichever developmental path the Albanians ultimately take, it will doubtlessly remain distinctive. One can only hope, however, that in its wake, improved access to data and Albanian decision makers will follow.

REFERENCES

Adhami, S. & K. Zheku (1981) Kruja dhe monumentet e saj 8 Nëntori, Tiranë

Albturist (1986) Republika Popullore Socialiste e Shqipërisë: hartë fiziko politike Hanid Shijaku, Tiranë

Alia, R. (1984) The correct line of the Party - the source of our victories 8 Nëntori, Tiranë

_____ (1986) Report to the 9th Congress of the Party of Labour of Albania 8 Nëntori, Tiranë

Anon. (1971) Atlas gjeografik i Shqipërisë Hamid Shijaku,
Tiranë
_____ (1972a) 'City planning and civic beauty', New Albania,
26(4), 12-13
_____ (1972b) 'The Bajram Curri City', New Albania, 26(4),
12-13
_____ (1973a) 'Rregullore mbi Mbrojten, Restaurimin dhe
Administrimin e Qytet-muzë te Gjirokastrës',
Monumentet, 6, 211-217
_____ (1973b) 'Rregullore per mbrojtjen dhe restaurimin e
ansambleve dhe ndertimeve te tjera me vlere historikë,
arkitektonikë dhe urbanistikë te qytetit te Korçës',
Monumentet, 6, 219-222
_____ (1977) Albania Norman Bethune Institute, Toronto
_____ (1981) 35 years of socialist Albania 8 Nëntori, Tiranë
_____ (1983) 'A new appearance for the ancient city',
New Albania, 4, 8-9
_____ (1984) 40 years of socialist Albania 8 Nëntori, Tiranë
_____ (1985a) 'Albanian economy in 1984 and forecasts for
1985', New Albania, 1, 4
_____ (1985b) 'How much is the rent for an apartment?'
New Albania, 6, 22-23
Banja, H. & V. Toci (1979) Socialist Albania on the road to
industrialization 8 Nëntori, Tiranë
Basha, H. (1986) Tirana et ses environs 8 Nëntori, Tiranë
Begeja, J. (1984) The family in the PSR of Albania 8
Nëntori, Tiranë
Berxholi, A. (1986) Durrës et ses environs 8 Nëntori, Tiranë
Besemeres, J. (1980) Socialist population politics
M.E. Sharpe, New York
Biber, M. (1980) 'Albania alone against the world',
National Geographic Magazine, 158, 530-557
Biberaj, E. (1985) 'Albania after Hoxha: dilemmas of
change', Problems of Communism, 34(6), 32-47
Bollano, P. (1984) 'The limitations of wage differentials',
Albanian Life, 29, 21
_____ (1988) Popullsia e Shqipërisë 8 Nëntori, Tiranë
Borchert, J.G. (1975) 'Economic development and population
distribution in Albania', Geoforum, 6(3/4), 177-186
Çarçani, A. (1986) Report on the 8th Five-Year Plan (1986-
1990) 8 Nëntori, Tiranë
Carter, F.W. (1986) 'Tirana', Cities, 3(4), 270-281
Cavolli, I. (1987a) 'Autumn days in villages', New Albania,
5, 16-17
_____ (1987b) 'Berat - a living museum', New Albania, 4, 6-9
_____ (1987c) 'Town in the centre of Albania'
New Albania, 3, 4-7
Dede, S. (Ed.) (1983) The earthquake of April 15, 1979 and
the elimination of its consequences 8 Nëntori, Tiranë
_____ et al. (1985) Enver Hoxha 1908-1985 Institute of
Marxist-Leninist Studies at the CC of the PLA, Tiranë

DeS (Drejtoria e Statistikes) (1962) Annuari stastikor i
    R.P.S.H. 1961 DeS, Tirane
_____ (1968) Vjetari statistikor i RPSH 1967 dhe 1968 DeS,
    Tirane
DPS (Drejtoria e Pergjitheshme e Statistikes) (1974)
    30 vjet Shqiperi Socialiste DPS, Tirane
Geco, P. (1970) 'L'accroisement de la population urbaine de
    la R.P. d'Albanie et sa repartition geographique',
    Studia Albanica, 2, 161-182
_____ (1973) 'Rendesia e qyteteve te medha ne popullsine
    qytetare te R.P. Shqiperise', Studime Historike, 1,
    53-71
Gurashi, A. & F. Ziri (1982) Albania constructs socialism
    relying on its own forces 8 Nentori, Tirane
Hall, D.R. (1981) 'Town and country planning in Cuba',
    Town & Country Planning, 50(3), 81-83
_____ (1984a) 'Foreign tourism under socialism :
    the Albanian 'stalinist' model', Annals of Tourism
    Research, 11, 539-551
_____ (1984b) 'Tourism and social change : the relevance of
    the Albanian experience', Annals of Tourism Research,
    11, 610-612
_____ (1985) 'Problems and possibilities of an Albanian-
    Yugoslav rail link'. In Ambler, J. et al. (Eds.)
    Soviet and East European transport problems
    Croom Helm, London, 206-220
_____ (1986) 'New towns in Europe's rural corner',
    Town & Country Planning, 55(12), 354-356
_____ (1987a) 'Albania'. In Dawson, A.H. (Ed.) Planning in
    Eastern Europe Croom Helm, London, 35-65
_____ (1987b) 'Albania's transport co-operation with her
    neighbours'. In Tismer, J.F. et al. (Eds.) Transport
    and economic development - Soviet Union and Eastern
    Europe Duncker & Humblot, Berlin, 379-399
_____ (1989) 'Cuba'. In Potter, R.B. (Ed.) Urbanization and
    development in the Caribbean Mansell, London
Hall, D. & A. Howlett (1976) 'Neither east nor west',
    Geographical Magazine 48, 194-196
Haxhiu, P. (1986) 'Personal property in Albania!'
    New Albania 4, 20-21
Hoxha, E. (1977) Report submitted to the 7th Congress of the
    Party of Labour of Albania 8 Nëntori, Tiranë
_____ (1981) Report to the 8th Congress of the PLA
    8 Nëntori, Tiranë
Karaiskaj, G. & A. Bace (1975) 'Kalaja e Durrësit dhe
    sistemi i fortifikimit përreth ne kohen evone antike'
    Monumentet, 9, 5-33
Kende, P. & Z. Strmiska (1987) Equality and inequality in
    Eastern Europe Berg, Leamington Spa
Klosi, M. (1969) 25 years of construction work in socialist
    Albania Naim Frashëri, Tiranë

Housing Policy in Albania

Kosinski, L.A. (Ed.) (1977) Demographic developments in Eastern Europe Praeger, New York

Kostanick, H.L. (1977) 'Characteristics and trends in South-eastern Europe: Romania, Yugoslavia, Bulgaria, Albania, Greece and Turkey'. In Kostanick (Ed.) Population and migration trends in Eastern Europe Westview, Boulder, 11-22

Kuke, S. (1983) 'High rates in housing construction', New Albania 2, 14-15

Lika, Z. (1964) Disa ceshtje mbi permiresimin e mëtejshem të planifikimit... Naim Frashëri, Tiranë

Lleshi, Q. (1977) 'Tirana', Geografski Horizont, 23 (3-4)

Marmullaku, R. (1975) Albania and the Albanians, C. Hurst & Co

Mason, K. et al. (1945) Albania Naval Intelligence Division, London

McIntyre, R.J. (1975) 'Pronatalist programmes in Eastern Europe', Soviet Studies, 27(3), 366-380

Meksi, V. (1987) 'Ample drinking water for all the villages and inhabited centres', New Albania, 5, 13

Mihali, Q. (1984) 'The Albanian economy 1983-1984', Albanian Life, 29, 34-37

Misja, V. & Y. Vejsiu (1982) 'De l'accroisement demographique en RPS d'Albanie' Studia Albanica, 19(10) 3-30

Mosko, S. (1984) 'Town planning in the service of the People', Albania Today, 77, 24-26

Muka, A. (1985) 'Karakteri i banesës popullore në fshatrat e Librazhdit', Librazhdi Almanak, 1, 61-91

Mullaj, B. (1984) 'Our ideal - socialism', New Albania, 5, 5

Nuri, F. (1982) 'Achievements in land reclamation and irrigation', Albania Today, 67, 24-26

Pajcini, I. (1983) 'New addresses in a city', New Albania, 6, 14-15

Papagjorgi, H. (1982) 'Peopling the countryside and extending the working class to the whole territory of Albania', Albania Today, 63, 14-19

Përgagitur Nga Shtëpia E Propagandës Bujqesore (1982) Bujqesia ne Republiken Popullore Socialiste te Shqipërisë Shtypshkronja e Re, Tiranë

Popovic, V. & B. Milic (1981) 'Montenegro after the catastrophic earthquake', Yugoslav Survey, 22(2), 27-48

Prifti, P.R. (1968) Albania's cultural revolution MIT Centre for International Studies, Cambridge, Mass.

_____ (1975) 'Albania - towards an atheist society'. In Bociurkiw, B.R. & J.W. Strong (Eds.) Religion and atheism in the USSR and Eastern Europe Macmillan, London, 388-404

_____ (1978) Socialist Albania since 1944 MIT Press, Cambridge, Mass.

Housing Policy in Albania

Qemo, G. & E. Luci (1983) 'The improvement of planning for the territorial distribution of productive forces and problems which emerge in this field'. In Institute of Marxist-Leninist Studies The National Conference : on problems of the development of the economy in the 7th five-year plan 8 Nëntori, Tiranë, 233-239

Qerimi, V. (1981) Chronicle of another battle won (April-October 1979) 8 Nëntori, Tiranë

Riza, E. (1971) 'Banesa e fortifikuar gjirokastritë' Monumentet, 1, 127-148

_____ (1975) 'Studim per restaurimin e nje banesë më cardak në qytetin e Krujës' Monumentet, 9, 107-125

_____ (1978) Gjirokastra : museum city 8 Nëntori, Tiranë

_____ (1981a) 'Banesa popullore në qytetin-muze të Beratit' Monumentet, 21(1), 5-34

_____ (1981b) 'L'architecture populaire urbaine en Albanie aux XVIIIe-XIXe siècles', Culture Populaire Albanaise, 1, 87-111

_____ (1983) 'The Institute for the Restoration of Monuments of Culture', New Albania 1, 24-25

Schnytzer, A. (1982) Stalinist economic strategy in practice : the case of Albania Oxford University Press, London

Selala, P. (1982) 'Well-being and its continuous uplift in the PSRA', Albania Today, 2(63), 20-24

Shehu, M. (1971) Report on the 5th 5-year Plan (1971-1975) Naim Frasheri, Tiranë

_____ (1976) Report on the 6th Five-Year Plan (1976-1980) 8 Nentori, Tirane

Shkodra, G. & S. Ganiu (1984) The well-being of the Albanian people and some factors and ways for its continuous improvement 8 Nëntori, Tiranë

Skarco, K. (1984) Agriculture in the PSR of Albania 8 Nëntori, Tiranë

Skenderi, K. & Y. Vejsiu (1983) 'The development of demographic processes and the socio-economic problems which emerge'. In Institute of Marxist-Leninist Studies The National Conference : on problems of the development of the economy in the 7th five-year plan 8 Nëntori, Tiranë, 155-164

_____ (1984) 'The demographic processes are inseparable from the socio-economic development', Albania Today, 77, 3-35

Smith, M. (1979) 'The earthquake in Albania', Albanian Life, 15

Strazimiri, G. (1987) Berati : qytet muze 8 Nëntori, Tiranë

Stretton, H. (1978) Urban planning in rich and poor countries Oxford University Press, Oxford

Suli, L. (1982) 'Tipologjia dhe arkitektura e banesës së vjetër të zonës së Konispolit', Saranda Almanak, 2 100-107

Housing Policy in Albania

Sulko, T. (1988) 'Tirana is rejuvenated every year',
    New Albania, 1, 10–11
Toci, V. (1971) 'Amfteatri : Dyrrahit' Monumentet, 2, 37–42
Vejsiu, Y. (1982a) 'The natural movement of the population',
    New Albania, 2, 12–13
_____ (1982b) 'What does the shift in population show?'
    New Albania, 6, 20–21
_____ (1987) 'The policy of the Party in the development of
    the demographic processes of the country', Albania
    Today, 3(94), 31–37
Wachsmuth, A. (1987) 'Das Länd der Skipetaren lasst die Welt
    hinein' NordWest Zeitung, September
Xholi, Z. (1985) For a more correct conception of national
    culture 8 Nëntori, Tiranë
Zanga, L. (1983) 'Albanian population growth' Radio Free
    Europe Research, RAD/253, 31 Oct
_____ (1987a) 'A new phase in Greek-Albanian relations'
    Radio Free Europe Research, RAD/233, 9 Dec
_____ (1987b) 'Albania to attend Balkan Conference'
    Radio Free Europe Research, RAD/192, 21 Oct
_____ (1987c) 'Establishment of diplomatic relations between
    the FRG and Albania' Radio Free Europe Research,
    RAD/165, 18 Sept
_____ (1987d) 'Glasnost in Albania' Radio Free Europe
    Research, RAD/200, 28 Oct
_____ (1987e) 'News media coverage of events in Kosovo'
    Radio Free Europe Research, RAD/223, 20 Nov
_____ (1988a) 'Albania expands international participation
    and co-operation' Radio Free Europe Research, RAD/106,
    10 June
_____ (1988b) 'Albanian foreign minister sets forth
    Albania's new-look diplomacy' Radio Free Europe
    Research, RAD/36, 4 Mar
_____ (1988c) 'Albania's contacts with the outside world'
    Radio Free Europe Research, RAD/87, 20 May
_____ (1988d) 'Albania's pace of making foreign contacts
    quickens' Radio Free Europe Research, RAD/150, 3 Aug
_____ (1988e) 'Albania's rejection of the Soviet experiment'
    Radio Free Europe Research, RAD/145, 1 Aug
_____ (1988f) 'Are Yugoslav-Albanian relations improving?'
    Radio Free Europe Research, RAD/185, 12 Sept
_____ (1988g) 'Changes in Albanian cadre policy' Radio Free
    Europe Research, RAD/65, 15 Apr
_____ (1988h) 'Criticism in Albania of conservative
    attitudes toward young people' Radio Free Europe
    Research, RAD/122, 1 July
_____ (1988i) 'Fifth Plenum reflects a new mood' Radio Free
    Europe Research, RAD/52, 24 Mar
_____ (1988j) 'Foreign languages and books win favour in
    Albania' Radio Free Europe Research, RAD/163, 22 Aug

_____ (1988k) 'Kosovo, Albania, and the Balkan
Conference' Radio Free Europe Research, RAD/20, 16 Feb
_____ (19881) '"Leftists" criticized for opposing innovation
in Albania' Radio Free Europe Research, RAD/14, 5 Feb
_____ (1988m) 'Major decision in Albania on cadre policy'
Radio Free Europe Research, RAD/116, 27 June
_____ (1988n) 'Satire against conservatives in Albania'
Radio Free Europe Research, RAD/97, 31 May
_____ (1988o) 'The birthrate declines in Albania' Radio Free
Europe Research, RAD/200, 5 Oct
_____ (1988p) 'The new Yugoslav-Albanian cultural agreement'
Radio Free Europe Research, RAD/31, 26 Feb

Chapter Nine

HOUSING POLICY IN YUGOSLAVIA

Shaun Topham

TRENDS IN HOUSING CONDITIONS AND CONSTRUCTION:
1945-1985

Housing has always been a major problem in
post-war Yugoslavia. As much as 75 per cent of
the total housing stock was damaged during the war
and over three million people were left homeless.
The new Yugoslavia inherited a housing stock of
3,400,000 units of a generally poor quality -
working out at nine square metres per head of the
population (Bežovan and Kirinčić 1987, p. 82).
The housing situation is in a great period of
change and considerable upheaval is expected in
the next few years. Yugoslav society is very open
and tremendous debate has occurred in recent years
on social, economic and political issues and
housing is no exception. However, difficulties
emerge in analysis due to the immense complexity
of the Yugoslav system and the problems occurring
in implementing policies in such a diverse
federation of republics.
In the period of reconstruction immediately
following the war, the major goal was
industrialisation and the building of new housing
was neglected. Housing was regarded as a
secondary problem which would resolve itself.
Most effort went into renovating and refurbishing
existing housing stock. The situation was made
more serious by the rapid urbanisation of the
country.
Yugoslavia in the post-war period brought
about a radical transformation of its social
structure. One of the most important indicators
of this transformation was the transfer of
population from the country to the town. The pre-
war agricultural structure was turned to industry

402

and no thought was given to its location other than political objectives. The proportion of the population that was agricultural had dropped from 66 per cent in 1948 to 50 per cent in 1961 and to 19 per cent in 1981. The proportion of urban dwellers increased from under 20 per cent in 1948 to nearly 50 per cent in 1980 (Singleton, 1976, p. 244). Housing conditions could not catch up in either quantitative or qualitative terms, with this rapid and largely uncontrolled urbanisation. This process in itself generated the housing crisis.

Until 1966 all industrial policies favoured the building of centres in each of the six republics, partly out of a desire to make full use of the existing infrastructure and productive reserves. Thus Yugoslavia's housing problems are those of the large cities, particularly Belgrade, Zagreb and Skopje. Ljubljana in Slovenia suffers less due to it being a less dominant centre of activity for Slovenia.

Further pressure was created by the trend away from extended families living together, causing a greater demand for housing. The post-war situation was further complicated by the tremendous regional variation in Yugoslavia with added differences between town and country.

Ideologically, the years following the war were years of Stalinism. All buildings containing more than three dwellings were brought into social ownership. The period 1945-1954 was characterised by centralised administration and budget allocation. Local Authorities (opstina) controlled the funds and built dwellings from this allocated budget and by 1950 the emphasis had been shifted to building new dwellings. By 1954, 310,000 dwellings were added to the initial stock of 3,400,000 which had survived the war (Jakšić, 1983, p. 45).

In the immediate post-war era, the town was not an attractive place to live. Despite policies aimed at industrialisation, to the detriment of agriculture and hence a worsening position of the countryside, propaganda had to be employed by the Communist Party to entice villagers to live in the towns. The urban population increased at treble the overall growth rate in the period 1948-1953 (Puljiž, 1977).

The period 1956-1965 saw a transition from central budgetary allocation. In 1957 a social fund for the construction of new dwellings was introduced. All work organisations paid a fixed percentage contribution to it from the personal income created by the workers. The early sixties witnessed what is referred to as the "First Housing Reform" which consisted of a series of laws aimed at introducing economic relations into this sector of the economy. Credit facilities were improved. Housing co-operatives, formed by groups of individuals, began to emerge on a small scale as a successful alternative to local authority provision of rented accommodation. The construction of private dwellings increased in importance with rising incomes but was dependent on the availability of credits from employers, bank loans and personal savings. It was hoped that these reforms would put the housing sector "on a more healthy basis for it had been until now a very neglected branch of our economy" and would "open it up for further development on a basis of self-management by tenants" (Kardelj, 1964, p. 398). In this period, 838,160 new dwellings were built (Jakšić, 1983, p. 46).

The years between 1966-1975 included the "Second Housing Reform". Housing enterprises were established. They were, like the rest of the Yugoslav economy by this stage, self-managed with the aim of providing greater autonomy to the enterprises. Workers were to make the decisions and the market was to be used to allocate resources. Construction enterprises competed for those funds allocated by the workers' councils to satisfy their housing requirements. The republican and municipal funds were abandoned with business banks taking their place with credit potential. A further 1,319,515 dwellings were constructed in this period (Jakšić, 1983, p. 47).

After 1976, major changes in the Yugoslav system were reflected in the housing sector. A new form of organisation emerged - "a self-managing community of interest". Its function was to bring together all interested parties in any particular activity, embracing such services as health and housing. Enterprises providing funding, local authorities, public services, planners, building enterprises and consumers would meet together and come to legally binding agreements. Working people could influence the location, quality, quantity and price of dwellings being constructed. Building enterprises would remain in market competition with each other. Between 1976-1983, a total of 1,140,029 dwellings were completed. This showed no improvement over the previous decade (Statistički Godišnjak Jugoslavije (SGJ), 1985).

The total housing stock increased according to census figures from 3,420,000 in 1951 to 4,082,000 in 1961 and from 5,043,000 in 1971 to 6,130,000 in 1981 (SGJ, 1987).

It could be said that there has been a housing crisis throughout Yugoslavia's post-war development. Even at the beginning of the 1960s when Yugoslavia was ranked high in Europe in terms of investment in housing as a percentage of total investment, there was a failure to provide a correspondingly large number of dwellings. The blame for this has been laid on the absence of long-term planning of industry, urbanisation and house building on a wider scale and on the unharmonious work of project organisers, investors and contractors. Inadequate land policy, low productivity, lack of integration and a tendency to a monopoly position of some building enterprises along with credit policies unfavourable to housing, all contributed to the failure to get satisfactory results. This is despite the fact that Yugoslavia has an excellent construction industry and that Yugoslav companies are sought after around the world, contributing to the inflow of foreign currency.

A consequence of this inability to meet demand has been a sacrifice in the quality of dwellings constructed in favour of quantity. Existing regulations relating to the ratio of larger units to smaller two- and three-roomed units have been ignored in order to supply as large a number of units as possible. Thus some new building in the social sector falls short of model standards. Similarly, planning and building regulations are flouted in the private sector. There exists the phenomenon of illegal "wild" private building around the larger towns, with many consequent faults in terms of drainage, services and roads.

Table 9.1 illustrates recent trends in the total dwelling stock, dwelling sizes and occupancy rates.

According to the Social Plan of Yugoslavia for 1986-1990, the growth in the rate of house building was to be 3.2 per cent. This would result in a housing stock of 650,000 dwellings - an increase of 10 per cent. Similarly, it is planned that the average floor area of a dwelling

Table 9.1: Yugoslavia: Total housing stock according to censuses

	1951			1961			1971			1981		
	Total	Urban	Other	Total	Urban	Other	Total	Urban	Other	Total	Urban	Other
Dwellings (000s)	3490	889	2601	4082	1167	2915	5043	2129	2914	6130	3048	3082
Average area (sq.m)	41.3	48.3	39.0	44.8	46.4	44.1	49.6	50.3	49.0	60.7	59.1	62.3
Area per head	8.7	11.5	8.3	10.0	10.3	9.8	12.2	13.8	11.2	17.0	17.6	16.2
Persons per dwelling	4.7	4.1	4.8	4.5	4.5	4.5	4.1	3.7	4.4	3.6	3.4	3.8

Source: SGJ, 1987, p. 200

Table 9.2: Yugoslavia: Housing stock by size and facilities available

Year All Locations	Dwellings (000s)	Rooms per dwelling					Dwellings installed with			
		1	2	3	4	5+	Electricity	Water	Central Heating	Bathroom
1971	5043	1189	2142	797	285	-	4430	1693	227	1238
1981	6130	1395	2444	1477	535	268	5865	4155	849	3127
1983	6539	1450	2604	1602	581	291	6274	4543	1011	3496
1984	6665	1467	2651	1641	596	299	6400	4663	1057	3612
Urban Locations										
1961	1167	539	418	128	61	-	1094	525	-	284
1971	2129	759	880	362	128	-	2096	1338	214	1054
1981	3048	753	1229	720	228	112	3006	2724	-	2298
1983	3282	788	1330	792	251	121	3240	2959	-	2529
1984	3353	800	1358	812	257	124	3311	3030	-	2597

Source: SGJ, 1987, p. 300

would increase to 62 square metres and space per head up to 10 square metres (Službeni List SFRJ, 31.12.1985.)

The housing stock in Yugoslavia is quite young by Western European standards due to the effect of the war. More than 70 per cent of the total stock has been built in the period 1945-1984.

The regulations governing area of new buildings are as follows: a "one-roomed" dwelling comprising of drawing room, kitchen, space for storage, food store, bathroom with WC and corridor must have a minimum area of 32 square metres.

A two-roomed dwelling should have a living room, double bedroom, space for storage, kitchen food store, bathroom with WC and corridor with minimum area of 45 square metres.

A three-roomed dwelling would have an extra double bedroom with a minimum area of 67 square metres and an additional separate WC whilst the four-roomed dwelling would have three double bedrooms, and extra WC and be of a minimum area of 85 square metres.

They also adopt half sizes by adding a "single bedroom" to each of the above, giving areas of 39 (1.5), 57 (2.5), and 78 (3.5) square metres (Zakona o Stambenim 1985).

Large numbers of flats are still regarded as sub-standard and it is calculated that there are 388,000 of a low enough standard to be regarded as a health risk. Three quarters of the flats provided with special welfare funds for the old, the unemployed and other disadvantaged groups are one- or two-roomed. The many inhabitants of these "solidarity" flats have to endure sub-standard conditions. The intense building of small units for quantity simply transforms the problem and does little to solve the housing crisis. The resulting total structure becomes increasingly inappropriate and unsuited to the needs of Yugoslav families. Table 9.2 illustrates how less than half the dwellings have a bathroom.

The overall land tenure in Yugoslavia, following a rapid acknowledgement of the failure of collectivisation programmes in the immediate post-war period, has essentially remained in private ownership. Land was re-allocated in 1945 with maximum ownership limits. At the height of collectivisation in 1950, 78 per cent of Yugoslavia's agricultural land remained in private hands (Singleton, 1976, p. 118). However, since 1958, when an Act of Nationalisation of Urban Dwellings and Building Plots was passed, all land in urban areas was nationalised. Existing owners only have the right to use land. They do not have the right to dispose of it. However, dwellings can be bought and sold.

Yugoslavia has always had a high proportion of home-ownership and the trend to more private ownership is illustrated in Figure 9.1. The trend initially coincided with increasing affluence but in recent years it also reflects changing political attitudes. Private house-building has always accounted for the largest proportion of new building - over 60 per cent in recent years.

Figure 9.1: Dwellings constructed between 1975 and 1984

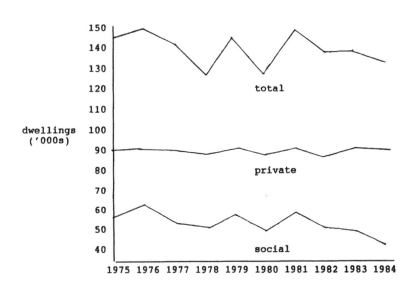

Source:SGJ.1987.p299.

From Figure 9.1 it can be seen that the private sector has remained steady whilst fluctuations can be attributed to fluctuations in social sector completions. A possible reason for this is the longer building period often associated with private building. Figures for 1984 for example show that although 130,000 dwellings were completed, a further 362,000 were in the process of being constructed with the vast majority of these being in the private sector. Houses are often built with the aid of family and friends and stop and start according to available finance for materials and so on.

Furthermore, the private sector, which mostly builds two-storey dwellings, tends to build larger houses. This is out of necessity, to satisfy a family's immediate housing need. But also this is seen by the authorities, in tourist areas, as the solution to the shortage of tourist accommodation, given the high investment costs required in providing beds in hotels rather than in private houses. However, since 1975 the social sector has built a slightly higher proportion of larger houses, although this has been far short of the required proportion.

Table 9.3: Yugoslavia: Dwellings constructed by size
(as percentages)

	Year	1 room	2 room	3 room	4 room	5+ rooms
social	1975	26	50	21	2	0.4
sector	1983	21	46	27	4.8	0.6
	1984	23	48	25	3	0.3
private	1975	16	42	28	8.6	5.0
sector	1983	16	33	32	15.7	8.7
	1984	10	32	33	15.4	9.4

Source: SGJ, 1987, p. 299

Yugoslavia is particularly rich in raw materials for the construction industry and this sector contributed almost 8 per cent of the gross social product in 1984. At one stage, in 1980, over 400,000 workers were employed within Yugoslavia whilst Yugoslav enterprises were able to compete around the world, employing a further 20,000. However, the industry has deteriorated in the 1980s, being badly hit by the policies aimed at "stabilisation" and since 1980 the number of workers employed in this area has fallen by 15 per cent (Economic Intelligence Unit, 1988, p. 19). This is reflected in the number of housing completions shown in Table 9.3. Of the total housing stock, given the modernness of most of it, only a small proportion is regarded as being constructed of unsuitable materials.

The quality has also improved slightly in terms of infrastructure to particular developments. It must be noted, however, that the private construction of housing is not out of luxury but out of necessity in most cases, as the only means of satisfying a family's immediate housing need.

Co-operatives have not played an important part in recent years (in terms of house supply) but a more substantial role is now envisaged for them as part of the way out of the housing crisis. In the period 1947-1960, 14.6 per cent of new building was by co-operatives but this had fallen after 1965 to 1.5 per cent in the period 1971-1979. However, in the Republic of Croatia for example 8,676 dwellings - 6.2 per cent of the total were built by co-operatives in the period 1981-1985 whilst this total is planned to double in the next planning period. Some of the larger enterprises solve their workers' needs through co-operatives - the so-called "closed" co-ops - accounting for half their number. The others are co-operatives open to all in a particular area (Bežovan and Kirinčić, 1987, p. 100).

The stock in the social sector is rented out through a variety of criteria set out by the enterprises for whose workers the housing has been constructed and by the municipalities in the case of that proportion of the stock which has been set aside as "solidarity" housing for those underprivileged or young members of society.

Yugoslavia's housing problem is closely connected with the development of the large towns and much of the post-war building in the social sector has taken the form of large overspill developments to these towns with no movement towards building new towns. All the large towns have had high-rise developments with dense populations. New Belgrade, across the river from Belgrade, is now a city of over one-quarter million inhabitants. Similar large developments have grown around Zagreb, Split, Skopje, Sarajevo etc. Since the 1960s the height of the blocks of flats have steadily diminished from some of 25-30 storeys to more modest low-level blocks of 3-4 floors. They are generally built in self-contained areas around their own facilities of shops, cafe, cinema, post office, vegetable market etc. and are well linked by bus to the centre of town.

There is now an imbalance between the shortage of housing in the towns and in the countryside, where there is excess housing stock in some areas. Similarly, given Yugoslavia's problem of having a developed north and an under-developed south, the housing situation is generally worse in the poorer republics. Table 9.4 illustrates the regional distribution of the housing stock and indicates the variations in standards between republics.

Problems associated with the low-income groups have been greatly alleviated since the introduction of the Solidarity Housing Fund to provide housing for the more deprived members of society. The shanty towns which had developed on the edges of the big cities have been cleared. There remains the problem of housing the migrant workers who have moved to the north, particularly the Albanians who work in Slovenia. Migrant workers often have to stay in barrack-type dormitory accommodation, whilst many who travel shorter distances to the big towns keep "one leg in the country" and travel back to their farms at weekends and put up with cramped conditions during the week. Such effects of the housing shortage clearly affect the mobility of labour.

The essential element in the Yugoslav social, economic and political system is the belief in self-management and the need to ensure a worker is not alienated from the results of his/her labour. Thus the enterprises are directly involved in financing housing construction and workers contribute directly through their earnings, part of which is allocated through their workers' council to housing. All interested parties combine together and organise themselves as a "self-managing community of interest" to provide housing. Housing can be constructed on behalf of the enterprise or bought by the enterprise directly from construction companies on the open market.

Table 9.4: Yugoslavia: Regional distribution of housing stock

	Total (000s) (1984)	Urban (%) (1981)	Socially owned (%) (1984)	Bathroom (%) (1984)	Average area per head 1951	1961	1971	(sq. m) 1981
Yugoslavia	6665	49.72	22.8	54.2	8.7	10.0	12.2	17.9
Bosnia	1130	40.98	19.8	48.4	5.8	7.1	9.0	14.6
Montenegro	149	53.43	26.8	53.0	6.7	7.8	9.9	15.3
Croatia	1488	52.71	25.7	59.9	9.8	11.1	14.1	20.5
Macedonia	481	57.56	13.7	51.7	6.9	7.5	10.5	16.5
Slovenia	636	52.73	32.9	73.3	10.5	11.5	15.7	21.6
Serbia	1814	54.69	24.3	51.2	8.3	10.4	12.2	18.1
Kosovo	259	38.59	11.2	26.1	6.8	7.3	7.9	10.6
Vojvodina	707	54.70	18.1	54.4	13.0	13.4	15.7	23.1

Source: SGJ, 1987, pp. 428, 553

Note: There are also strong regional variations within republics

In addition to this direct source of finance, workers contribute 3 per cent of their earnings to the "Solidarity Fund" to enable the under-privileged to be housed and to subsidise the rents in needy cases. Similarly, the developed republics are expected to show "solidarity" with the underdeveloped and assist them. There is a federal fund set at 1.8 per cent of the product from the social sector with another fund set at 0.8 per cent targeted at improving social services. There has been particular help with electrification. Enterprises can help those building their own houses but generally they must seek credit from the banks and use their own savings. Much new private housing is financed through the earnings of returning gastarbeiter. At one stage over half a million Yugoslavs were employed in West Germany.

Yugoslavia has received international aid on two recent occasions to tackle the immense devastation caused by earthquakes. The first was in Skopje in 1963 which left many dead and 150,000 homeless, and more recently the Montenegran earthquake of 1979. Much of the devastation which affected huge areas of Montenegro is still being rebuilt.

With cheap labour and a rich source of exportable basic building materials, it would be expected that the cost of construction of a new flat would be low. In fact the price of a flat is very high and prohibitive to many citizens. This can be attributed in part to poor organisation, particularly in view of the fact that private dwellings can be cheaper than those which have had the benefit of specialists with equipment and a trained workforce with the cheaper access to building and other materials. Further factors causing the high prices are the infrastructure costs, land being expensive due to a shortage of serviceable sites, and a variety of taxes which have to be met. It is estimated that the average flat costs the equivalent of five years work for the average worker.

As a contrast to the high cost of new housing, rents are very low. Rents in 1981 accounted for only 3.8 per cent of a family's total household expenditure. This is put into perspective by the fact that 4.3 per cent was spent on tobacco and drink. Indeed, as Table 9.5 shows, rent has in fact been falling in real terms since 1969.

Table 9.5: Yugoslavia: Average expenditure of a four-member
family

Year	Percentage of total expenditure spent on a) Accommodation	b) Tobacco and drink
1969	5.4	4.2
1970	4.5	4.2
1971	4.4	4.3
1972	4.1	4.4
1973	4.3	4.6
1974	4.2	4.6
1975	3.9	4.6
1976	3.9	4.4
1977	3.6	4.3
1978	3.7	4.0
1979	3.6	4.0
1980	3.4	4.0
1981	3.8	4.3

Source: Sekulić, 1986, p. 359

In Croatia, rent paid was 3.09 dinar per
square metre in 1976. In 1985 this had risen to
27 dinars per square metre - an increase of nine-
fold. But during the same period prices had risen
fifteen-fold and personal income thirteen-fold.
Thus rents have nominally fallen. There is wide
variety in rent paid in differing areas and from
town to town within a particular republic. Rents
in Croatian towns in 1986 varied from the 19
dinars per square metre in Split and 24 in Zagreb
to the 40 and 65 dinars paid in Varaždin and
Osijek. Rents in the more developed republics and
areas were approximately double those of the
underdeveloped (Saveznogo Zavoda za Statistiku,
1987).
However, the rent for a flat in the private
sector can be 5-10 times higher than in the social
sector. There is insufficient data available as

in many cases the flats are let illegally. Drives to eradicate this problem of illegal renting have had the effect of driving up rents of those let legally by reducing the supply of flats. An individual may own two or three dwellings according to size although they are restricted to the use of one social dwelling. It is the illegal letting of social dwellings by individuals having alternative accommodation which causes problems.

Table 9.6: Yugoslavia: Rent: Dinars per square metre on 30.6.1986

Yugoslavia (average)	48
Slovenia	64
Vojvodina	60
Croatia	52
Serbia	50
Montenegro	40
Bosnia	33
Kosovo	31
Macedonia	26

Source: Saveznogo Zavoda za Statistiku, 1987

THE ROLE OF HOUSING INSTITUTIONS AND ORGANISATIONS

It must be questioned whether an accurate description of the Yugoslav system is possible. The socio-economic and political system has been in constant change over the last four decades. It has been described as "one of the most fascinating social projects of our time - an attempt to create a genuine democratic socialist society, to change the whole fabric of economic, political and cultural life, to implement the change rapidly, on a mass scale and through the existence of an enormous ethnic and national diversity" (Marković, 1978, p. 134).

There has been a whole series of constitutional changes, the latest taking place in 1988. The basic pillars on which the Yugoslav version of socialism is founded are those of self-management evolving from workers' control and the use of the market to reduce administrative intervention to a minimum.

Housing Policy in Yugoslavia

The aim of the 1974 constitutional changes
was to provide a framework in which the majority
of citizens could directly participate in the
management of society, including its housing
needs. The result was a set of organisations
unique to the Yugoslav constitution. The basic
organ- isations of associated labour are the
smallest legal entities distinguishable as
producers of a marketed or marketable output.
They have their own income statement and balance
sheet and can be an enterprise in their own right,
but normally an enterprise (or organisation of
associated labour) consists of several basic
organisations coming together.

The delegate system is used throughout and
these basic organisations link together in "self-
managing communities of interest". These
organisations are of great importance in the
sphere of public utilities, welfare, education
etc. In a particular commune there would normally
be "communities of interest" covering education,
science, culture, health and welfare with others
covering housing, transport and public utilities.
Representatives of both providers and users of
services are brought together and these
organisations have an equal status with the
communal bodies in allocating resources. The
participants make "self-managing agreements" to
achieve their stated objectives.

Thus the area of housing is covered initially
at the basic organisational level. Enterprises
are obliged to pay a percentage of each worker's
earnings to a housing fund organised at the local
commune level to partly pay for "solidarity"
housing for the underprivileged and partly to form
subsidies. The workers' council also allocates a
minimum amount to providing housing for the
workers employed. This amount can be as high as a
particular enterprise can afford. Thus the more
successful an enterprise is, the more likely are
the employees to provide adequate housing
facilities.

The self-managing community of interest
brings together the planners, the consumers (in
the form of delegates from the workers' councils)
residents, construction firms, utility suppliers
etc. - indeed anyone who has an "interest" in
the housing process. The construction
organisations, their suppliers, the utility
suppliers etc. also have their self-managing
organisation. A common lubricant to the whole

system is the involvement of the socio-political organisations - naturally the League of Communists but also the Socialist Alliance - a broader-based organisation, the unions, veterans organisations etc. The banks which are involved to provide credit are similarly controlled by their customers - the enterprises using the services. Self-managing communities of interest in housing cannot, as a rule, operate at a loss. All their resources are owned by members and they do not have their own resources. Everything is carried out to a basic financial plan and they are only able to spend what has been promised to them. Clearly such an elaborate system requires strong participation all round to make it work.

A further requirement is that as much of the economic judgement revolves round the existence of a market. A strong country-wide market is a pre-requisite, whilst the healthier any economy is, the more likely its system is to perform as intended with the minimum of intervention.

We will examine later some of the problems which have been encountered in solving Yugoslavia's housing crisis through this institutional framework.

Certainly one problem is that the majority of the housing construction is outside of the social sector and has been more difficult to control given the low level of resources which can be spared to tackle this problem.

The planning departments have been instrumental in introducing schemes aimed at urban renewal in the older town centres using communal funds from the Housing Fund. Their contribution has been more significant in the centres of tourism.

A further institution to be mentioned are the housing co-ops which are once more being turned to as a possible solution to some of the housing problems.

HOUSING MARKETS, ALLOCATION MECHANISMS, RELATION-SHIPS BETWEEN SUPPLY AND DEMAND

The Yugoslav Constitution refers to the worker's right to a dwelling, but there is no explicit definition within the framework of the Yugoslav socio-political system of what this right entails and what should satisfy "housing need". What is usually turned to as a guide is square

metres of living space per head. The Long-term
Stabilisation Programme of 1982 set out the task
of overcoming the housing crisis and improving
conditions of housing and communal standards. The
target to be aimed at had to depend on 'real needs
and abilities'. This obliged all socio-political
communities to establish stages for assessing
their aims and instruments (Komisija Saveznih
Društvenih Saveta, 1982, p. 130).

It is argued from many sides that the present
simple definition of need - falling into simple
categories of those who have and those who do not
have a dwelling - is inadequate and that too much
emphasis is given to the protective function as a
shelter. Greater emphasis should be put on the
quality of life implied and that humans ought not
to be regarded as a "mechanical element" in the
housing process. But, in such a crisis situation,
shortage of dwellings dominates (Čaldarović, 1980;
Živković, 1981).

The position, however, is not one of simple
shortage. Indeed the position has reversed
between the censuses of 1971, when there was a
total deficit of 332,000 dwellings in relation to
5,375,000 households, and 1981, when the number of
dwellings actually exceeded the number of
households by 135,000 (6,321,000 flats - 6,186
dwellings) (SGJ, 1987.) It is argued that a
simple administrative redistribution of the excess
capacity to those in need could cure the problem
at a stroke. The problem, however, lies in the
location of these dwellings, for they are not in
the large towns where they are desperately needed;
the excess is predominantly in the rural areas
(Sekulić, 1986, p. 365).

There are similar disparities between the
republics. Why the administrative solution could
not work may be illustrated using Croatia as an
example.

The surplus of 9,752 dwellings in rural
Croatia is of little comfort to the 20,271
families seeking adequate accommodation in Zagreb.
Overall it is estimated that Croatia requires
another 100,000 flats. In 1986 the Socialist
Alliance claimed that taking the whole of
Yugoslavia, half a million workers had not solved
their housing needs, echoing the Croatian
situation. When the definition of need is
extended to spatial requirements the scale of the
problem of overcrowding becomes clear. The figure
of 20 square metres per head is taken as the

target to achieve whilst 10 square metres is taken as the socially agreed minimum. Again, taking the Croatian figures, the inhabitants of some 28,757 dwellings have less than 6 square metres per head whilst the inhabitants of a further 137,158 dwellings have 6-10 square metres per head. Thus there are 817,000 people living in 166,000 dwellings in Croatia which fall below the accepted pathological threshold at which the likelihood of mental disease increases (Antić and Jakšić, 1983, p. 20-24).

Table 9.7: Yugoslavia: Shortage or surplus of dwellings in relation to number of households in Croatia

Year	1961	1971	1981
Total	-129,566	-100,843	-42,428
Urban	- 90,796	- 90,382	-52,180
Rural/Mixed	- 38,790	- 10,461	+ 9,752
Zagreb	- 41,971	- 35,324	-20,271

Source: Sekulic, 1986, p. 365

A further 685,000 people living in 222,000 dwellings have only 10-14 square metres per head. This is below the critical standard, according to Lauwe, sufficient to ensure the balance between individual and family life (Lauwe, 1959, p. 108). This means that in Croatia there are 1.5 million inhabitants in around 388,000 dwellings who have a very low standard of housing, sufficient to damage their health. And as Table 9.4 points out, Croatia with an average 20.5 square metres per head of space is above the Yugoslav average of 17.9 square metres per head. Thus the problem is worse in other areas - notably Kosovo, with an average of only 10.6 square metres per head.

Figure 9.1 illustrates the source of supply of new housing, with the private sector providing 60-70 per cent of the total compared with the social sector's 30-40 per cent. The amount spent on new housing in Yugoslavia in 1986 amounted to

3.2 per cent of the national income. However, this figure does not include all the private housing given the unofficial nature of a proportion of it. The construction sector in total accounted for 10.6 per cent of the national income.

The distribution of the housing stock poses one of the more serious problems which the Yugoslavs are faced with. This is worsened by the acute shortage - giving less to distribute - but also by the fact that Yugoslavia has legal income differentials reducing the spread by which wages can differ. Thus, the more egalitarian an income structure, the more apparent and serious any discrepancies in other areas become. Certainly this is the case with housing.

There are clear gainers and losers in the distribution process. Clearly those unable to solve their housing problems at all are at the bottom of the order. At the top are those workers and officials who fulfil the criteria for being allocated a socially owned flat. Enterprises which have more success are able to command higher housing funds and are able to provide more flats. The shipbuilding industry, for example, in 1986 could allocate 324,000 dinars per worker to solving the housing problem, wholesaling - 310,000, oil and gas industry - 254,000, whilst at the other end of the spectrum the building material industry and the drink producers could only afford 6,000 and 12,000 dinars per worker respectively (Bežovan and Kirinčić, 1987, p. 150). Thus clearly where a worker works can have a significant influence on his chances of getting social housing. Similarly, the ability to provide accommodation is a factor which strongly influences the mobility of labour. Agreements have now been made to prevent the more affluent enterprises from 'over-spending' on opulent flats in the most desirable locations.

Despite the distortions caused by inflation, the following figures for Croatia illustrate the advantage of working in a successful enterprise or sector of the economy.

There are similar problems with the workers employed in the non-productive sector. For example, only 57 per cent of the police force could claim to have adequately satisfied their housing needs whilst the Federal Government found itself short of 1,480 dwellings for the federal officials in 1987 and had to order 606 new ones at a cost of 5 billion dinars (Politika, 22.5.1986).

Another example is that of a group of 14 doctors at a medical centre who had waited 4 years for suitable accommodation who have turned to the co-operative structure to solve their problems (Bežovan and Kirinčić, 1987, p. 105).

The problem of the underprivileged has to a great extent been solved by the "solidarity" housing. It is a large section of the middle strata who suffer most. They have either to rent privately or build their own homes. It is this category which accounts for the large amount of private building - turning to this as a "second-choice" solution to their problems, being unable to acquire a "social" dwelling. A further consequence is that those who fail to get a social flat are subsidising those who have managed to acquire one as all workers still have to pay the contribution to the Housing Fund.

The key problem with private house building is its socially undefined position. It is neglected in planning and hence expansion is socially uncontrolled. Social intervention is exclusive to "collective" building whilst private building is treated as marginal, but in reality it accounts for 60-70 per cent of the total building in any year. It is marginalised without any clear social status and as we see is often illegal.

The housing co-operative sector as we know it in the West has had a chequered history in Yugoslavia, despite the rest of the system being based on self-management and co-operation. The period between 1956-1963 was particularly rich with 1400 well-financed co-operatives thriving in a suitable political climate with access to credit. However, by the 1970s this number had fallen to only 52 after the reforms of 1965. Whilst 14.6 per cent of all dwellings were built by co-operatives in the period 1947-1960, only 1.5 per cent were built between 1971-1979. These figures are included in the totals for the private sector (Bežovan and Kirinčić, 1987, p. 100).

Housing co-operatives are still shaking off the poor reputation they earned in the past, their activities verging on the criminal in some cases. Poor building, corruption, the building of week-end houses and commercial enterprises passing themselves off as co-operatives to be allowed the favoured access to building land were some of the problems. The whole area was generally weakly regulated.

Table 9.8: Yugoslavia: Allocation to housing by individual foreign trade enterprises in Croatia (million dinars)

		1982	1983	1984	1985
"Kupotrade"	income per worker	5.26	11.51	12.54	27.49
	allocated to housing per worker	0.75	0.96	0.58	2.00
	number of workers employed	33	31	34	40
"Astra"	income per worker	2.11	3.22	6.53	15.37
	allocated to housing per worker	0.19	0.20	0.34	0.46
	number of workers employed	90	91	88	87
All foreign trade enterprises	income per worker	1.43	2.33	3.78	7.89
	allocated to housing per worker	0.11	0.14	0.20	0.37
All Croatia	income per worker	0.47	0.63	1.02	1.89
	allocated to housing per worker	0.02	0.02	0.04	0.05

Source: Bežovan and Kirinčić, 1987, p. 151

Further, the principles of solidarity were not always extended to co-operatives, as they were regarded as working mainly for the short-term interests of the members.

However, the housing co-operative is now seen as one possible aid to solving the present problems with many municipalities encouraging the formation of housing co-operatives. In Slovenia particularly they have started to develop - in 1987 there were 71 in 34 municipalities whilst there were 76 in Croatia. The main problem they now face is in getting land and often it is their relationship with the municipality which determines it. Indeed, some of the larger enterprises such as Jugoturbina have turned to the co-operative as the solution to their problem and because of their size and influence on local municipalities, have not had problems in this area. Jugoturbina, an enterprise employing more than 18,000, have been able to build 15-20 per cent cheaper than through the communities of interest (Bežovan and Kirinčić, 1987, p. 102).

There is still little organised help to back up the co-ops and often they are more costly than other forms of building. Generally those workers likely to build privately do not have much information on the possibility of building co-operatively. It is regarded as one of the tasks of the unions to educate in this area for this direction has been taken when the opportunity has been presented. It is the aim that each municipality should have at least one well-qualified housing co-operative, although until now they have mainly been in the large towns. Tax concessions on materials have been promised to help this and it is hoped that the banks will form a model to help supply credit and that suitable land should be made available. There is an argument concerning the provision of subsidies in providing land in this case as it could be viewed as a penalty to those having to finance this subsidy and unfair on those whose housing is not being subsidised in this way.

At the moment Yugoslavia clearly has two forms of distribution: one based in principle on ideas of social justice and equalitarianism and the other on principles of supply and demand. Many difficulties of a consistent housing policy stem from the problem of the two systems having to co-exist - where a dwelling has two realities - as a means of private consumption or as a socially owned resource merely being used by an individual according to need.

The criteria for allocation are set out by the workers' council in a work organisation and incorporated in the rule book. This is an area where the unions play an important part, being much more involved in the "welfare" of the workers than in the more defensive aspects of their western counterparts. A survey of the criteria adopted in the rule books of twenty work organisations showed considerable variation (Bežovan and Kirinčić, 1987, p. 153). Some organisations took as few as 3 criteria, others as many as 20. The majority, however, employed 6 or 7, most including:

    a) existing housing position and status;
    b) relationship to work - e.g. essential
       worker, qualifications, position etc;
    c) total working experience;
    d) socio-economic position of the worker
       and his/her family.

With a large number of work organisations, the existing housing position of the applicant accounted for 70 per cent of the allocated "points". With others it was noticeable that the criteria of expertise in carrying out a job was an important factor. However, some work organisations do not have rule books covering this and others are illegal, whilst sometimes management arrogantly settles the problem.

The allocation of housing remains one of the areas which causes most conflict in work organisations and annually increasingly large numbers of cases are brought before the "court of associate labour". The number of cases rose to 2461 in 1983 and to 2656 in 1985, accounting for 20 percent of cases in this particular court, while such cases accounted for 35 per cent of those unresolved and sent up to a higher court. Similarly in the first eight months of 1986, the number of cases dealing with housing examined by the Federal Trade Union organisation accounted for 40 per cent of the total. There is nothing in the rule book to ensure the re-allocation of social flats which are too big for families whose children have grown up (Bežovan and Kirinčić, 1987, p. 154).

PROBLEMS AND POLICY ORIENTATIONS

Essentially the problems faced by the housing sector are linked with the serious economic situation which Yugoslavia has been struggling to overcome in recent years. Yugoslavia is involved in a harsh programme of what it refers to as "economic stabilisation" which commenced in 1982. The root of the problem lay in heavy foreign debt and the necessity to both save and earn foreign currency coupled with very high inflation rates of several hundred per cent. Whilst the foreign earnings have improved, the inflation has worsened. The problems of price controls and local monopolies have affected the development of a Yugoslav-wide market on which much depends in terms of efficiency. It would remain to be seen how much more successful the Yugoslav system would be if it was working in a healthy economic climate.

Clearly the major housing problem is the shortage of housing in urban areas with a correspondingly unsuitable size structure to meet the requirements of Yugoslav families. The high cost and difficulty in getting credit prove major obstacles. Some social groups and professions are in such a position that although they occupy a relatively high social position in the division of labour and carry out a socially important role, e.g. doctors and teachers, they are not in a position to secure adequate housing for themselves. One health centre was having problems in carrying out an efficient service for none of its 8 doctors could find suitable accommodation and reported that "not even in ten years would we be able to acquire a single flat from the centre's income" (Borba, 1986). Other workers contribute to the housing fund all their working lives without receiving anything back from it. For example, the newspaper Borba reported the case of some pensioned miners, some from the best mines, "who in 75 years had never solved their housing problem and were still living in wooden shacks without ever a chance of getting a flat. Paradoxically the mine from which one miner came had 180 flats at its disposal" (Borba, 1985).

Given the low standards achieved, the first priority for most Yugoslavs is a simple improvement in their circumstances. However, there are data available showing their preferences. For example, research in Belgrade has revealed that 93 per cent of residents feel that a balcony or terrace is an essential component of a dwelling. The use to which it would be put is indicated by preferences given for its situation. Fifty-seven per cent thought it ought to be next to the kitchen - implying its use as an extension to the working area - whilst 30 per cent thought it ought to be next to the living room and 10 per cent next to the bedroom. There was a great desire for collective space for children (41 per cent), young people (26 per cent), old people (9 per cent), whilst clearly the 10 per cent of respondents wishing for pram-storage facilities indicated a problem (Čaldarović, 1980, p. 27-29).

Preferences through Yugoslavia showed that 22 per cent of residents preferred a single-storey dwelling, 52 per cent preferred 2-4 floors, 10 per cent favoured 5-8 floors and 5 per cent above 8 floors. However, preferences vary from town to town. For example, in Novi Sad, 51 per cent preferred a single-floored dwelling, 28 per cent favoured 2-4 floors with 15 per cent choosing 5 and above (Čaldarović, 1980, p. 30).

Generally, the infra-structure, services, public transport, location and shopping facilities are taken care of in most developments and local "community organisations" (mesna zajednica) bring together at a very local level, citizens and suppliers of all services typically consumed in a neighbourhood. This is not always the case in the private sector, with an inadequate inspectorate to fully control new buildings. Private building can also take decades to complete, being in many cases dependent on savings or funds sent from abroad to purchase materials etc. to continue the work. Indeed many of the gastarbeiter saving money abroad left primarily to be adequately and immediately housed and to provide the opportunity for solving their own problem at a later date (11 percent of those asked gave the opportunity of an immediate flat as the reason). Thus the enormous amount of building work which can be seen is deceptive as many houses will take years to complete. Clearly, turning to private building should not be seen as a luxury (Mežnarić, 1977).

A further problem associated with the private sector is the phenomenon of "wild building" in the suburbs of the large towns where the facilities are minimal and contravene every urban plan. It is estimated that there are a third of a million

such illegal houses, some of which are part of very large settlements such as Kaluderica near Belgrade with an estimated 10,000 illegal dwellings. Those who have solved their housing problems in such a way have given the following reasons for doing so:
- 44 per cent because of the size of family and inability to solve their problem;
- 27 per cent because of the expense of sub-letting and its status;
- 12 per cent because their work organisation was unable to help;
- 5 per cent because the land was cheaper;
- 4 per cent inherited the land;
- 3 per cent because they wanted to own a house;
- 2 per cent because of the cost of the documentation for building legally
(Čaldarović and Richter, 1975).

It must be noted also that the illegal builders were the most unlikely to receive a flat through "normal" channels and were generally of a lower social standing and educational level. Fourteen per cent of the builders in one illegal area (Zagreb-Dubrava) had no formal school qualifications whilst only 3 per cent had high qualifications. But, in the corresponding "elite" suburb (Medvejčak), this figure was only 4.8 per cent unqualified and 20 per cent with the high qualifications. Seventy-five per cent of the illegal builders came from an "ordinary worker" background compared to the overall 55 per cent for Zagreb (Čaldarović and Richter, 1975). This leads us to examine the stratifying effect of existing allocation policies.

As Vujović points out,

> Despite the large number of dwellings built since the war, analysis of data concerning areas such as the illegal "wild" building areas, the status afforded sub-tenants, the distribution of flats which are socially owned, the housing status of workers employed in low accumulating work organisations, the status of owners of "weekend" houses, the neglected housing areas and the socially segregated housing areas together reveal that the housing problem is one of

the major manifestations of social stratification in Yugoslavia. (Vujović, 1979, p. 465)

The basis of segregation in housing stems from the distribution of social dwellings on criteria of position in the work place rather than upon need for housing. Although the intention of the socialisation of the dwelling policy was exactly that of cancelling out social stratification in that field, the policy has in fact produced one of the most important stratifications in urban Yugoslavia. It has introduced a composition of groups which support the status quo. What is interesting is the fact that the structure of these groups does not correspond with the traditional view of "high" and "low" strata. The most influential protect their position by claiming to defend those of "low" strata. Essentially it is a matter of exploitation by those who have socially owned dwellings over those who do not have them. The following examples clearly show the extent of social segregation in housing and it is an area which is discussed and criticised in the Yugoslav press. It is regularly accepted as a problem with no apologists defending the situation.

The groupings in Yugoslav society are allocated by sociologists into political rulers, managers, intellectuals of "system" labour, intellectuals of socially necessary labour, clerks, skilled workers, unskilled workers, private craftspeople and individual agricultural producers. Using these groupings, it emerges that the likelihood of having a socially owned flat decreases with social status - 57 per cent of politicians have socially owned flats, 54 per cent of managers, 37 per cent of intellectuals and 25 per cent of clerks. However, this privileged status is afforded to only 16 per cent of the skilled and unskilled workers (Lay, 1986, p. 23). Similarly the number of workers living with relatives falls according to how well educated they are. Table 9.9 indicates some quality differences in housing amongst different social groups.

Possession of a weekend house was found amongst 21 per cent of managers, 15 per cent of intellectuals, 9 per cent of clerks and 3 per cent of workers. Similar figures can be found showing a correlation between education and housing

Table 9.9: Yugoslavia: Quality of housing by social group

	Social group			
	Politicians	Managers	Qualified workers	Unqualified
Less than 15 sq.m. per head (%)	21	18	43	52
With 2 or less rooms (%)	32	29	47	55
With no water supply (%)	1	–	18	36
With no flushing WC (%)	1	–	16	34
With no bathroom (%)	1.5	1	19	36
With no parquet floor (%)	5	7	35	56
Sample size (total 2079)	401	384	406	402

Source: Lay, 1986, p. 24

standards (Lay, 1986, p. 25).

Segregation according to location has also developed in Yugoslavia. Segregation is undesirable and research shows that not only does social zoning exist, but new housing developments reproduce it. Areas occur in most cities where collections of homogenous groups gather; (Lay 1984; Čaldarović, 1975, 1986 and 1987; Sekulić, 1986.)

What is of great concern, stemming from the allocation of social dwellings, is that those with flats do not need to spend on private housing and can hence direct their spending to luxury goods - cars, weekend houses, foreign holidays etc. Clearly the demand structure brought about by this social inequality has further consequences on the economic system and particularly on the demand for foreign currency for travel and imported luxury items.

Table 9.10 shows the correlation between educational structure, likelihood of gaining a flat and reasons for wanting to be considered for a flat.

It must be pointed out that those in the highest categories were still possibly justified in requesting a larger flat.

The official answer to the failure of the 1981 to 1985 Croatian plan was:

> The basic causes of such unsatisfactory results in house building are primarily not enough building land, a poorly developed commune infra-structure, the high price of dwellings, weak 'communities of interest', not enough dwellings being built for the open market, bad organisation in the preparatory stages, slow development of socio-economic relationships and, finally, insufficient and unforthcoming financial means, with often changing conditions of credit and increasing rates of interest. (<u>Delegatski Vjesnik</u>, 21.03.1985, p. 45)

The unions have taken up issues concerned with planning, seeking to erase the country/town differentials and to rationalise the use of town land. With Zagreb alone accounting for a quarter of Croatia's housing shortage, any developments in the smaller centres must ease the pressure on the

Table 9.10: Yugoslavia: Reasons for requesting a change of dwelling related to educational qualifications (1978)

Reason for request	Educational Qualification							
	VSS	VSS	VKV	SSS	KV	NSS	PKV	NKV
no dwelling	43.5	42.0	45.3	43.4	43.4	48.0	46.6	47.2
sub-tenant	5.5	7.3	7.0	7.6	6.4	8.0	8.7	8.3
living with relatives	21.1	19.5	22.2	18.8	15.5	18.7	15.5	14.8
unhygienic	2.9	6.9	6.9	12.0	10.3	11.3	16.1	13.1
dwelling too small	27.0	24.3	18.6	18.2	24.4	14.0	13.1	12.5
income per worker index (social sector=100)	178.5	132.8	120.7	106.5	91.3	82.1	74.1	65.4
access to a socially-owned dwelling index	2.53	1.96	1.16	1.20	0.74	1.17	0.58	0.66

Source: Bežovan and Kirinčić, 1987, p. 147

Note: The educational levels referred to reflect the current Yugoslav usage, ranging from the highest (VSS) to the lowest unqualified level (NKV)

larger centres.

Horvat points out that land which takes two months to change hands in the USA can take as long as eight years in Zagreb and makes it an ordeal to attempt to move from one city to another, with no organisation for marketing land or information service. The land may constitute up to half the total building costs whilst the documentation can take 2-3 years. There is an inbuilt delay of three months due to the formality of first offering the land for sale to the <u>opstina</u>. The deposit for land is huge - 30-50 per cent - whilst inflation makes it difficult to plan building cost (Horvat, 1982, p. 8).

Clearly there are deep problems with shortage of land and the existence of a dual system where town land is nationalised and that outside is in private hands. The problem occurs where the two systems meet as towns spread and new roads radiate out and land changes hands in anticipation of "urban sprawl". Bearing in mind that the rational exploitation of an individual's plot of land is not always best for the community, it is argued that a third "border" category should be established with regulated protective zones round towns to prevent speculation and ensure rational use (Miličević, 1987, p. 29).

The communities of interest were criticised for the failure of the plan. It is claimed that there is a) excessive administration and a bureaucratic mentality, b) lack of interest causing inquorate meetings and c) a slow, complicated decision-making process with a predominance in decision making by workers (within the community of interest organisation) who are out of touch with those they are representing. It is argued that more common services ought to be shared by groups of communities of interest (<u>Komisija za zivotne i radne uvjete radnika</u>, 1987, p. 200).

However, the communities of interest have complained that they have been thwarted in carrying out schemes to build large numbers of flats for the market due to land problems and the failure of the banks to be able to provide credit for such initiatives. A model for such marketing initiatives needs developing.

A shortage of dwellings clearly allows opportunities for exploitation in the classic sense and in response campaigns have been run to return illegally let flats to social use. A recent campaign - "You have a house - Return your flat" - has had a success given the estimate that between 8-10 per cent of the total stock is being misused or the real owners are unknown in certain areas. A call has been made to take an inventory of the whole housing stock with the unions involving the most interested parties - the homeless - in the exercise, rather than this being carried out by administrators (Kompasi, 1987, p. 193).

Bežovan has argued that since 1965 there has not been a housing policy for the economic reform of 1965 did not treat the problem of polycentric development, the urbanisation of the countryside, conservation issues nor the problem of illegal building (Bežovan and Kirinčić, 1987, p. 168). It was believed that the successful performance of a system of self-management would eradicate these problems. A harsher viewpoint was that "in the case of Yugoslavia it is the complete absence of concepts and models and proposals for solutions which furnishes the urban crisis with a touch of the apocalypse" (Mlinar, 1984, p. 110).

PROSPECTIVE DEVELOPMENTS

Future demographic trends, particularly in the under-developed regions, offer no solution to the problem. Increasing pressure is being put on the large towns. Families are getting smaller, putting increased pressure for housing for any given level of population. The half-million Yugoslavs who have been working abroad are slowly returning with their skills and savings and accommodation needs. The government has introduced a scheme to provide credit for 30,000 of these returning workers to house themselves.

Yugoslavia is once again in a position of drastic reform to its system with constitutional changes being introduced. The most recent Housing Reform has been proposed by the Federal Authorities (November 1987) to be discussed by the Republics as part of the Stabilisation Programme. It states as basic measures:

    i) the introduction of economic criteria in order to secure housing and communal funds
    ii) the transfer of the burden of financing from the social to the private sector
    iii) to provide better credit facilities for individuals. (Politika, 1987.)

Implicit in these proposals are a series of very significant policy switches. Throughout the post-war period, a dwelling has been treated as a good of social consumption and housing problems as social problems. Economic values have not been fully applied to housing. As a consequence, damage has been caused to maintaining and reproducing housing stock. The principle of division according to the results of labour has been broken, introducing social differences in housing and in society in general. Resources have been diverted, productivity hindered and inflation stimulated. Thus it has been concluded that "social ownership of our housing stock does not guarantee socialism or benefit for our society. Indeed it is the reverse, for by advocating social ownership what is meant is privilege for those who have as opposed to those who do not. They are in favour of political monopoly rather than the logic of the market" (Bajt, 1986, p. 72). It is regarded as an illusion to refer to social ownership when a flat can only be used in a "private" way.

To enable economic criteria and the market to be employed:
  i) Rents are to rise to provide adequate amortisation.
  ii) Social dwellings are to be sold.
  iii) Money subsidies are to be increased to poorer families.
Tax changes are designed to help those repaying credit and to exempt them from taxes on building materials and land during the building period. Some municipalities have already started selling socially owned flats, particularly older property, with revenue raised being returned to the housing fund.

The increase in rents to a "real" economic level has commenced. 1990 is the date envisaged by which adequate rates of amortisation will be fully realised with increases being made in stages. Dwellings are being revalued and rents are to be set at 2.5 per cent of the value of the dwelling. Thus rents are now rising dramatically ahead of inflation, although due to the low base start they still stand at only 1 per cent of the value of the dwelling whereas they should be much higher if the 1990 target is to be reached. This is all being carried out administratively by "direct social control" rather than through the mechanism of self-management in order to directly

affect the socio-political structure.
It is accepted that the market will bring new stratifications but these will reflect the economic power of the household, rationalising the structure of consumption and providing resources and a better supply of housing. Economic prices imply a variety of dwellings by size and standard but also the additional threat of further segregation into good and bad and elite quarters. It is argued that planning can alleviate such phenomena and that the best parts of town with the nicest outlook ought not to have the better conceived and more expensive developments (Bjelajać, 1982, p. 51).

The proposals accept that there is a shortage of 640,000 dwellings and that it is necessary to build 116,000 a year up to the year 2000 and that as private building has already reached 60-70 per cent of the total construction, the greatest part of funding in future - 80 per cent - should be secured by individuals (Politika, 19.11.1987.)

However, attention is already being drawn to possible problems which could arise with the implementation of such measures, particularly in terms of developing sufficient funds at a time of such high inflation. For example, in the Republic of Serbia, total housing funds amounted to 190 billion dinars in 1986. In the first six months of 1987 they amounted to 220 billion dinars - a nominal growth of 15 per cent. However, over the same period, the cost of housing construction grew at a rate of 430 per cent! (Politika, 19.11.1987).

A 1986 anti-inflationary Law on Income and Revenue caused a 20 per cent drop in housing construction and it is feared that the attempt to switch emphasis to the private sector will create a similar stagnation. The plan also stipulates that in return for extending amortisation periods from 56 to 100 years, rents should include a contribution to the housing fund. The overall effect on rents is that they would increase by approximately 310 per cent. However, in some municipalities this could vary from 120 to 1,000 per cent (Politika, 19.11.1987). Clearly, with falling real incomes as a result of the 'stabilisation' programme such increases could add to the pressures being manifest in the street demonstrations witnessed in Belgrade in 1988.

Other topics not covered by the plan include the status of illegal building, the increasing prices of building materials, and a land policy:

Housing Policy in Yugoslavia

so it has been criticised as being uncomprehensive.

In conclusion, it appears that there is no easy way out of solving Yugoslavia's housing crisis. The post-war principle that housing should be publicly owned has been abandoned and replaced with the view expressed by Kiro Hadjivasilev at a recent Federal Social Council meeting that

> Socialism will not lose anything if it has private ownership. Housing can be privately owned for it falls into the realm of private spending. It should be the concern of the individual with the help of society with provision still being made for the under-privileged. (Politika, 1988).

REFERENCES

Antić, L. and Jakšić, I. Stanovanje i stanbena izgradnja u okviru stambeno-komunalne privrede. In Znanstvene osnove dugorocnog razvoja SR Hrvatske, Ekonomski institut, Zagreb, 1983

Bajt, A. Alternativna ekonomska politika. Globus, Zagreb, 1986

Bežovan, G. and Kirinčić Stambena politika i stambene potrebe. Radnicke novine, Zagreb, 1987

Bjelajać, S. Kako se stvara socijalni zoning grada. Gledista 1-4, Beograd, 1982

Borba 30 May 1985

Borba 25 April 1986

Čaldarović, O. Neki aspekti socioloskih istrazivanja stanovanja u nas. Gledista 9-10, pp. 20-33, Beograd, 1980

_____ Izgradnja novih naselja u nas kao socioloski problem drustvenog koncepta naselja. Stambena i komunalna privreda, br.5-6, Zagreb, 1982

_____ Neocekivane posljedice planiranja razvoja gradova u Jugoslaviji. Sociologija, vol. XXVIII br.4, pp. 561-571, 1986

_____ Socijalna pravda i nejednakosti -prilog razmatranju socijalne stratifikacije u urbanim sredinama. Revija za Sociologiju, vol. XVII (1987) br.1-2, p. 105

Čaldarović, O. and Richter, M. Socioloska studija juzne Dubrave-Poljanica u Zagrebu. Urbanisticki zavod grada Zagreba, Zagreb, 1975

Housing Policy in Yugoslavia

Caldarovic, O. and Loncar, N. Drustveni okvir fenomena individualne stambene izgradnje u nas. Stambena i komunalna privreda, br.3-4, Zagreb, 1985
EIU. Quarterly Review; Yugoslavia, London, 1988
Gledista, br.1-4, pp. 3-11, 1982
Grupa autora. Stanovanje, aktualne problemi teorije i prakse. Savremena administracija. Beograd, 1975
_____ Drustveni aspekti individualne stambene izgradnje. Institut za drustvena istrazivanja, Zagreb, 1975
Horvat, B. Drustveno-ekonomske karakteristike stambenih potreba, Beograd, 1982
Jakšić, I. Izdaci za stanovanje kao uzrok stvaranju socijalnih razlika u oblasti stanovanja. Socijalni rad, 1/1974, Zagreb
_____ Stan-dobro osobne ili drustvene potrosnje. Ekonomika, br.10, Zagreb, 1977
_____ Stanogradnja i stanovanje u Zagrebu te njihov doprinos razvoju grada. Doktorska disertacija, Ekonomski fakultet, Zagreb, 1982
_____ Retrospektiva stanogradnje u Zagrebu, Hrvatskoj i Jugoslaviji. Bilten zavoda SKD, Zagreb, 7/1983
_____ Stanarina. Ekonomski pregled, 1-2, Zagreb, 1984
_____ Financiranje stambene izgradnje kao funkcije tretmana stana. Financijska praksa, br.4, Zagreb, 1985
_____ Drustveno-ekonomska priroda i makroekonomska struktura cijene stana. Stambena i komunalna privreda, br.9-10, Zagreb, 1985
Kardelj, E. Problemi nase socijalisticke izgradnje, knjiga V. Kultura, Beograd, 1964
Komisija Saveznih Društvenih Saveta, Beograd, 1987
Komisija za zivotne i radne uvjete radnika, Beograd 1987
Kompasi, M. Neki aspekti stambene politike i zadovoljavanje stambenih potreba s prijedlogom mjera i akcije organizacija sindikata. In Bezovan, G. above pp. 187-195
Lauwe, C. de. Famille et habitation, II. CNRS, Paris, 1959
Lay, V. Loncar, N. and Seferagic, D. Upotrebna vrijednost naselja kolektivne stambene izgradnje. Urbanisticki zavod grada Zagreba, Zagreb, 1984
Lay, V. Kvalitet svakidasnjeg zivota drustvenih grupa. Revija za Sociologiju, Vol. XVI, br.1-4, pp. 19-29, 1986
Marković, M. Praxis: Critical social philosophy in Yugoslavia. Boston Studies in the Philosophy of Science, vol. 36, Synthese Library, 1978
Mežnarić, S. Motivi odlazenja slovenskih radnika na rad u Njemacku, produzavanja boravka i vracanja u Sloveniju. Revija za sociologiju, 1-4, pp. 20-42, 1977
Miličević, G. Mere zahvatanja gradske rente u nasim uslovima. In Bezovan, G. above pp. 27-41

Housing Policy in Yugoslavia

Mirowski, W. and Mlinar, Z. Urban social processes in Poland
and Yugoslavia: theoretical and methodological issues.
Polish Academy of Science, USP, Warsaw, 1984.
Mlinar, Z. Socioloski aspekti kucne samouprave na primjeru
Nove Gorice. Komuna, br.9, Beograd, 1984
Pjanic, Lj. Stambeno pitanje u gradovima FNRJ. Ekonomski
institut FNRJ, Beograd, 1954
Politika, 11 November 1987
Politika, 20 February 1988
Puljiž, V. Eksodus poljoprivrednika. Zagreb, 1977
Richter, M. and Caldarovic, O. Socioloska analiza naselja
Knezija, Kalinovica, Srednjaci u Zagrebu.
Urbanisticki zavod grada Zagreba, Zagreb, 1974
Rogic, I. Sociologijske odrednice stambene reprodukcije.
Dometi II, Rijeka, 1980
Seferagić, D. Problemi kvalitete zivota u novim stambenim
naseljima. Doktorska disertacija, Filozofski
fakultet, Zagreb, 1985
Sekulić, D. Putevi i stanputice stambene politike.
Sociologija, vol. XXVII br.3, pp. 347-371, 1986
Stambeno zadrugarstvo u Jugoslaviji. Savez stambenih
zadrugara Jugoslavije, Beograd, 1980
Singleton, F. Twentieth Century Yugoslavia. Macmillan,
London, 1976
Stojanovic, D. Marxova teorija rente i razvoj komunalno
zemljisne ekonomike u nasim oslovima. Komuna, br.4/
1982
Vujović, S. Stambena kriza i ljudske potrebe. Sociologija
XX, 4 pp. 441-465, Beograd, 1979
Zivkovic, M. Prilog jugoslavenskoj urbanoj sociologiji.
Beograd, 1981

Documents

Aktualna pitanja stambene politike i zadaci SSRN. Beograd,
1979
Analiza dokumenata Saveznih drustvenih saveta o Dugorocnom
programu ekonomske stabilizacije u sektoru stambene
privrede, industrijalizovana stambena izgradnja. br.5,
Beograd, 1985
Analiza razvoja SR Hrvatske u razdoblju od 1981-1985.
Delegatski vjesnik, br.297, 21.03.1985
Delegatski vjesnik, 359, 1986
Drustveni dogovor o najvisem nivou standarda i normativna
stana zgrade i naselja koji se moze financirati
drustvenim sredstvima Republicki komitet za
gradevinarstvo, stambene i komunalne poslove zastitu
covjekove okoline, Zagreb, 1986

Dugorocni program ekonomske stabilizacije stambene i
komunalne privrede. Komisije Saveznih drustvenih
saveta za probleme ekonomske stabilizacije, Dokumenti
komisije, CRS, Beograd, 1982
Informacija o problemima iz stambene oblasti na koje radnici
ukazuju u predstavkama dostavljenim Vecu SSJ i
njegovim organima. VSSH, Beograd, 1986
Saveznogo Zavoda za Statistiku, Beograd, 1987
Službeni List SFRJ. Beograd, 31.12.1985
Stambena politika u udruzenom radu. Radnicke novine,
Zagreb, 1981
Statistički Godišnjak Jugoslavije. Beograd, 1987
Zakona o stambenim odnosima donosenjem Drustvenog dogovora.
Narodne Novine, Zagreb, 1985

Chapter Ten

HOUSING POLICY IN HUNGARY

Gábor Locsmándi and John Sillince

INTRODUCTION

Hungary has made great strides in housing provision since the end of the war, during a time when war damage and migration resulting from rapid industrialisation created enormous difficulties and shortages. Initially new housing was not of high quality, but later quality improved, as did the amount of demolition of old stock. The first big surge in housing construction came during the 1960s with the development of industrialised methods of building. This was followed by a vast increase in mortgage funds for owner occupation in the second acceleration in the 1970s, during which period shortages were reduced considerably, allowing very progressive allocation and subsidy policies to be introduced in 1971 for the needy. These policies were partially withdrawn in 1983 as a result of the country's economic difficulties, since when state funding for housing has been reduced, and since when the private sector has been encouraged. In January 1989 the mortgage rate for new mortgagees was raised to 18 per cent; there were also considerable rent rises.

TRENDS IN HOUSING CONDITIONS AND CONSTRUCTION 1945-1986

In 1945 Hungary was still a relatively backward agrarian country whose housing stock was largely in poor condition or damaged by the war. The period until 1960 saw no great improvement in housing problems. Investment was concentrated in heavy industry in Budapest and the North. The largest number of new dwellings (mostly privately built) was in the villages, yet the village population fell by 11 per cent from 1949 to 1960. And despite 100,000 new dwellings in the towns (excluding the capital, Budapest) the number of town inhabitants per dwelling rose by 21 per cent due to large-scale migration.

Throughout the late 1950s, especially in the VII, VIII, and IX Party Congress meetings, there was criticism of the lack of housing in Hungary. In 1960 a 15-year housing plan was launched which aimed at 50-80,000 new dwellings per year, an ambitious stepping-up of scale from the average 26,000 new dwellings per year of the previous 11 years. Moreover, investment funds shifted away from heavy industrial investment toward a geographically more even spread of investments in light industry and infrastructure. Also, the difficulty of controlling Budapest's growth was acknowledged - the 15-year plan accepted the need for a quarter of a million more dwellings in the capital (i.e. a quarter of all planned provision).

The target of a million new homes in 15 years was reached. With demolitions, the total net increase 1960-1975 was 816,000 new homes. Despite this the number of dwellings with no basic amenities fell very slightly, and by 1970 still two thirds of the nation's housing stock was without basic amenities.

Although the situation is still unsatisfactory, there was a dramatic improvement between 1949, when 9.2 million people lived in 2.467 million dwellings, and 1986, when 10.64 million people lived in 3.846 million dwellings. This improvement is illustrated in Table 10.1.

The largest qualitative advance has been since 1966, when demolition rates rose markedly, to a peak in 1976-1980 (99,200 demolished) after which they declined (to 76,000 in 1981-1985), and when construction rates rose, to a peak in 1976-1980 (452,700 built) after which they declined (to 369,000 in 1981-1985) (KSH, 1985b, pp. 8-9).

However, the problem of quality has not been merely due to history prior to 1945. Much new build until about 1970 was of poor quality. For example, in the period 1961-1965, 37.7 per cent of new dwellings were without a bathroom, 71.4 per cent without flush toilet, 7.4 per cent without electricity, 51.8 per cent without mains water, and 80.6 per cent without mains gas. The improvement here has been considerable: comparable

Housing Policy in Hungary

1981-1985 figures are 0.8 per cent, 1.2 per cent, 0 per cent, 0.9 per cent and 63.9 per cent (KSH, 1985b, p. 12).

Table 10.1: Hungary: Housing quality

	1949	1986
% of one-room dwellings	70.5	19.5
% of two room dwellings	24.6	48
% of dwellings without mains water	83	22.8
% of dwellings without flush toilet	87.4	34.3
% of dwellings without mains gas	93	69.8
% of dwellings without sewerage	N/A	21.4
% of dwellings without bathroom	89.9	27
Average density (persons per 100 rooms)	265	126
Average density (persons per 100 dwellings)	373	277

Source: KSH, 1985a, p. 277

Notes: N/A is Not available
All 1986 figures for 1st January

There has been a gradual increase in dwelling size, from 52 square metres in 1971-1975 to 54 square metres in 1980-1985 for state-built dwellings, and from 69 square metres in 1971-1975 to 80 square metres in 1981-1985 for private sector dwellings. The national average (state and private sectors) has grown from 63 to 74. The Budapest average has grown from 55 to 60. Differences between the two sectors are most marked for dwellings of three or more rooms. For example, in 1981-1985, the average size for the state sector was 32 square metres (one room), 55 (two), 75 (three), while for the private sector it was 38 (one), 63 (two), 102 (three) (KSH, 1985b, p. 15). There has been a gradual rise in the maximum permitted dwelling size, from 100 square metres to 140 square metres (about 5 habitable rooms). The 1988 maximum is 10 habitable rooms - a very generous limit.

The problem of low housing quality has been more a rural than an urban one. In 1949 the percentage of dwellings without amenities was 56 in Budapest, 86.4 in the towns, and 97.6 in the villages. Despite subsequent advances, the rural-urban bias still remains. The 1970 figures are 33 for Budapest, 47 for the towns, 67 for the villages (EGSZI, 1979; KSH, 1978, p. 8), and the bias still persisted in reduced form in 1987 (Békesi, 1987.)

In 1985 74.5 per cent of dwellings were privately owned. Tables 10.2 and 10.3 show how much private and state sector new dwellings were constructed in recent years. The fall in construction since 1980 has affected the towns most, then the villages, and Budapest least.

Table 10.2: Hungary: New state and private sector construction of dwellings

	1961-1965	1976-1980	1981-1985
New state sector	104,100	162,000	81,500
New private sector	178,200	289,000	289,000

Table 10.3: Hungary: New construction of dwellings in Budapest, the towns and the villages

	1976-1980	1981-1985
Budapest	85,600	79,400
Towns	201,500	159,800
Villages	165,600	135,500

Note: With some variations, "towns" are settlements outside Budapest larger than 8,000 population

Table 10.4 shows the large effect of government attempts (by providing cheap mortgages via the state building society and credit company OTP) to encourage home ownership since 1970. The category of 'shared house and freehold flat' is that of ready-built OTP flats (as well as a pre-existing number of old subdivided houses) which grew rapidly. The category "own house" includes inherited, paid-off, and self-built houses. The first three groups (i.e. the private sector) rose by 34.2 per cent in 1970-1985 compared with the state sector rise of 17 per cent. The table also shows that the number of dwellings in villages fell in 1970-1980. Also, the proportion of state dwellings is highest in Budapest.

In 1985, 34,356 detached houses were built (47.4 per cent of the total of 72,506): 56,564 (78 per cent) were of 4 floors or less (including detached houses), 12,312 (17 per cent) were of 5-10 floors, 1,657 (2.2 per cent) were of 11 or more floors. Besides these, 3,527 holiday homes were built. Whereas in the 1960s and 1970s most high-rise blocks were 10 floors or more, since 1980 the emphasis has switched increasingly to 5-storey walk-up blocks.

In terms of growth rates of settlements of different sizes, it is the towns between 15,000 and 20,000 which have grown fastest between 1972 and 1982 (93 per cent). The next fastest growth has been in the big towns of 100,000 to 500,000 (a growth rate 1972-1982 of 59.4 per cent). The small towns have grown despite low state housing and low infrastructure spending (Sillince, 1988). But they do have very high investment per head in industrial companies, presumably expanding where labour is most available (KSH, 1982b, pp. 205, 207, 209). The large towns have grown fast because of high levels of housing and infrastructure investment.

Low income groups tend to have worse and more expensive accommodation, both in the state and private sectors (Szelényi 1983, Sillince 1985a, Dániel 1985). Partly because of this, after 1975, special tenants' rebates and preferential mortgages (see later) were made available for families with 3 or more children. A well-publicised, though numerically small number of small, "temporary state" flats were also made available for young couples. However, these dispensations have slowly been withdrawn in recent years - more reliance is now given to waiting time for allocation of state flats. Homelessness and squatting do not occur.

Table 10.4: Hungary: Ownership of the housing stock (in thousands), 1970, 1980, 1985

	1970				1980				1985
	Budapest	towns	villages	all	Budapest	towns	villages	all	
Own house	176	486	1446	2108	168	555	1434	2157)	
Shared house and freehold flat	34	19	4	58	70	123	17	210)	2828.5
Self-build group	19	29	1	49	61	106	5	173)	
Government and other owners	389	269	161	819	410	347	119	877	958.5
Total	619	804	1611	3034	709	1132	1575	3417	3787

Source: KSH, 1982b: 1985a

Table 10.5 shows some indicators of housing size and quality for the richest and the poorest individuals. The size and quality rise steadily from the poorest to the richest individuals. (For other predictors of housing size and quality see Dániel, 1985.)

Table 10.5: Hungary: Dwelling amenities, by individual monthly income, per 100 households, 1979

	Less than 320 forints	More than 2240 forints
Rooms	121.4	228.2
Bathrooms	12.2	73.8
WCs	10.2	71.8
Hot water facilities	11.2	71.8
Modern central heating systems	2.0	51.5
Telephones	1.0	35.9
Modern dwellings	8.2	41.7

Source: KSH, 1982c, p. 336.

Table 10.6 shows that single-parent families are not noticeably worse off - they have comparable accommodation to that of childless couples. Indeed the table shows that flat size and quality are generally distributed appropriately, at least with regard to family status.

Retired people generally have larger and higher-quality dwellings than their incomes would suggest. This is undoubtedly due to low residential mobility (Dániel, 1985), rather than to policy.

New house- and flat- building by the state sector is only a fifth of that in the private sector (see Table 10.7). Within the private sector, the state building society, OTP, plays a prominent role both in finance and by direct

works. As far as building method is concerned
(see Table 10.7) most construction using
traditional methods is private sector
construction.

Table 10.6: Hungary: Size and quality of housing by family
status, 1979

Family Status	Per 100 households					
	Rooms	Bath rooms	WC	Hot water	Modern central heating	Modern flats
Non-family household	156.3	36.2	36.3	33.1	21.1	17.2
- one individual	152.3	34.9	35.1	31.8	20.3	16.3
- more individuals	181.7	44.1	43.5	40.9	25.8	22.6
One family household (without relative)	195.7	54.3	49.9	49.8	26.4	29.0
- couple without children	179.9	46.3	42.2	42.2	21.2	30.3
- couple with children	207.3	59.5	54.8	55.1	29.3	28.7
- single parent with children	179.1	48.1	46.9	43.8	26.8	26.1
One family household (with relative)	230.5	59.5	52.2	53.2	24.0	33.9
- couple without children	214.4	51.7	50.0	48.3	20.0	30.0
- couple with children	235.9	62.2	52.9	54.8	25.3	35.2
Average	194.1	52.0	47.6	47.3	24.8	27.6

Source: KSH, 1982c, p. 332

Table 10.7: Hungary: Numbers and building methods of newly
constructed dwellings (rate per 1000
population in parentheses)

	1985	Yearly Average 1981-1985
New council tenancies	8,162 (0.77)	9.717
New council built dwellings for sale	2,936 (0.28)	4,609
Other new state dwellings	1,858 (0.17)	1,971
Total new state sector construction	12,956 (1.22)	16,297
New OTP - built dwellings	18,469 (1.73)	17,631
Other new dwellings using OTP finance	40,102 (3.76)	38,555
Other new dwellings not using OTP finance	980 (0.09)	1,484
Total new private sector construction	59,551 (5.59)	57,640
Total new dwelling construction	72,507 (6.8)	73,937
Building Method:		
- prefabricated panels	24,144 (2.26)	26,000
- concrete cast in situ	4,858 (0.42)	4,600
- bricks on reinforced concrete frame	1,450 (0.14)	2,780
- other modern methods	798 (0.07)	760
- traditional methods	41,619 (3.9)	39,800

Source: KSH, 1985a, p. 278

The majority of current construction is dwellings with 3 or more rooms. In 1985 56.7 per cent had 3 or more rooms (3.86 per 1000 population), 36.6 per cent had 2 rooms (2.49 per 1000), and 6.8 per cent had 1 room (0.46 per 1000). About half of all new construction is of detached houses (34,356 were built in 1985, a rate of 3.2 per 1000 population). Very few semi-detached new houses are built (only 1424 in 1985, a rate of 0.13 per 1000). Multi-dwelling buildings (mostly state sector-built) form the other large group - for example in 1985 there were 25,115 new dwellings grouped together with more than 25 dwellings per building (KSH, 1985b, p. 42).

The financial changes in recent years can be shown by comparing state and private investment. In 1976-1980 state housing investment was 75.1 billion forints (43.3 per cent of the total); private housing investment was 98.4 billion forints (56.7 per cent of the total). In 1980-1985 state housing investment was 55.1 billion forints (24.8 per cent of the total) while private housing investment was 167.3 billion forints (75.2 per cent of the total). Considering all investment (all sectors) in 1976-1980 total national investment was 1029.7 billion forints, of which housing was 173.5 billion forints (16.9 per cent). By 1980-1985 total national investment was 1120.3 billion forints, of which housing was 222.4 billion forints (19.9 per cent). This reflects the stability of housing investment relative to the cutbacks in productive investment after 1980. Considering the state housing sector only, where there have been cutbacks, total national state sector investment in 1976-1980 was 924.8 billion forints, of which housing was 75.1 billion forints (8.1 per cent) while the figures for 1981-1985 are 469.8, 55.1 (11.7 per cent).

In 1985 the average new housing cost was 12,000 forints per square metre, a rise of 53.8 per cent on 1980. There were regional variations. Budapest was highest at 13,000 forints (a rise of 51.2 per cent), Györ was lowest at 10,300 forints (a rise of 68.8 per cent).

The costliest type of dwelling in 1985 was the detached state-built house (neither a local council tenancy nor built by a local council for sale), averaging 98 square metres and costing 1,842,500 forints to build. The cheapest was the high-rise flat bought with an OTP mortgage, averaging 57.2 square metres and costing 656,500 forints to build.

To buy a dwelling, a person can borrow up to 180,000 forints from OTP at 2 per cent interest per annum over 25 years. For sums between 180,000 and 400,000 forints 3 per cent must be paid (Szarka and Székély, 1987). Bank loans are available for larger sums, but interest is 8 percent per annum over ten years. For families "in need", OTP gives interest-free mortgages. However, the concept of "in need" has been drastically narrowed since the late 1970s. To buy, rebuild, or extend a family (detached) houses, one can borrow, since 1983, up to 90,000 forints, 70 per cent of the purchase price at 3 per cent interest per year from OTP. Otherwise, family houses can be financed by OTP ten-year loans at 8 per cent interest a year. An extra 140,000 forints mortgage is available at 3 per cent interest for modernising central heating (away from oil) and for insulation. To change a one- storey into a two-storey house (to encourage the efficient use of land) one can borrow 50 per cent of the cost up to 140,000 forints at 3 per cent over 25 years and a bank loan up to 50,000 forints at 6 per cent over ten years. For an extra dormer room (in the attic) the money is available subject to the necessary local council permission. OTP loans are not available for large dwellings (of more than 5 habitable rooms).

Thus for the great majority of private housing construction, capital is made available at 3 per cent rate of interest. Up to 45 per cent of the total dwelling cost can be paid for by social policy allowances, i.e. in state lump-sum subsidies. On top of this the prospective owner occupier may benefit from his or her employer's help, which is usually in the form of a zero-interest loan (Ijmkers, 1985 p. 349). This suggests that government help to owner occupiers is far more considerable than that suggested by de Felice (1987, p. 119).

Since the onset of Hungary's current economic difficulties in about 1979, house prices have tended to rise fast. For example in Budapest dwellings in the worst (best) districts in 1980 were valued at 5,000 (10,000) forints per square metre. In 1987 values were 10,000 (40,000) forints per square metre. The 1987 Budapest average was 25,000 forints per square metre, and outside Budapest, in the provinces, it was 15,000. Large villas in the fashionable parts of Budapest attracted (in 1987) 15,000 dollars private rental per year (mostly to Austrians and West Germans). These values compare with an average individual income of 7,000 forints per month (1988).

The trend since 1980 has been for local residents to bear an increasing proportion of the costs of infrastructure provision. Prior to piped gas being installed, residents are asked to vote for or against it. If it is installed in the street, those residents who do not want it still must pay 12,000 forints, while it costs 70,000 forints to be laid into the house. Sewerage, where this is provided, costs 40,000 forints, and a telephone line costs 60,000 forints.

THE ROLE OF HOUSING INSTITUTIONS AND ORGANISATIONS

There are three ways of acquiring home-ownership. One can build a house oneself. One can join a housing co-operative. Or one can buy a dwelling. In 1985, out of a total of 59,551 private sector completions, 37,675 were self-built, 19,629 were bought, and 2,247 were built by a housing co-operative.

In the vast majority of cases where one builds oneself, the couple and family members, friends, and neighbours do nearly all the work. Often a private builder provides an inexpensive overseer role. Dwellings are predominantly simple detached bungalows or 2-storey dwellings. This method offers maximum choice of dwelling type, and involves about a quarter of the cost of buying a ready-made dwelling. Employing a builder to do all the work speeds up the job but makes it more expensive.

With housing co-operatives, payment is in instalments as and when it can be afforded. The person can provide all the labour himself or pay extra. Some groups are completely self-build, others provide a cheap shell for the owner to complete.

Ready-built dwellings are provided by local councils (2,936 in 1985) and by OTP (41,082 in 1985). Their number has expanded since the growth in mortgage finance in the 1970s.

Besides home-ownership, there is reliance on state housing provision. In 1985 8,162 extra local council tenancies were created, and there

451

were 1,858 other new state flats. This latter group are built for members of the army, police and so on, and are not distributed by local councils. Allocation is by means of a points system, administered by local councils. The point system is applied not only to council flats, but also to co-operative and OTP flats. Thus even housing which finally ends up as privately owned passes through some form of state-mediated allocation process. The councils must publish waiting lists with points, but do not need to justify the points given. A proportion of council flats are given to families whose homes have been demolished; such an allocation does not use the points system. Government and Party officials obtain the best state flats, as do some instances of inducing "key workers" (e.g. doctors, teachers) to migrate to where they are needed. There was some criticism of the lack of equity in the system, and some experimentation in the middle seventies, giving greater weight to "need" in allocation of points. These experiments have receded now, as a reaction arose for example, to the spectacle of large families getting good state flats at low rent, and "selling" them (the exchange price can be half to three-quarters of what the flat would fetch if it were private), afterwards moving back to cheaper accommodation. Moreover, state provision is declining and little accommodation is now available for such noble purposes.

The social function or role of state housing has changed in an important way in the last 3 decades. Up to 1970 state housing was the cheapest and was of highest quality. But after 1970 the situation changed, and the relative quality of state flats has fallen.

State flats have relatively very low rents. Tenants can stay without condition for a lifetime, and the occupancy right can be inherited. The low rents have induced an illegal market to grow up - the tenancy right of the flat is sometimes sold to the highest bidder. For new tenants, payment of a deposit, or entrance fee, offsets the bargain character of state-built flats to let.

The institutional structure of the state building industry has many implications for design, building development, management, and rehabilitation (Mihályi 1978). In the 1960s eleven huge building corporations were established to produce factory-built industrialised housing.

These enjoy a powerful monopoly position. They make a narrow range of products which are difficult or imposible to adapt to non-high-rise housing types. Their infrastructure (large cranes, lorries, etc) are only suitable for high-rise building. They suffer from a problem of over-capacity, created in the 1960s when the fashion was for very large plants and companies. In the 1960s the idea was to build the majority of flats with state resources: something like a ratio of 70-80 per cent state, 20-30 per cent private, was aimed at. This, now that government resources devoted to housing are starting to be reduced (along with greater encouragement of private home-ownership and self-build) is leading to a problem of over capacity. Because of over-capacity most of these large house-building companies are in a position of imbalance. This results in problems of a lack of new investment, and of widening their range of products. These problems are aggravated by the large numbers of unskilled workers which they employ. Such large numbers cannot easily be diverted to other sectors. They have relatively low efficiency compared to the industrial average. The companies are beset by problems of rising energy and raw material prices. On the other hand, private builders tend to take away the better-trained workers from the state building firms, so that the labour market position of the large state building firms is not advantageous. It is under these circumstances understandable that state building firms try to keep their monopolistic position by lobbying national, county, and city politicians to get more orders and to get the necessary state subsidies. But state resources are limited - often when construction costs rise local councils react by cutting back on housing targets rather than by raising the budget allocation. This further aggravates the problem of over capacity and creates a vicious circle.

The problems of the lack of cheaper methods of building became strikingly apparent after the shift in the ownership distribution of newly-built flats in the late 1970s and early 1980s. Experts hoped this would lead to a radical change in the attitude of the big building firms, and to a more flexible approach to the needs of potential owners. Though many of them tried to work out new schemes - building terraces for example - the prices of them could be afforded by only a small minority of potential owners, particularly among those on the waiting lists. The whole procedure tended to prove the relative cheapness of high-rise building and paradoxically led to a reinforcement of the monopoly position of the state building companies. They have been able to freely dictate architectural and planning solutions because of this monopoly. This is why a nationwide series of television debates on high-rise schemes, on the effect of such environments on people's ways of life, and on escalating housing costs, subsided after a promising start in the early 1980s. Politicians agreed to further subsidise state building firms, and to do little to reform them.

These institutional problems underlie many questions associated with urban renewal and rehabilitation. Opinion among architects and planners has followed opinion elsewhere and become much more favourable towards rehabilitation. But the state building firms have neither the skills nor the interest in such work. Until now rehabilitation, besides that in tourist areas, has remained a political question rather than having a certain future. Renewal is becoming more important - the 1975-1990 housing plan greatly increased the planned demolition rate. Recently the demolition rate declined, due to financial reasons rather than to a rise in the political fortunes of the rehabilitation lobby.

POLICY ORIENTATIONS

Housing policy has been used as an instrument in attaining other policy objectives. It has been intricately connected with the problem of restraining Budapest's growth, the problem of rural depopulation, the problem of rewarding bureaucrats in the state and party organisations, and the problem of low population fertility.

Budapest has had special problems (Sillince 1985b, Hegedüs and Tosics 1983), partly due to the weakness of planning controls over immigration into the capital, and partly due to the large housing shortages there. If we consider the proportions of housing shortage calculated from waiting lists for state dwellings, Budapest rose from 3 per cent in 1960 to 39 per cent in 1975, whereas in the towns it was much lower (21 per cent in 1960, 25 per cent in 1975) and in the villages it fell noticeably (41 per cent in 1960, 35 per cent in 1975) (Sillince 1985a). In 1975 the estimated housing shortage was still 470,000, of which 65 per cent was in Budapest and the towns. Besides stricter measures in Budapest, the current (1975-1990) fifteen-year housing plan envisages much reduced housing investment in the towns to discourage movement away from the rural areas (VATI 1971, Barath 1981). This involves a target housing stock change in 1980-1990 of 21 per cent for Budapest (compared with a change of 16 per cent 1970-1980), a target housing stock change 1980-1990 of 25 per cent for the towns (compared with 41 per cent for 1970-1980), and a target housing stock change 1980-1990 of 3 per cent for the villages (compared with a negative change 1970-1980 of -3 per cent) (KSH, 1973, p. 34; Council of Ministers, 1975; KSH 1982a, p. 289).

There has been a realisation that keeping people in the villages prevented the development of further housing problems in the towns. This realisation is now enshrined in the Long Term Plan for Settlement Development (Boros and Lacko, 1986). Also, it was found that the average size of a village family was higher than that of a town family (2.95 for Budapest, 3.05 for towns, 3.35 for villages, for 1975: EGSZI 1979, p. 26), representing a potential source of further over-crowding in the towns. Moreover, the environmental factors which influenced lower birthrates in the towns were considered detrimental to the national policy of encouraging fertility (Cseh-Szombathy, 1979; Gyözöné, 1978).

Partly because of Budapest's large size compared with the national population (it has about 20 per cent of the population) and partly to spread economic development more equally across the country, a policy of decentralisation from Budapest has evolved. The problem of the growth of Budapest has been more complicated than was first realised. The regulations in the 1950s and 1960s were that those who had worked in the capital for a minimum of five years or who could show evidence that they had lived there in poor housing for five years had a right to be put on the various Budapest housing waiting lists. The first regulation was frequently circumvented by commuting for a period of five years (often from very far afield or by weekend commuting). The second could be got round by getting relatives in the capital to say one had lived with them. There was also shared private accommodation available - often at exploitative rents. The problem for the planners was that Budapest was one of the engines of economic growth in the 1950s and 1960s. So the 1960-1975 fifteen-year housing plan accepted this and proposed a quarter of all new provision for Budapest. Attempts were made, therefore, to change the balance away from Budapest by other methods. Expanding Budapest economic enterprises were tempted by government money to set up branches outside the capital. Also, more new towns were designated with special government aid, to offset Budapest's hypertrophy slightly, and to attract potential migrants to the capital away to other areas. Decentralisation has occurred but mainly for other reasons, mainly those associated with enterprises' search for labour (Sillince, 1988).

However, in the 1975-1990 housing plan the planned distribution of over-crowding favours Budapest against the villages, a pattern unhelpful to decentralisation. For 1990 the planned number of individuals per hundred rooms are: 100 for Budapest, 109-115 for the towns, and 118 for the villages. The 1990 planned numbers of individuals per 100 flats are 222 for Budapest, 227-228 for the towns, 235 for the villages (Council of Ministers, 1975). However, relatively small flats are now being built in Budapest, an average in 1981 of 57 square metres compared with the national average of 70 square metres; also, only 26.7 per cent of 1981 constructions in Budapest had three or more rooms, compared with a national average of 45.6 per cent, suggesting a less favourable policy towards Budapest (KSH, 1982b, p. 131).

Proposed in the 1975-1990 fifteen-year housing plan is an increase in total housing stock from 3,573,536 to 4,030,000 dwellings, after allowing for demolitions. The total planned new dwellings amount to 1,200,000, of which 550,000-600,000 is the net increase after allowing for demolitions. This represents a rate of 80,000 new dwellings per year, a continuance of the highest rate during the previous fifteen-year plan period, and a rate of 40,000 demolitions per year, a doubling of the highest previous rate. The 1990 target is likely to be met, although partly because demolitions have proceeded much slower than planned. Annual numbers have been falling -

the 1985 figure of 12,490 demolitions was the lowest since 1963.

There are a multitude of often conflicting objectives involved in policy. The allocation of state housing is often used as a wage supplement. A private electrician or plumber can earn 40,000 forints per month (in 1985) while a company director gets about 20,000, a government minister 25,000, an average Party worker at his workplace gets nothing (besides his usual income) for his Party activities. While such examples may not be entirely typical, they do fit in with public perceptions of the rewards of different occupations. One understandable reaction by government organisations with influence upon housing allocation decisions is to seek to use housing as a means of attracting and retaining useful personnel. Therefore the government perceives a need to use housing to compensate for the deficiencies of the system of rewards. Why the system of rewards cannot instead be changed is an interesting political question. Suffice to say that the system has become intolerant of major changes. Also, the allocation of state housing is used to encourage the migration of key workers. For example, village primary teachers earn only 4,000 forints a month and village doctors only 6,000 forints. Pickvance (1986, 1988) asks interesting questions about employer influence in housing allocation, although the rapid geographical decentralisation of Hungarian industry suggests that such influence has not been very successful in attracting labour (Sillince, 1988). Also, another objective of state housing is to redistribute incomes and help the needy. Parallel to similar principles in the private sector, has grown up the principle of permanent rights. State flats can be occupied for life. Occupants other than heads of households acquire the ability to inherit such rights. This principle may conflict with helping the needy. It reduces the mobility of dwellings. There is also the objective of encouraging fertility, but long waiting lists are no great encouragement. The most recent objective is the encouragement of private finance. It conflicts with many previous objectives.

Housing Policy in Hungary

MAJOR PROBLEMS

There are essentially two housing markets - the state and the private sectors. Within the state sector there is a chronic housing shortage (Kórnai, 1980) characterised by average waiting times which have remained at about 8-10 years since the early 1970s. However, some deserving cases receive state housing relatively quickly while others must wait much longer. The private sector is characterised by high prices. Both markets are "imperfect" in the sense that there is friction between buyer and seller: in the state sector those waiting do not know how long they need to wait, nor whether it is worth their while making alternative arrangements; in the private sector information is not freely available about what is for sale and at what price.

The most important question which relates to segregation is whether or not the 300,000 or so gypsies are forming ghettos as seems to be occurring in District VIII in Budapest. With the exception of the early 1970s, gypsies are usually the least skilled and poorest and are therefore given the worst housing, a policy likely to lead to segregation. Official data do not exist about gypsies. With regard to social segregation of a more general nature, little is known and then only with regard to Budapest. The motivation for official interest came after Joszefváros (in Budapest) was redeveloped and mutilated. Partly the interest arose from the renewal versus rehabilitation debate. Several studies were made in the late 1970s. One of them (Ekler, et al. 1980) revealed the continuation of out-migration from parts of the central core. Some small sub-areas showed a loss as high as 25 per cent over ten years. Moreover, the pace of out-migration was found to be increasing. Though the overall tendency throughout Great Budapest was found to be a gradual levelling out of ecological differences in the different social groupings (confirmed by the 1950-1980 period by Sillince 1985b and for Prague between 1930 and 1970 by Musil 1987) many of the older denser areas were found to show a fall in social status and a rise in average age. This approach to studying ecological differences on a district-by-district basis has been criticised by Csanádi and Ladányi (1985). Their Budapest study argues that the rich may live in good flats facing the street, and the poor may live in bad flats facing the backyard (or light well); both rich and poor may occupy the same street, yet significant segregation is still occurring.

Another aspect of segregation is that while half of the manual industrial workforce lives in villages, only a quarter of clerical and professional workers do (Konrád and Szelényi, 1971; Danta, 1987) although recent trends noted by Sillince (1988) may be reducing these differences.

The following are the results of a series of sample surveys of 2,500 representative individuals throughout the country in 1970-1976 by the Central Statistical Office (Hoffmann, 1981). No matter what their present housing circumstances, the family (detached) house was preferred by all, moderated somewhat by experience in a high-rise flat. This preference increased with age. Council high-rise flats were consistently unpopular. Questions on attitudes revealed that people are preoccupied with shortages and the difficulties of getting a dwelling. The strength and persistence of these attitudes has been established recently (Farkas and Pataki, 1987).

Szelényi (1983) has shown how in several socialist countries including Hungary, housing is treated as a kind of wage supplement. For his 1971 case studies of Pécs and Szeged, the best and most subsidised housing, he showed, was administratively allocated to the highest occupational groups. 1979 national data (Magyar Szocialista Munkaspart, 1980) show that among occupational groups, the intelligentsia (non-self employed and middle class in "mental" occupations) had by far the largest proportion of the best government tenancies (50 per cent compared with the average per occupational group of around 20 per cent) and the smallest proportion of owned houses (only 34 per cent compared with the average of 74 per cent). This occupational group comprises key workers who get preferential treatment on waiting lists. It is also concentrated more in Budapest, which has the largest proportion of government tenancies, especially of the high quality ones. This suggests two things. First, there is a significant amount of privileged access in the Hungarian housing market (see also Konrád and Szelényi, 1971; Szelényi 1972, 1983 and for comparisons with USSR, Di Maio, 1974; and Hoffmann, 1981, p. 77). For further more recent evidence of inequity in Hungarian and other Eastern European housing see Szelenyi (1987), Ciechocinska (1987), Musil (1987), Dangschat (1987), and Hegedüs (1987). Second, despite the widespread popularity of some ownership an important sector (and that which is highest paid) of the population considers it to be unnecessary. (However, this elite preference is changing towards purpose-built owned flats.) An indication of the good bargain which the (controlled) rental sector represents is given by the fact that between 1954 and 1975, only in three years (1956, 1972, 1973) did income from council rented property exceed maintenance expenses (Council of Ministers, 1975). Another indication is that between 1959 and 1982, 70,000 council dwellings were offered for sale to sitting tenants in Budapest. Only 2,300 were bought by private individuals. However, this situation may be less true for state-built dwellings since 1970 (see above). In the uncontrolled sector, there is a small but important sector of illegal tenancies at high rents 1,200 forints a month for one room in Eger in 1983 (Falus 1983).

Sillince (1985a, Table 9) shows that rents and mortgages are relatively cheaper the higher the person's income. He shows that for the occupational group which rents more than it mortgages (the intelligentsia) a high income is combined with a rent of 4 per cent of income. Whereas, for example, for the occupational group which is most heavily involved in home ownership, the labourers in co-operatives (94 per cent home ownership), the necessary mortgage repayment is 9 per cent of income. Szelényi (1983, pp. 59–68) shows similar data for rents and deposits. Mortgages, like rents, are massively subsidised. Thus high-income groups are subsidised more than low-income groups in relative and absolute terms.

In overall terms, the housing of the top occupational group is the best quality on a range of indicators while co-operative labourers are provided with the worst housing (Sillince, 1985a, Table 10; KSH, 1982c, p. 49).

For those in rented flats and houses, the intelligentsia, junior non-manual workers, and foremen have equal quality housing, whereas, again, co-operative labourers occupy the poorest accommodation. For those in mortgaged flats and houses, the intelligentsia, junior non-manuals, and the retired occupy the best quality housing, and manual workers occupy the worst accommodation (Magyar Szocialista Munkaspart, 1980). Sillince (1985a, Table 11) shows some housing quality indicators and income for paid-off owners. Some of the poorest families are in the worst accommodation. This is easily understandable - the majority of owned property is rural and of lower quality. For mortgagees the picture is not so clear-cut (Sillince 1985a, Table 12). There is only a vague association between income and housing quality. Besides amenities within the dwelling, size and number of rooms are other important indicators. Sillince (1985a, Table 13) shows that in terms both of numbers of rooms and of density of people per room, the richer families do better than poorer ones. Dániel (1985, Table 11) shows that how much housing subsidy a person receives is positively related to income. For example, those individuals on less than 12,000 forints per year receive 9.6 per cent subsidy; those on above 55,000 forints receive 13.0 per cent subsidy. For other Hungarian data see de Felice (1987).

PROSPECTIVE DEVELOPMENTS

The high priority of housing policy in Hungary up until the mid-1970s dated from discontent which surfaced after the Uprising and political changes of 1956. In the 1960s, as a result, a modern, large-scale house-building industry was created. In the 1970s volumes were further increased by means of an expansion in mortgage credit. Numbers, size, and quality of dwellings all dramatically increased. But outside the private self-build sector, state officials (mainly in local councils) continued to influence allocation. This led to unequal access: better-off people got the best housing almost free.

This situation was criticised by Szelényi (1983) and Dániel (1985). Szelényi followed market-oriented ideas by Liska (1969) and suggested that the Central Planning Office should cut housing construction and relax controls on private house building. Since those criticisms were written in May 1972 those suggestions have been put into practice (although state rents are still very low). The relative quality of state-built dwellings has fallen. But only 17 per cent of families now receive state-built housing, so that only a minority of new dwellings are allocated by the state. A far more important recent source of inequality is family help in the acquisition of home ownership. It is convenient for the government to overlook this, however, since it wishes to encourage home ownership.

Summarising the current state of knowledge, it would be fair to say that today the fact that Hungarian and other East European housing is inequitably distributed is unquestioned. What is in doubt is (a) whether inequalities are increasing or decreasing, and (b) whether (although a purely private sector housing system would lead to the greatest inequalities) a mixed private sector and state sector system reduces the inequalities of a purely state sector system. On (a) there are differences of opinion. Musil (1987) has argued for a trend to less social class segregation and has put forward data on the 1930–1970 period in Czechoslovakia, while Szelényi (1987) argued that these data and other available East European data were not detailed enough prior to 1970 to reach a conclusive answer. Other criticisms have been advanced by Csanádi and Ladányi (1985). On (b) with particular regard to Hungary, both Tosics (1987) and Hegedüs (1987) argue that the market cannot counteract inequalities created by state allocation processes. Instead they argue for "intervention of the society into the state and into the market". Szelényi, who has argued instead for the development of a more mixed welfare state economy in Hungary (see Szelényi and Manchin, 1986) comments that this "is an attractive idea (and a very fashionable one in eastern Europe today) but it has to be further elaborated to be believable" (Szelényi, 1987, p. 4). The crucial point of disagreement concerns what happened in Hungary during the 1970s, Tosics and Hegedüs arguing that state allocation methods became more egalitarian, and Szelényi arguing that the much larger construction volumes enabled such changes to occur. Certainly, unequal access to housing is one important measure of inequalities in socialist countries, and Hungary seems to share this problem with other East European countries. Whether more steering towards the market would reduce these inequalities must remain an open question. A much more certain conclusion is that periods of expansion in construction volumes such as the 1970s allowed experimentation towards greater

egalitarianism. The harsher economic climate
since 1980, however, has meant that other themes
of market-orientation have instead been emphasised
- the need to make prices reflect costs more
faithfully, and the encouragement of the already
important private sector. The latest (1988) move
to encourage the private sector has been the
relaxation on the size of private companies, now
able to employ up to 500 workers. The costs of
state-built housing construction will continue to
be on the political agenda. They continue to rise
- private firms are now three times as productive
of manpower.

THE NEW HOUSING POLICY IN 1988

Hungarian housing policy is at an interesting
turning point. During 1987 and 1988 there has
been much debate (often ill-informed and
consisting of stereotyped images) and a general
expectation that drastic change is on the way
(Ernst, 1988).

At the highest level of rhetoric, the
government maintains an optimistic attitude: there
is almost no problem. Average figures such as
persons per room and square metres per person are
often referred to in these contexts. Yet there
are still 8 per cent of households without their
own flat (Dániel, 1985, p. 408). The housing
shortage is a real one. Despite as yet many
confusing signals, the politicians always say two
things: we can't spend more on housing (meaning,
probably, that less will be spent), and we haven't
solved the problems yet (meaning, be patient).

In the public mind, four problems are
paramount. First, relatively rich and important
people have large, good quality flats at low rent:
most people resent this. Second, some lucky
enough to have the money have built for themselves
large mansions - often of 200 square metres, but
some of 300 square metres exist too - huge even by
Western standards. At the other end of the social
scale there is a growing number of people who
cannot solve their housing problem themselves -
most notably young couples, who must remain with
parents. Fourth, people ask why the cost of
building houses is so high, and why the government
cannot do something about it. It is these four
problems (stated above in the vivid stereotyped
form of most public debate in Hungary, where such

openness of debate is as yet very new) which the government has clearly signalled that the new policy will address. It is interesting that design problems, such as the unpopularity of high-rise housing, are a long way behind in the public's mind.

So it is relatively clear what the public wants. But a policy meeting all these objectives is unlikely, partly because the money simply is not available, and partly because the social differences and class tensions (these taboo concepts would be expressed very differently in Hungary) opened up by the new policy debate are very dangerous, politically.

The optimism of the government and the use of Panglossian averages are a way of saying that the problem has been solved so that housing can be downgraded relative to the needs of the economy. For here is the crunch: Hungary is going through troubled economic times. There is a huge foreign (Western) debt, real incomes are falling due to the government's stabilisation programme. Hungarians must now pay income tax: for the middle class - not rich, not poor - this means roughly a halving of second incomes.

However, the problem has not been solved. Many dwellings are shared by more than one household (the government, euphemistically, refers to families rather than to households). Many dwellings are old, small, and of poor quality. The house-building and house-repairing industries are both expensive and inefficient. The better-off, the intellectuals, in general have more square metres of housing space, in lower-density environments, than do workers. The budget burden on the state is considered by the government to be excessive. More alarmingly, there is now a significant minority - perhaps 10 per cent of the population - with Western values and expectations. This group will in the future set the pattern of values and expectations for the remainder. Thus any discussion of Hungarian housing must consider the large and growing gap between Western and Hungarian housing conditions. In the light of this, earlier policy mistakes, such as the ten-storey developments of the 1970s (lacking the open space of similar high-rise developments of the 1960s), may prove costly in the future. Even now, such flats are proving difficult to sell, despite their moderately good interior size and quality.

So what are the options currently under consideration by the government? In Budapest and the big cities, where maintenance costs of the existing stock are roughly two thirds the cost of new state construction (Mihályi, 1978, Figure 2) where city councils want to reduce their maintenance costs, there are ideas to try to strengthen the "ownership attitude" of residents, by forming new, smaller organisations representing each city block. There is also the idea to reprivatise state flats. There is also a strong body of opinion pushing for a "market solution", for raising rents and mortgages to economic levels (that is, to cover building costs). Alongside this body of opinion is a more enlightened group arguing that a safety net system must be set up - an open subsidy or rebate for the poor - on the lines of, for example, the UK. Besides these "demand side" measures (all the above ideas involve reducing the state's burden, an ominous sign) there is the idea to siphon off private money through "special" banks which would play a role similar to that of building societies in the UK. Let us take these ideas one by one.

At present the maintenance of the state housing stock is carried out by huge city council organisations. Where possible, money is clawed back from residents, to cover costs of repairs. But such costs, for example in Budapest, may run to 700,000 forints (roughly £7,000 or 100 times the average monthly salary). If a resident were to pay this money, the value of his refurbished flat would not be increased by such a large amount, because the density of old city blocks is so high - traffic noise and fumes, difficulty of parking, lack of green outlook, lack of daylight. It would be better if such rehabilitation schemes involved small-scale demolition (for gardens, for daylight) and decanting (for amalgamation of smaller into larger flats). But this would necessitate some residents being moved to totally new housing on the city periphery, and high-level political resistance to this exists on grounds of cost. So far two experimental schemes have been in operation. These are at Mayakovskiy Street in District VI and in Block 15 in District VII, both in Budapest. A majority initially voted for rehabilitation, knowing they would have to pay for it. But the cost estimates rose, and many poorer residents did not move back in because they could not afford to pay the additional mortgage or extra rent involved.

In future the government would prefer each city block to be represented by an organisation having some "social ownership" status (either a co-operative, or a new type of organisation having the same role as a private property developer, or a "community of owners" - <u>tarsaházi közösség</u>) which would strengthen residents' ownership attitudes and channel repair activities. But there are legal problems. These new private enterprises would require the same legal rights - to sign bills, have trading accounts, have special tax rights etc. - as state enterprises have. So far the government has never granted this equal right to private enterprises. The experience of the organisations most similar to these envisaged organisations - the "community of owners" - has been mixed, involving corruption and speculation. At the heart of the problem is that if a totally new management system is to come in, one must avoid the situation where an owner, or small group of owners, controls a building with a large number of flats. Hence the need for some new type of organisation which is ideologically removed from private property development.

But the skirting of ideological obstacles will not remove all social problems. The poorest people live in the smallest, worst accommodation. If the government really wants to reduce the state's repair bill, how will such city blocks be improved? Will the middle-class minority dictate, or manoeuvre, rehabilitation schemes which force poor people to leave?

Another idea for change is reprivatisation. A state tenant has the right to purchase his flat from the state. At present most of the old city blocks are state owned while about 5 per cent are in mixed ownership. In new post-war developments state and private ownership varies on a block by block (or stairway by stairway) basis. Reprivatisation will thus create a trend towards mixed blocks, making the role and composition of small management organisations more complex and difficult. Moreover, reprivatisation is not proving popular - only good flats in small blocks with a good environment are selling well.

Besides the strengthening of "ownership attitudes" and reprivatisation, there is the idea to force up mortgage rates and rents. Until now state tenants have enjoyed a favoured position. The state is prepared to pay a lump sum if tenants leave a flat. In Budapest the sum reaches 500,000 forints - large compared to the 50-100,000 forints entrance fee paid by new tenants. Yet despite this incentive few choose to leave. Hence, goes the argument, why not raise rents? Moreover, many rich and important people live in low rent, good quality flats. Surely, it is argued, they should pay more. It looks likely now that rents will rise considerably, perhaps with the creation of a rebate system to help poorer people. A strong raising of rents at least above a certain critical level (Dániel and Semjén, 1987) will surely give a boost to reprivatisation and to tenants moving in return for a lump sum. For private owners, it seems almost certain that the mortgage rate will jump from 3 to 15 per cent. Despite initial government support for retrospective legislation and strong lobbying for it by the Youth Organisation, there is now a public government promise (Kossuth Radio, 1988) that the rise will not affect existing mortgages. (In April-May 1988 a fierce middle-class reaction to proposals to make the rise retrospective was successful.) The middle-class argument is that mortgages are too high, and that rents are too low: the argument goes "we built the flat with our own hands" or "we paid for it from our own earnings". Thus the strains that are developing are revealing class tensions, with no means of publicly admitting this fact. Moreover, the official view is that there is no residential segregation: "there are many different people on our housing estates". But segregation does exist, in Szeged and Pécs (Szelényi 1983), and more visibly between the gypsy and prostitute District VIII where residents have an average of 12 square metres per person and the mansions in the Buda Hills in District II of Budapest where the average living space is 35 square metres per person. Such segregation may be accentuated by local management organisation to control repairs: a poor majority may force out a rich minority and vice-versa. Yet it is prohibited to speak about segregation and class interest. In 1988 on Hungarian TV a sociologist spoke about the fact that the middle class has had its income drastically reduced by the new tax system and that any rise in mortgages will leave them unable to pay. His point was that only the top ten percent - the rich - can afford any such rise. Yet such language - speaking about subgroups of the population rather than about averages - was met by embarrassed silence. Such taboos discourage informed debate. For example, a crucial variable in deciding whether or not to moderate the raising of rents and mortgage rates by a rebate system for the poor, will be the cost of this system to the government. Yet such a figure has not been publicly disclosed. To summarise, this "market approach" - the raising of rents and mortgages to the economic level - has won uncritical support, without sufficient acknowledgement of either the difficulties this will cause for poor people, or the fact that potential savings to the state budget will be much diluted by the necessary rebate system for the poor. At least the government will be satisfied - it will achieve its main objective (along with the new taxes) of cutting back demand.

Among experts, some rebate system seems impossible to avoid, and it does have rational properties too. At present everything is done from the centre. Money goes from one level of the government housing bureaucracy to the next - often by means of political deals and compromises which have little to do with housing problems. If a rebate system were instituted the needs of each local authority would become easily visible.

Within the Hungarian economy there is a great deal of spare cash - saved by the relatively rich and to a lesser extent by the middle class from their second incomes. Now, with the setting up of the new tax system people know their savings and hence their incomes will be investigated, and so there has been a rush to buy cars, jewellery, and property. Property prices have risen fast everywhere in the last ten years but particularly fast between 1986 to 1988, and also particularly fast in Budapest. One pragmatic political reaction to this has been the suggestion that incentives (tax concessions) should be provided to induce savers to invest in special banks, which would then make available funds for others to get mortgages. However, as long as property speculation is so lucrative relative to banking the money, such a policy does not seem very likely to succeed.

What is the timetable for the new policy? "Theoretical ideas" - the collection of the main ideas, the working out preliminary material and many alternatives - were ready by September 1988 for a committee chaired by Laszlo Marotti, the

Environment Minister. One Ministry of Construction source has indicated that there is a fixed amount of money available for housing, and that the main question is how to get the money that is missing, from higher rents and mortgages, given that perhaps a rebate system is needed. In January 1989 the mortgage rate for new mortgagees was raised to 18 per cent; there were also considerable rent rises.

Unfortunately, the policy outlook is not encouraging. There are three areas where debate contains simplifications and unanswered questions. First, a purely market system will create new problems, because poor quality city flats need to be improved somehow and because many cannot afford higher rents or mortgage repayments. Second, more flats and hence more money is needed than the politicians think, and tough decisions must be made to distinguish priority areas from less needy areas, using relevant statistics rather than political deals: the big cities need more per capita than the villages, and the inner cities likewise. Third, there has been the red herring of new-build versus rehabilitation, the former supported by a shrinking, yet still overweight state building industry, the latter by changing architectural fashion: ironically the argument that the city stock is large enough and that rehabilitation only is required, can be used by politicians who don't want to spend more on housing.

From the viewpoint of the average housing consumer the improvements in average housing quality and space represent a change for the better over the last 20 years. But for the coming generation of housing consumers the outlook is not good. In 1968 the price of the average new flat was 320,000 forint; most was provided as credit by the State Bank. Only 10,000 forint was required in cash - the equivalent at that time of half an average annual salary. By 1988, however, the average new flat cost 900,000 forint, of which only 400,000 was provided by the State Bank as credit. The remaining 500,000 forint cash needed represented 5 years' average salary.

Housing Policy in Hungary

REFERENCES

Barath, E., (Ed.) (1981) Orszagos Teruletrendezesi Terv
   Konceptio. VATI 11, Budapest: Regionalis Irona
   _____ (1983) "Decentralizalasi tendenciak a magyar
   településhalozat féjlesztéseben a 70-es évtizedben",
   Városépítés, Vol. 26, No. 2, 26-30
Bekesi, L., (1987) "Lakaspolitikai celok es
   telepulesfejlesztes", Városépítés, XXIII, No. 2, 1-40
Boros, F., and L. Lacko, (1986) "New trends in settlement
   policy in Hungary", Foldrajzi Kozlemenyek, Vol. XXXIV,
   No. 3, 205-221
Central Committee, (1970) A Magyar Szocialista Munkáspart
   Kozponti Bizottsaga. Minutes of Meeting, 10 March
   1970, Budapest: Kossuth Könyvkiádó.
   _____ (1978) A Magyar Szocialista Munkaspart Központi
   Bizottsága. Minutes of Meeting, 12 October 1978,
   Budapest: Kossuth Könyvkiádó
Ciechocinska, M., (1987) "Government interventions to
   balance housing supply and urban population growth:
   the case of Warsaw", International Journal of Urban
   and Regional Research, Vol. 11, No. 1, 9-26
Compton, P.A., (1979) "Planning and spatial change in
   Budapest", Ch. 16 in French, R.A. and Hamilton, F.E.I.
   (Eds.) The Socialist City ; Spatial Structure and
   Urban Policy, pp. 461-492, New York : Wiley
Council of Ministers, (1975) Elóterjesztés a Miniszertanacs
   részére a lakásépítés és gazdálkodás 1990 - ig szolo
   térvról. Budapest: Council of Ministers
Csanádi, G., and Ladányi, J., (1985) "Budapest - a
   városszerkezet történetének es Különbözö társadalmi
   csoportok városszerkezeti elhelyezkedésének
   nemökológiai viszgálata". Unpublished manuscript
   referred to in Szelényi 1987
Cseh-Szombathy, L., (1979) A csaladpolitika célja,
   tartalmi Kore, eszkoz rendszere. Tarsadalmi
   terveszési fuzetek III Köt, 20-21, Budapest :
   Országos Tervhivatal Tervgazdasági Intézete
Dangschat, J., (1987) "Sociospatial disparities in a
   'socialist' city. The case of Warsaw at the end of
   the 1970s", International Journal of Urban and Regional
   Research, Vol. 11, No. 1, 37-60
Dániel, Z., (1980) "Igazsagos vagy igazsagtalan
   lakaselosztas", Valóság, Vol. 4
   _____ (1981) "A lakasszektor reformja", Valóság, Vol. 12
   _____ (1983) "Public housing, personal income and central
   redistribution in Hungary", Acta Oeconomica, 31, 87-104
   _____ (1985) "The effect of housing allocation on social
   inequality in Hungary", Journal of Comparative
   Economics, Vol. 9, No. 4, 391-409

Dániel, Z., and A. Semjén, (1987) "Housing shortage and rents: the Hungarian experience", Environment and Planning, Vol. 21, No. 1, 13-29

Danta, D.R., (1987) "Hungarian urbanisation and socialist ideology", Urban Geography, Vol. 8, No. 5, 391-404

Dienes, L., (1973) "The Budapest agglomeration and Hungarian industry - a spatial dilemma", Geographical Review, 356-377

Di Maio, A.J., (1974) Soviet Urban Planning : Problems and Politics, New York: Praeger

Dolescsko, K., (1983) "Hol tart a budapesti lakásgazdalkozas?" Népszabadság, July 30, 9

EGSZI, (1979) Épitésgazdasági és Szervézetési, 4884/5 Tanulmany, Budapest : Épitésgazdasági és Szervézesi Intézet

Ekler, D., Hegédus, J., Tosics, I., (1980) A városépités alkalmazott társadalmi-gazdasági modelljének elméleti és módszertani kérdései : a városféjlödés társadalmi-térbeli összefüggései. "Budapest példáján I és II Kötét". Budapest : Budapest Városépitési Tervezo Vallalat

Ernst, G., (1988) "Patthelyzet a lakásfronton", Valóság, No. XXXI, Vol. 7, July, 71-88

Falus, G., (1983) "Garzonház Egerben", Népszabadság, March 26, 3

Farkas, K., and Pataki, J., (1987) "The 8th year of seven lean years", Jel Kep, No. 2, 54-157

de Felice, M., (1987) "Housing", in Kende, P., and Z. Strmiska, Equality and Inequality in Eastern Europe, Berg, Leamington Spa, pp. 111-127

Ferge, Z., (1979) A Society in the Making, Harmondsworth, Penguin

Gyözöne, K., (1978) A falusi lakokörnyezet alakulasrol. Budapest : Szövetkezeti Kutató Intézet Közlemények

Hamilton, F.E.I., (1970) "Aspects of spatial behaviour in planned economies". Papers of the Regional Science Association, 25, 83-105

_____ (1979) "Spatial structure in East European cities", in R.A. French and F.E.I. Hamilton (Eds.), The Socialist City. London, Wiley

Hegedüs, J. and Tosics, I., (1983) "Housing classes and housing policy : some changes in the Budapest housing market". International Journal of Urban and Regional Research, Vol. 7, No. 4, pp. 467-494

_____ (1987) "Reconsidering the roles of the State and the market in socialist housing systems", International Journal of Urban and Regional Research, Vol. 11, No. 1, 79-99

Hoffmann, I., (1981) Lakáskörülmények. Budapest : Kossuth Könyvkiádó

_____ (1983) "A folyomatos árszinvonal-emelkedés
hatása lakásépítésre", Közgazdasági Szemle, 30,
608-612
Hörcher, N. and Vajdovich-Visy, E., (1985) Structural
changes in the Hungarian settlement structure,
Budapest XXVth European Congress of the Regional
Science Association, August 27-30
Horváth, T., (1983) "Morognak a lakók", Magyarország, 20, 35
Ijmkers, F., (1985) "Changes in the Hungarian housing
financing system - some theoretical remarks", in
International Federation for Housing and Planning,
Papers and Proceedings, International Congress,
Budapest, pp. 349-352
Konrád, G. and Szelényi, I., (1969) Sociological Aspects
of the Allocation of Housing. Budapest : Sociological
Research Group, Hungarian Academy of Sciences
_____ (1971) "A késleltett városfejlödés tarsadalmi
konfliktusai", Valosag, Vol. 15, No. 12, 19-35;
(appeared as "Social Conflicts of Under-Urbanisation",
in M. Harloe (Ed.), Captive Cities, Wiley, 1977)
Kórnai, J., (1980) Economics of Shortage. Amsterdam, North
Holland
Kossuth Könyvkiádó, (1976) Az Életszinvonal Alakulasa
Magyarországon 1950-1975. Budapest : Kossuth
Könyvkiádó
Kossuth Radio, (1988) Interview with Mr. Petrovai, Deputy
Minister of Housing, 8.30 p.m., August 2
Kovacs, I., (1984) "Development of living standards in
Hungary from 1975-1983", Acta Oeconomica, 33, 155-180
KSH, (1973) Népszamlalás 1970 26 Kötét. Budapest:
Központi Statisztikai Hivatal
_____ (1978) Lakásépítés és megszünés 1977. Budapest:
Központi Statisztikai Hivatal
_____ (1982a) Statisztikai évkönyv 1981. Budapest:
Központi Statisztikai Hivatal
_____ (1982b) Területi statisztikai évkönyv 1981.
Budapest: Központi Statisztikai Hivatal
_____ (1982c) Életkörülmények és lakásviszonyok. Budapest:
Központi Statisztikai Hivatal
_____ (1985a) Statisztikai Évkönyv. Budapest: Központi
Statisztikai Hivatal
_____ (1985b) Lakásstatisztikai Évkönyv. Budapest: Központi
Statisztikai Hivatal
_____ Ten Yearly. Népszamlalás. Budapest: Központi
Statisztikai Hivatal
_____ Yearly. Lakásépítés és megszünés. Budapest:
Központi Statistikai Hivatal
_____ Yearly. Lakásstatisztikai Évkönyv. Budapest:
Központi Statisztikai Hivatal
_____ Yearly. Demográfiai Évkönyv. Budapest: Központi
Statisztikai Hivatal

_____ Yearly. Területi Statisztikai Évkönyv. Budapest: Központi Statisztikai Hivatal

Liska, T., (1969) "A berlákás - Kereskedelem koncepciója", Valóság, Vol. 1, 22-35

Magyar Nemzet, (1983a) "Peldaul Somogy", Magyar Nemzet, 20 July, 3

_____ (1983b) "A lakás tulajdonosság Tatabanyában", Magyar Nemzet, 27 July, 2

_____ (1983c) "Problémák Építöiparnál", Magyar Nemzet, August 7, 1

_____ (1983d) "Házépítés sajat eróból?" Magyar Nemzet, August 7, 5

Magyar Szocialista Munkáspart, (1980) Magyar Szocialista Munkáspart 12 Kongresszusa. Budapest : Kossuth Könyvkiádó

Mihályi, P., (1977) "Történeti szempontok a magyarországi lakashiány értékeléséhez", Valóság, Vol. 5

_____ (1978) "A typical waste of a centrally planned economy : unsatisfactory maintenance of council houses in Budapest", Economics of Planning, Vol. 14, No. 2, 81-95

Musil, J., (1987) "Housing policy and the sociospatial structure of cities in a socialist country - the example of Prague", International Journal of Urban and Regional Research, Vol. 11, No. 1, 27-36

Népszabadság, (1983) "Több ezer új lakás keszül el a fövárosban", Népszabadság, September 2, 1

Nök Lapja, (1982) "Lakás Lakás", Nök Lapja, Vol. 34, No. 33, 3-5

Pickvance, C.G., (1986) "Economic organisation and housing change in the pattern of state socialist redistribution". Paper for the XI World Congress of Sociology, New Delhi, 18-23 August 1986

_____ (1988) "Employers, labour markets, and redistribution under state socialism : an interpretation of housing policy in Hungary 1960-1983", Sociology, Vol. 22, No. 2, May, 193-214

Sillince, J.A.A., (1985a) "Housing as social problem versus housing as historical problem : the case of Hungary", Environment and Planning C Government and Policy, Vol. 3, 299-318

_____ (1985b) "The housing market of the Budapest Urban Region 1949-1983", Urban Studies, Vol. 22, 141-149

_____ (1988) "Regional policy in Hungary : objectives and achievements", Transactions of the Institute of British Geographers, Vol. 12, 451-464

Szarka, I., and I. Székely, (1987) "A lakásépítés pénzügyi támogatása és a tanácsok lehetöségei", Városépítés, Vol. XXIII, No. 2, 6-8

Housing Policy in Hungary

Szelényi, I., (1972) "Tarsadalmi struktura es lakásrendszer Budapest Kandidatus értekezés. (Doctoral thesis)
_____ (1977) "Social conflicts of under-urbanisation". In M. Harloe (Ed.), Captive Cities, London, Wiley
_____ (1978) "Social inequalities in state socialist redistributive economics", International Journal of Comparative Sociology, 19, 63-87
_____ (1982) "Inequalities and social policy under state socialism", International Journal of Urban and Regional Research, 6, 121-127
_____ (1983) Urban Inequalities under State Socialism. Oxford: Oxford University Press
_____ (1987) "Housing inequalities and occupational segregation in state socialist cities: Commentary to the special issue of IJURR on East European cities", International Journal of Urban and Regional Research, Vol. 11, No. 1, 1-8
Szelényi, I. and Konrád, G., (1969) Az uj lakótelepek szocialógiai problémái. Budapest: Akademiai Kiado
Szelényi, I., Manchin, R., (1986) "Social policy and state socialism". In Esping-Anderson, G., Rainwater, L., Rein, M., (Eds.) Stagnation and Renewal in Social Policy, White Plains, Sharpe
Tomassy, I., (1983) "A magánlakásépítés helyzete Nográd Megyében". Városépítés, Vol. 19, No. 2, 8-10
Tosics, I., (1987) "Privatisation in housing policy: the case of the western countries and that of Hungary", International Journal of Urban and Regional Research, Vol. 11, No. 1, 61-78
Varga, E., (1983) "A telekalakitasi eljarásrol". Városépítés, Vol. 19, No. 3,11
VATI, (1971) "Orszagos Településhalozat-Féjlesztési Konceptio". Budapest: Épitésügyi és Városféjlesztési Miniszterium

CONCLUSION

John Sillince

The quantity and quality of a nation's housing is justly seen as an important indicator of its standard of living. At a time when Western Europe has reached a rough equivalence between the demand for and supply of housing (despite large subgroups whose special needs are not met), and when quality of housing and access to it is being seen there as more important than quantity, in Eastern Europe and the Soviet Union quantitative shortages still continue to be the dominant problem. These shortages are partly due to poor quality and under-investment in housing relative to industry before 1975, and in stagnant or declining investment totals after 1975. Since the late 1970s Eastern Europe and the Soviet Union have been affected by the delayed recession caused by the oil price shock of 1973/1974, and this has led to attempts in several countries of the region to shift the financial burden of housing away from the state to the individual. This recent reaction against such a full intervention by the state in housing has been a pragmatic reaction to economic recession, since the degree of the reaction in the West and the East has closely matched the degree of each country's economic dependence on oil. The persistence of housing shortages in Eastern Europe and the Soviet Union is in a way remarkable, because the problem of war damage, the extent of rural-urban migration and the birth rates which gave rise to post-war shortages have all declined. In order to discover what shortages exist, official statistics must be used comparatively across the region to gauge the scale of the housing crisis, and corroboration must be obtained from press reports in those countries.

Conclusion

Although the clearest measure of housing shortage is excess of households over dwellings, statistics on households are not everywhere available. The next best indicator of new households is the number of marriages, and this in fact is the method used above to arrive at estimated housing shortage. This measure must be supplemented by comparisons of construction rates, dwelling sizes, average living space, and so on. Quality is also important as suggested above, but here only anecdotal evidence exists.

The housing shortage in Eastern Europe compares oddly with the relative plenty in the West. Yet in the most advanced industrialised countries there are tremendous mismatches in particular housing sub-markets (and some evidence that homelessness is beginning to resurface after a period of under-investment and low construction rates). For example, in 1975 while the stock of 5-room dwellings in the FDR exceeded demand, there was a shortage of 3-room dwellings - such demand will increase by 25 per cent by 1990 together with a 40 per cent rise in demand for 2-room dwellings. These large deficiencies exist at the same time as an aggregate level matching of supply and demand. The same sort of changes are also, for example, affecting the USA, where there was a seven-fold increase in the numbers of young people living alone between 1960 and 1978, and where now a third of single households are elderly.

The effects of housing shortages are many and varied. Homelessness is a direct result, but it is difficult to define, depending upon how sympathetic or unsympathetic is the housing policy towards the homeless - upon how much families are expected to pile up generations within the same dwelling, as happens in Eastern Europe today. Ultimately the search for quantitative measures of homelessness is a futile one. It is easier to judge how many families suffer overcrowding, how many young couples are forced to live with parents, how long on average households wait to get a dwelling. Along with this are personal and press accounts and court cases of the frustrations resulting from the house allocation system, and stories about illegal sublets. One can only imagine what the effect upon quality of life must be. There are inevitably tensions between relatives, and probably there is a lowering effect upon the birth rate.

Housing policy in Eastern Europe has progressed through many phases. It is tempting to suggest a model of policy stages, from minimal interference such as exists in the USA where the state has virtually withdrawn from public support for low-income families, and where state housing accounts for only 2 per cent of the stock, through social housing oriented towards equality, to comprehensive provision and management. Yet such a model would be very misleading. First, if it were suggested that governments progress chronologically from less to more interference, a number of objections could be raised. For example, such a progression has gone into reverse since the oil price shock made governments in the West and the East cut back on housing. One of the interesting aspects of this cutback has been the pragmatism of the move - the lack of effect of political ideology, and the similarity between East and West in cutting back on housing, which has fewer resistances to cuts than have education, health, and income maintenance, whatever the political milieu. Moreover, in some countries, such as the USA, governments set their starting position at minimum interference, and make deviations from this when under pressure, while in others, such as in Eastern Europe, governments have begun at comprehensive provision and management and then modified their position when problems arose. Second, any attempt to correlate the range represented by the model with the Right-Left political spectrum would come to grief when one considered politically centrist nations like Sweden, whose housing system veers more towards comprehensive provision and management than those of some of the countries discussed in this volume. Since 1945 in Sweden 45 per cent of all new housing has been constructed by the state and co-operatives account for 20 per cent - only 35 per cent is private; government financial assistance strongly favours poorer people; uniquely among Western countries the middle class spends a higher proportion of its income on housing than does the working class; working class and middle class housing areas there are virtually identical; 65 per cent of housing is post-1945; the ratio of dwellings per capita and the quality of dwellings are very high.

In housing studies it has become a commonplace to say that housing policy serves many different (often conflicting) goals. This is

Conclusion

certainly true in Eastern Europe. Besides the
need to provide each household with its own
dwelling, there are many other aims, pursued by
the different agents of housing supply. The
governments want to solve the housing shortage, to
encourage women to have larger families in those
countries like Hungary and the Russian part of the
USSR where birthrates are low, to attract
qualified labour to areas where it is scarce, to
prevent rural depopulation, to reward war
veterans, to protect the old, the sick, and the
poor and to provide employment for large numbers
of unskilled building workers. Yet they are also
(in Yugoslavia already, in Hungary very soon, in
the USSR maybe) wanting to raise rents and
mortgage rates considerably in order to reduce
state housing budgets. On the other hand,
economic enterprises, in such labour-intensive
economies as those of Eastern Europe, want to
maximise labour supply, either directly by their
own dwelling allocation policies or indirectly
when consulted by housing allocation officials in
city councils and central government who often
control allocation of houses for sale as well as
for rent. Housing co-operatives too have partisan
goals: in Hungary, Yugoslavia and the USSR, for
example, they are middle class and conscious of
property values: their members want to maximise
investment potential; they do not want young
families with noisy children, or families with
social problems. Also, there are private self-
builders, usually in smaller towns and villages,
usually inside some mutual-aid group comprising
relatives and friends. And there are the building
workers who work illegally, on the black market,
as a second job usually: all these want to get raw
materials, usually illegally, and building permits
from often reluctant local councils. They are
usually untrained and often do poor quality work.
Then there are the large, industrialised, building
companies which want work for their factories and
workforces, which lobby local and national
politicians to get work, and which have
monopolistic powers to fix high prices if
political connections are not successful. Because
there are advantages in starting many projects
simultaneously, buildings take a long time to
build, resulting in poor quality finishes. Also,
there is the vast collection of industries
supplying raw materials to the building industry,
whose products are often of poor quality which

leads to poor finishes on completed buildings.
Also, there is the architectural profession which
is critical of simple, crude, industrialised
designs, and which lobbies for greater
environmental sensitivity in new buildings and for
a larger role for rehabilitation.

All these actions and their conflicting aims
and pathological methods are different from those
in advanced Western countries, which have their
own conflicts and pathologies. Yet amazingly
there are many similarities between perceptions of
governments and other actors within these two
systems. In both systems there is an increasing
use of cost-effectiveness studies, in the West at
an advanced stage, in Eastern Europe materialising
in a criticism of existing policies, a criticism
of shortages and poor quality, a criticism of the
more exotic forms of housing inequality resulting
from administrative allocation. Also in both
systems, again stronger in the West than the East,
there is a growth in popularity of rehabilitation,
more government encouragement to homeowners to
carry a greater share of housing costs, and less
willingness by governments to support state
housing and state subsidies. While in many
Western countries an important part of government
subsidies go to middle-class individuals with the
aim that these will vacate accommodation allowing
poorer individuals to rise in the housing chain -
the filtering concept - in many countries of
Eastern Europe it is again the middle class which
is also taking on a crucial role in the newly
emerging policies such as the sale of state
housing, and the encouragement of the co-operative
and private construction sectors. Also, while the
West has traditionally relied on financial policy
tools, Eastern European governments have
previously relied on administrative ones - the
decisions of government officials in allocating
housing for rent and for sale, using bureaucratic
criteria. But the current trend in Eastern Europe
- of long standing in Yugoslavia and Hungary - is
towards the increasing use of financial policy
tools (the varying of rents and mortgages) and the
decreasing use of administrative ones (such as
allocation). The clearest recent (1987) indicator
of this is from the trend-setter in the new
policies - Yugoslavia. Housing funds for
enterprises, co-operatives and individuals are now
to be determined on economic criteria, i.e.
credit-worthiness, rather than social ones (such

Conclusion

as number of children, poverty, war service etc.).
And by 1990 it is planned that "economic" rents
will be charged in Yugoslavia. A safety net -
housing rebates - has been set up there to help
the needy. This "financial" approach seems likely
to be followed in Hungary where mortgage rates
have risen from 6 per cent previously to 15 per
cent together with large rent rises at the end of
1988. As yet it is unclear what changes will
occur in the USSR, although such radical measures
seem unlikely: more probable is a mixed financial-
administrative arrangement where a common rent is
charged for "standard" accommodation with much
higher rents to reflect better quality, location,
and larger size and so on, based on
administratively-determined assessments of housing
accommodation. What the outcome will be is
difficult to guess. Already strong tensions have
been created in Hungary, for example, and some
experts claim that large numbers of people - even
including the middle class - will be unable to
pay. However, this claim may be an example of the
fact that when expression of opinion becomes
freer, middle-class groups are the most skilled at
asserting their preferences. Certainly it seems
that the financial changes in Hungary and
Yugoslavia will lead to a cut in demand and hence
in house construction, at least in the short to
medium term. This does not seem to be the case in
the USSR, where government plans for expanding the
housing stock remain firm, and where it is more
likely that a second objective - to keep rents low
- will be dropped somehow. How the government of
the USSR will achieve this raising of rents is not
clear, however, bearing in mind the long history
there of very low and stable rents. Apart from
Yugoslavia, household expenditure on housing in
all the countries in the region after 1945 was
kept artificially low. In Hungary an attempt to
raise rents failed in the 1970s but those
pressures had reasserted themselves by 1983. And
in Poland since 1979 the proportion of household
expenditure devoted to housing has risen, although
rents are still far below "economic" levels.
These developments have as yet had little impact
upon housing policy in Czechoslovakia, Bulgaria,
Albania and Romania, where rents are still very
low, and where the economic recession in Comecon
countries since 1979 has made itself felt in ways
(such as stagnant wages, sharp increases in the
role of the black market) other than the inflation

which has ravaged Yugoslavia since the late 1970s and Poland since 1987 and affected Hungary more moderately since 1983.

Another aspect of the housing policy changes in Eastern Europe has been the move to encourage the private and co-operative sectors and reduce the state sector. This development has affected all countries in the region except for Romania and Albania. In Czechoslovakia the rise in the co-operative and private sectors began in 1980, while in Hungary and Bulgaria it happened somewhat earlier, in 1970. In Poland the co-operative sector has traditionally been the dominant one, but encouragement of the private sector only began during the last ten years. In the GDR the expansion in private construction came about after 1973. The USSR has only just begun to encourage its private and co-operative sectors. In Romania the trend has actually gone the other way — because of concern over speculation and housing inequalities there was a cutback in sales in 1986. About these policy moves several things need to be said. In the Balkan part of the region — Yugoslavia, Albanian, Hungary, Romania, Bulgaria — home-ownership has always in recent times been high even by Western standards and this has survived throughout the post-war period, so attitudes favourable to property ownership already exist. In the USSR on the other hand, property ownership and any demonstration of materially superior status engenders resentment and suspicion. Another thing is that the great bulk of private construction is at present self-built in rural areas. Also, the private sector (and co-operative sector in the USSR and Yugoslavia) encounter problems of obtaining raw materials and bureaucratic permits. Bureaucracy and legal processes of land acquisition delay development, and this affects small builders particularly hard: in Yugoslavia the laudable tradition of wide consultation and self-management unfortunately aggravates this problem. Moreover, local governments may not share the central government's enthusiasm for the new policy, as happens in the USSR. Also, at present the private sector has an ambiguous and politically unsupported position, as for example in Hungary and Yugoslavia, even when it accounts for the majority of house construction.

Conclusion

Some countries have begun to attempt to get to grips with the problem of scarce and poor quality building materials. In Hungary workers can collect all the profit from evening and weekend production of building materials. In the USSR there is the idea to set up special small brick factories. But the problem unfortunately goes much farther than this and concerns the need throughout the region to upgrade product quality and to introduce a means by which products go to where they are most needed. This affects existing factory working methods such as quality control, productivity measurement, incentive schemes, and trading agreement methods in the CMEA based on barter by weight by bureaucrats rather than a free market. These and other areas of production and distribution are in great need of reform and as yet it is unclear whether such reform will be successful.

The relative importance of economic enterprises within the housing system varies considerably between the countries of Eastern Europe. It does not vary for any labour market-related reason but seems to be due to historical accident. In the USSR economic enterprises have a very important role, partly because before 1917 industry had already begun a tradition of employer-provided, low-rent housing, and partly because local government had almost no decision-making powers prior to 1917 and hence no experience of housing administration to offer. In Yugoslavia also economic enterprises have developed a strong role in housing. On the other hand, in Hungary and Czechoslovakia, for instance, economic enterprises play a small and declining role. Their effect upon housing administration is on the whole not beneficial - the ability to get a dwelling through them depends on whether the enterprise is rich or poor, and (at least in the USSR) they are inefficient at housing administration. Reforms begun in the USSR aim at eventually transferring much control from enterprises to local government, at the same time reforming the at present unresponsive and inefficient local government system. These reforms are only in their infancy. They aim to reform by means of pressure from the electorate rather than by tinkering with bureaucracy, but at present ideas to reform the electoral system are decidedly modest.

The present position in 1988 for Eastern European housing policy is not a clear one. It is not certain that the changes begun in Yugoslavia, Hungary, Poland, and the USSR will succeed. Indeed it is not clear what the changes are since in most cases only the outlines of policy initiatives have been drawn. It is ironic that the pressures for change in housing policy in Eastern Europe are greatest where the political system is at its most pluralistic (relative to others within the region), yet this type of political system is the most likely to vacillate and compromise. Also, the current reaction against collectivist housing policies in Eastern Europe must not be seen as inevitable or irreversible. It is not the first time that collectivist policies suffered a setback - this happened also in Great Britain, France and Germany when utilitarian political ideas were temporarily dominated by Malthusian ones during the rapid growth of poor-relief costs between 1800 and 1830.

Will the current reforms create a dichotomy within Eastern Europe between those countries where the poor are seen as the main object of social policy help and those where everyone has the right to it? At present in Eastern Europe social policy is strongly "universalistic" in theory, with everyone enjoying equal rights to housing, education, health provision and so on. Although the practice is rather different, the principle of equal rights does remove the stigma which associates itself with the "particularistic" social policies in some Western countries - means-tested housing rebates and social security payments for example. In Eastern Europe cheap state housing is a sign of belonging to society. Stigmatisation is reserved for "parasites", criminals, prisoners, mental hospital patients and gypsies. Usually it is difficult or impossible to stand on a street and distinguish which housing is rented state housing, which is state-built for sale, and which is co-operative-built for sale. Living in a co-operative-built housing block may be important to a Moscow intellectual, but a worker in a nearby state-built block will not feel in an inferior social position because his flat was built by the state. Almost-free housing confers on East European societies the attributes of the gift relationship (moral obligation towards the giver - the state). It is almost as if, as in pre-monetary societies, the gift remains

Conclusion

spiritually a part of the donor. It is unclear
to what extent new attitudes towards economic
productivity spurred on by new policies and
incentives and other attributes of a more
individualistic culture will cause people to
regard their relationship with their housing
system as one of exchange (in which their position
is one of economic equivalence and equality).
Already it is possible to suggest that these new
attitudes are more likely to come about in, say,
Hungary rather than, say, the USSR. The USSR
approach to rent rises probably will be strongly
universalistic - a common price with higher prices
for better quality of housing, while the Hungarian
approach seems to be towards a rebate system,
necessarily means-tested, and necessarily
particularistic. This difference between the two
countries, used for illustrative purposes, seems
understandable bearing in mind differences in
nearness to the West (distance, trade dependence,
tourist orientation, ease of foreign travel),
length of communist experience, size of pre-
communist middle class, and recent liberalisation
experience.

Appendix : Exchange Rates to the Pound Sterling

Albania	10.0533	lek
Bulgaria	1.4956	lev
Czechoslovakia	26.12	koruna
GDR	3.25	ostmark
Hungary	94.9959	forint
Poland	976.60	zloty
Romania	14.781	leu
USSR	1.0891	rouble
Yugoslavia	11310.57	dinar

Source: Financial Times, 21 February 1989, p. 32

BIOGRAPHICAL NOTES

Dr Gregory Andrusz is Senior Lecturer in the Centre for Community Studies, Middlesex Polytechnic. He carried out postgraduate research in Warsaw, was a British Council scholar in Kiev in 1975-1976, and has made several study visits to the USSR. He is the author of Housing and Urban Development in the USSR (Macmillan, 1984), and of several articles on planning and housing in the USSR.

Professor Dr Hanns Buchholz is Head of Department, Department of Geography, Hannover University.

Dr Frank Carter is Lecturer in the Geography of Eastern Europe at University College and School of Slavonic and East European Studies, London University. He has published articles on the urban development of Sofia, regional development in Bulgaria, that country's environmental pollution problems, and preservation of its historic heritage, as well as articles on planning and geography of other parts of Eastern Europe.

Dr Andrew Dawson is Senior Lecturer in the Department of Geography in the University of St. Andrews. He has had a long-standing interest in Eastern Europe, and especially the relationship between the state and economic and settlement development in Poland. Among his publications about the area, he edited Planning in Eastern Europe (Croom Helm, 1987).

Dr Derek Hall is Head of Department, Department of Geography, Sunderland Polytechnic. He has been travelling to Albania since 1974. In recent years he has been a tour leader for study groups visiting the country. He has published research on development in peripheral socialist countries, social ecological processes, UK transport policy and urban development in India. He is a regular contributor on the socialist world for Town and Country Planning magazine.

Mr Gábor Locsmándi is Lecturer in the Department of Town Planning at the Technical University, Budapest (Városépítési Tánszék, Müszaki Egyetem, Budapest, Hungary). He is a practising architect-planner and besides many architectural designs for various local authorities in Hungary, he has published a report on housing segregation for Budapest City Council.

Mr David Short is Lecturer in Czech and Slovak Language and Literature at the School of Slavonic and East European Studies, University of London. He spent 1966-1972 in Prague and in 1969 married a Czech. He is the author of Czechoslovakia (Clio Press, 1986, World Bibliographical Series Vol. 68); of 'Czech and Slovak' in B. Comrie (ed.) The World's Major Languages (Croom Helm, 1987); of annual bibliographical surveys of Czech (1981-1988) and Slovak (1982-1985) language studies in The Year's Work in Modern Language Studies, (Modern Humanities Research Association); and of articles on a variety of Czech authors. He has contributed to Czech dictionaries and has translated several textbooks by Czech authors.

Dr John Sillince is Senior Lecturer in the Department of Computer Science, Coventry Polytechnic. In 1979 he married a Hungarian. He is the author of A Theory of Planning (Gower, 1986).

Dr Shaun Topham is Research Fellow at the Postgraduate School of Yugoslav Studies, Bradford University. His doctorate concerning advertising and socialism was researched at Belgrade University. He is currently a Calderdale MBC councillor sitting on the Housing Committee. His other main research interests are in tourism and self-management.

Dr David Turnock is Lecturer in the Geography Department, Leicester University. Since 1968 he has travelled widely within Eastern Europe and Romania. He has published many articles on Romania, in particular on modernisation there.

Index